SPECTROSCOPY AND KINETICS

COMPUTERS IN
CHEMISTRY AND INSTRUMENTATION

edited by

JAMES S. MATTSON   HARRY B. MARK, JR.   HUBERT C. MACDONALD, JR.

Volume 1: Computer Fundamentals for Chemists (IN PREPARATION)
Volume 2: Electrochemistry: Calculations, Simulation, and Instrumentation
Volume 3: Spectroscopy and Kinetics

SPECTROSCOPY AND KINETICS

Edited by

James S. Mattson
DIVISION OF CHEMICAL OCEANOGRAPHY
ROSENSTIEL SCHOOL OF MARINE
AND ATMOSPHERIC SCIENCES
UNIVERSITY OF MIAMI
MIAMI, FLORIDA

Harry B. Mark, Jr.
DEPARTMENT OF CHEMISTRY
UNIVERSITY OF CINCINNATI
CINCINNATI, OHIO

Hubert C. MacDonald, Jr.
KOPPERS COMPANY, INC.
MONROEVILLE, PENNSYLVANIA

MARCEL DEKKER, INC. New York 1973

Copyright © 1973 by MARCEL DEKKER, INC.

ALL RIGHTS RESERVED

No part of this work may be reproduced or utilized in any form or by any means, electronic or mechanical, including photocopying, microfilming, and recording, or by any information storage and retrieval system, without the written permission of the publisher.

MARCEL DEKKER, INC.
95 Madison Avenue, New York, New York 10016

LIBRARY OF CONGRESS CATALOG CARD NUMBER: 72-91433
ISBN: 0-8247-6058-1

Printed in the United States of America

INTRODUCTION TO THE SERIES

In the past decade, computer technology and design (both analog and digital) and the development of low cost linear and digital "integrated circuitry" have advanced at an almost unbelievable rate. Thus, computers and quantitative electronic circuitry are now readily available to chemists, physicists, and other scientific groups interested in instrument design. To quote a recent statement of a colleague, "the computer and integrated circuitry are revolutionizing measurement and instrumentation in science." In general, the chemist is just beginning to realize and understand the potential of computer applications to chemical research and quantitative measurement. The basic applications are in the areas of data acquisition and reduction, simulation, and instrumentation (on-line data processing and experimental control in and/or optimization in real time).

At present, a serious time lag exists between the development of electronic computer technology and the practice or application in the physical sciences. Thus, this series aims to bridge this communication gap by presenting comprehensive and instructive chapters on various aspects of the field written by outstanding researchers. By this means, the experience and expertise of these scientists is made available for study and discussion.

It is intended that these volumes will contain articles covering a wide variety of topics written for the nonspecialist but still retaining a scholarly level of treatment. As the series was conceived it was hoped that each volume (with the exception of Volume 1 which is an introductory discussion of basic principles and applications) would be devoted to one subject; for example, electrochemistry, spectroscopy, on-line analytical service systems. This format will be followed wherever possible. It soon became evident, however, that to delay publication of completed manuscripts while waiting to obtain a volume dealing with a single subject would be unfair to not only the authors but, more important, the intended audience. Thus, priority has been given to speed of publication lest the material become dated while awaiting publication. Therefore, some volumes will contain mixed topics.

The editors have also decided that submitted as well as the usual invited contributions will be published in the series. Thus, scientists who have recent developments and advances of potential interest should submit detailed outlines of their proposed contribution to one of the editors for consideration concerning suitability for publication. The articles should be imaginative, critical, and comprehensive survey topics in the field and/or other fields and which are written on a high level, that is, satisfying to specialists and nonspecialists alike. Parts of programs can be used in the text to illustrate special procedures and concepts, but, in general, we do not plan to reproduce complete programs themselves, as much of this material is either routine or represents a particular personality of either the author or his computer.

The Editors

PREFACE

This volume contains sections on the application of computers to problems in spectroscopy (Part I) and chemical kinetics (Part II). As chemical kinetics is a highly specialized subject, relatively complete discussions of the various aspects of computer applications to reaction rate studies are given in the four chapters of Part II of this volume. However, the various applications of computers to the numerous types of spectroscopic or optical methods for study of atomic and molecular structure, chemical reactions, chemical analysis, etc., are so large that both complete volumes and parts of volumes will be devoted to this subject. Thus, Part I of this volume, concerned with spectroscopic applications, is only a part of the material to be discussed on this subject in this series.

Chapter 1 is concerned with the application of the computer to magnetic resonance data analysis. This chapter also discusses the applicability of "on-line" use of large computers. As the general trend of on-line applications of computers is the use of small, built-in (specialized) computers, this chapter offers an alternative approach which may be more feasible in certain situations. It should be mentioned that several chapters in future volumes will be devoted to the subject of the design of on-line laboratory systems. Chapter 2 discusses problems encountered in the important area of determining the optical properties (especially thickness) of very thin films. Chapter 3 considers, in detail, the application of both digital and analog circuitry as well as on-line computers to the problem of obtaining highly accurate kinetic data. The instrumental advances discussed here have opened up a new dimension in the use of reaction rates in the quantitative study of reaction mechanisms and the analysis of complex systems. Chapter 4 is concerned with the application of analog computer simulation of complex chemical and biological reaction rates in order to determine reaction mechanisms and carry out kinetic-based analysis of such systems. Chapter 5 deals with potentialities of digital computer Monte Carlo based simulation of reaction rate equations, while numerical methods for the solution of complex kinetic differential equations is treated in Chapter 6.

The editors wish to thank Bonnie Koran for the excellent line drawings of many of the figures for this volume and several of their colleagues and/or students who have read the chapters and given constructive suggestions concerning the presentation and the subject material.

Cincinnati, Ohio
November 1972

J. S. Mattson
H. C. MacDonald, Jr.
H. B. Mark, Jr.

CONTRIBUTORS TO THIS VOLUME

HOWARD J. BUTCHER, Nadir Metallurgical Co., Elmont, New York 11003

S. R. CROUCH, Department of Chemistry, Michigan State University, East Lansing, Michigan 48823

JIŘÍ JANATA, Imperial Chemical Industries Limited, Petrochemical and Polymer Laboratory, The Heath, Runcorn, Cheshire, England

JOHN J. MANOCK, Chemistry Department, Western Carolina University, Cullowhee, North Carolina 28723

HARRY B. MARK, JR., Department of Chemistry, University of Cincinnati, Cincinnati, Ohio 45221

LOUIS NEWMAN, 10746 Camarillo Street, North Hollywood, California 91602

NORMAN C. PETERSON, Kirk Chemical Laboratory, Polytechnic Institute of Brooklyn, Brooklyn, New York 11201

ERIC N. C. RANDALL, Department of Chemistry, University of Cincinnati, Cincinnati, Ohio 45221

CONTENTS

INTRODUCTION TO THE SERIES   iii

PREFACE   v

CONTRIBUTORS TO THIS VOLUME   vii

PART I: SPECTROSCOPY

Chapter 1. APPLICATION TO MAGNETIC RESONANCE
SPECTROSCOPY: THE USE OF THE
LARGE COMPUTER IN CHEMICAL
INSTRUMENTATION   3
Louis Newman

I.	Introduction	4
II.	The Magnetic Resonance Experiment	5
III.	Data Acquisition — The Interface Device	11
IV.	ESR Data Handling and Analysis	21
V.	Data Processing — The Computer Program	32
VI.	Conclusion	40
References		41

Chapter 2. THE CALCULATION OF THE OPTICAL
PROPERTIES OF THIN METAL FILMS
FROM INTERNAL REFLECTANCE DATA   43
Eric N. C. Randall and Harry B. Mark, Jr.

I.	Introduction	44
II.	The Principles and Terminology of Internal Reflection Spectroscopy	47
III.	Thin Films and Thin-Film Optics	57
IV.	Experimental Apparatus and Techniques	60
V.	Method for Determination of Film Thickness and Optical Constants of Thin Films	64
VI.	Discussion and Summary	82
Appendix I.	Symbols and Abbreviations	86
Appendix II.	The Calculation of the Reflection Absorbances of an n-Phase System	88

ix

Appendix III. A Fortran IV Language Computer Program for
the Calculation of Reflection Absorbances of an
n-Phase System   90

Appendix IV. A Fortran IV Language Computer Program for
the Calculation of the Optical Properties of
Thin Metal Films from IRS Measurements 95

References 101

PART II. KINETICS

Chapter 3. APPLICATIONS OF COMPUTER CIRCUITRY
AND TECHNIQUES TO KINETIC METHODS
OF ANALYSIS 107
S. R. Crouch

I.   Introduction 108
II. Analytical Kinetic Techniques 108
III. Instrumentation for Kinetic Methods 138
IV. Small Digital Computers in Kinetic Analysis 190
V.   Future Perspectives 203
References 204

Chapter 4. ANALOG COMPUTER SIMULATION OF
KINETIC MODELS 209
Jiří Janata

I.   Introduction 210
II. Closed Kinetics 211
III. Open Kinetics 233
IV. Biochemical Applications 243
V.   Miscellaneous Applications 248
References 257

Chapter 5. THE APPLICATION OF THE MONTE CARLO
METHOD TO CHEMICAL KINETICS 267
John J. Manock

I.   Introduction 267
II. Simulation of Chemical Reactions 269
III. Conclusion 288
References 291

CONTENTS

Chapter 6. INTEGRATION OF COMPLEX RATE EQUATIONS
USING INFINITE SERIES 293
Norman C. Peterson and Howard J. Butcher

I. Introduction 293
II. Numerical Integration 295
III. Analytical Functions Solutions 297
IV. Infinite Series Solutions 299
References 310

AUTHOR INDEX 315

SUBJECT INDEX 321

SPECTROSCOPY AND KINETICS

Part I

SPECTROSCOPY

Chapter 1

APPLICATION TO MAGNETIC RESONANCE SPECTROSCOPY:
THE USE OF THE LARGE COMPUTER IN CHEMICAL INSTRUMENTATION

Louis Newman

10746 Camarillo St.
North Hollywood, California 91602

I.	INTRODUCTION	4
II.	THE MAGNETIC RESONANCE EXPERIMENT	5
III.	DATA ACQUISITION — THE INTERFACE DEVICE	11
	A. Storage Devices	13
	B. A Digital Tape System	15
	C. Gaussmeter Readings	18
	D. Recording of Signal Intensity	19
	E. Magnetic Tape Format	19
IV.	ESR DATA HANDLING AND ANALYSIS	21
	A. General Description	22
	B. Theoretical Description	23
	C. Subroutines	24
	D. Program Usage	27
	E. Input Cards	27
	F. Results from ESR Spectra Analysis	28
V.	DATA PROCESSING — THE COMPUTER PROGRAM	32
	A. The Experimenter	33
	B. Program Objectives	34
	C. Modular Construction	35
	D. A Spectroscopy Program — RAMAN	36
VI.	CONCLUSION	40
	REFERENCES	41

Copyright © 1973 by Marcel Dekker, Inc. All Rights Reserved. Neither this work nor any part may be reproduced or transmitted in any form or by any means, electronic or mechanical, including photocopying, microfilming, and recording, or by any information storage and retrieval system, without permission in writing from the publisher.

I. INTRODUCTION

Many instrumental tools are slow-scanning spectrometers which plot a response versus an independent variable such as time, wavelength, or magnetic field. The output from the instrument is usually in the form of a two-dimensional graph. The problem of attempting to decipher this thin line of ink wandering about a sheet of paper has wasted many man-hours and has often proved fruitless and frustrating. For recording results from an experiment, this method has many mechanical disadvantages:

1. The recorder may be nonfunctional.

2. The pen may not write correctly.

3. The paper output is inconvenient to handle and store.

4. Measurement of results is often inaccurate, time-consuming, and subject to human error.

A large computer with the proper environment can handle the data from an instrumental experiment in a more convenient and accurate fashion. A definite time delay between the performance of an experiment and the final output of the results must be acceptable, and a large digital computer must be available at reasonable cost. The computer can then be used to display the data graphically, to store them in a convenient compact form, and to perform the necessary measurements and calculations. The large computer method is most advantageous when the reduction of the experimental data to meaningful quantities is complex, or when a higher degree of accuracy is required. Often, the needed results can be obtained on pages suitable for publication within five minutes or an hour after the completion of the experiment.

In order to implement the computer, a system must be devised to transfer the data to the computer (the interface), and the computer must be given instructions for the processing of the data (the program). Once an interface has been constructed and a program written, the computer will be able to extract the desired information from the experiment in a more accurate and less time-consuming fashion than previously. However, in practice, the large computer is often not utilized in the most optimal manner.

The use of a large digital computer involves three groups of people: the systems staff who operate the computer, the designers of the interface and the program, and the technicians who perform the experiment. The political problems involved in dealing with these people are often so substantial as to preclude the effective use of the computer.

1. The Large Computer in Chemical Instrumentation

Case A: Dr. Dolittle had the results from his experiment punched onto paper tape and planned to analyze these results with a program written in PL/I which he received from a friend at State U. Result: His local computer could not read paper tape nor could it understand PL/I.

Case B: Professor Budget decided to have the results of an experiment placed on magnetic tape and processed via computer. Result: Parity errors on his low-cost tape unit destroyed much of his data, and his research grant was reduced to zero by excessive computer charges.

Case C: Assistant Footsore had to walk three blocks to the computer, once to get his spectral data transferred and again to submit his program on cards. The output was not available until two days later. Result: He measured his spectra by hand.

These problems can be solved either by understanding and accepting the computer facilities that are available or by convincing the appropriate staff members that the operational capabilities of the computer need to be increased. Troublesome and inefficient cases of computer usage need not occur if the chemist, electrical engineer, programmer, and technician are able to communicate with each other and decide in advance what the objectives of using the computer are. Alternatively, one person can learn to understand all of the necessary operations from experiment to result.

This chapter discusses the methods and difficulties connected with using a large computer to produce results from experiments. First, a specific magnetic resonance experiment is used as an example which is suitable for computer application. The possible methods of interfacing an experiment to the computer are mentioned and the data acquisition method chosen for the magnetic resonance experiment is described in detail. Next, the computer program which produces meaningful parameters from the magnetic resonance spectrum is outlined. Finally, more general considerations for spectroscopic applications are delineated, and a program is designed to take advantage of the facilities of a large digital computer for data handling and processing.

II. THE MAGNETIC RESONANCE EXPERIMENT

In general, the magnetic resonance experiment involves placing a sample in the center of a magnetic field and applying electromagnetic radiation to the sample. The magnetic field is then increased or decreased slowly while the electromagnetic radiation is monitored. (Alternatively, the magnetic field may be kept constant and the frequency of the electromagnetic radiation changed slowly.) Absorption of energy at a given magnetic field results from a change in orientation of particles in the sample. The resultant plot of the electromagnetic energy absorption versus the magnetic field (or frequency) is influenced by two factors in stable molecules. First, particles with different magnetic environments in the molecule may absorb energy at different magnetic field strengths. Second, magnetic particles near the absorbing particle will interact with the magnetic field, and change the position of the absorption. This interaction is independent of the magnetic field strength. The most common types of magnetic resonance are nuclear magnetic resonance (NMR), in which the particle is an atomic nucleus with odd spin, and electron spin resonance (ESR), in which the particle is an unpaired electron.

The specific experiment described here is the use of ESR on organic free radicals in solution. Since only one compound that forms a free radical was used in a given sample, and the radical is randomly oriented in solution, the magnetic environment is equivalent for all the molecules. Therefore, if other magnetic particles were absent in the molecule, only one resonance absorption (line) would be observed. However, a nucleus in the molecule with spin $\frac{1}{2}$ is oriented either with (50%) or against (50%) the magnetic field, and the interaction with the free electron results in the original line being split into two absorption lines. Another nucleus in the molecule with spin will again split each of the two lines into more lines. Magnetically equivalent nuclei have the same magnitude of interaction with the electron. This interaction is referred to as the hyperfine splitting constant or the coupling constant. After all nuclei have been accounted for, the resultant spectrum may have a very large number of lines and many of these lines may overlap each other. The analysis of such spectra by hand to find the hyperfine splitting constants can be time-consuming or next to impossible. The problem is further complicated by the usual arrangement of the ESR spectrometer which displays the first derivative of the absorption curve in order to improve the signal-to-noise ratio.

The ESR spectra of some 4,4'-dialkylbiphenylide radical anions are depicted in Figs. 1-8. The arrow in each figure is 2 G long and points in the direction of increasing magnetic field. The radical anions are formed by reducing the parent hydrocarbon with an alkali metal using tetrahydrofuran (THF) as the solvent [1]. The concentration of the bright green radical anions is about 5×10^{-4} M. The X-band spectra are taken at 100 kHz modulation frequency at low microwave power levels and modulation amplitudes. The unpaired radical electron occupies a molecular orbital which is delocalized around both benzene rings. This electron probably makes the rings coplanar, so the molecule has three twofold symmetry axes. The spectra were taken at a scanning rate of 1 G/min.

1. The Large Computer in Chemical Instrumentation

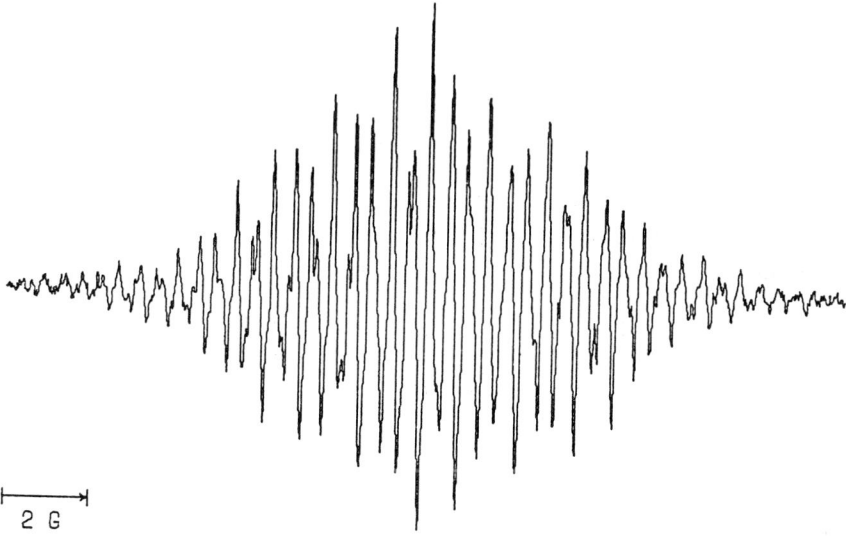

Fig. 1. The ESR spectrum of potassium 4,4' bis (trideuteromethyl) biphenylide in THF at 222° K.

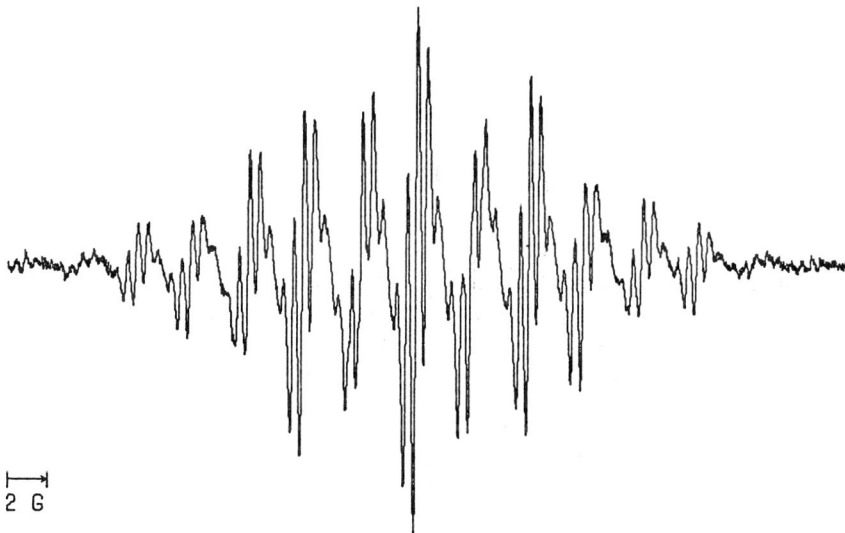

Fig. 2. The ESR spectrum of potassium 4,4'-dimethylbiphenylide in THF at 222° K.

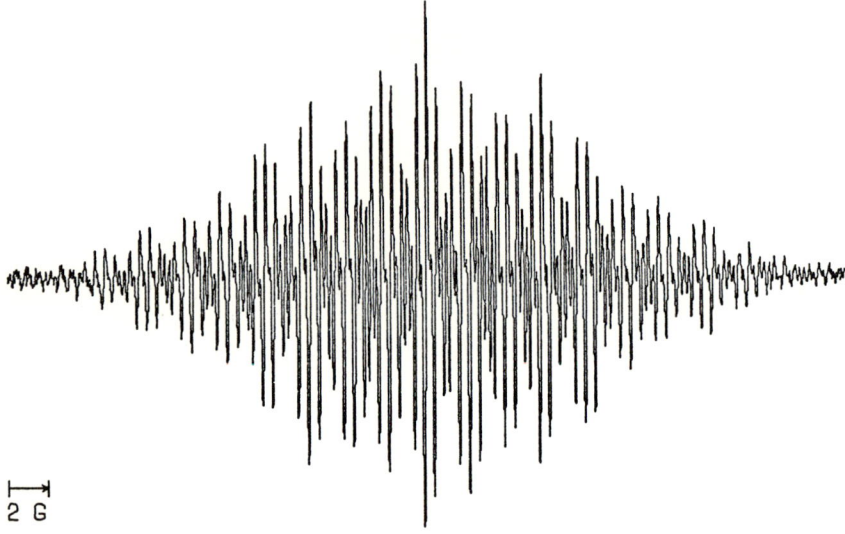

Fig. 3. The ESR spectrum of potassium 4-methyl, 4'-(methyl-^{13}C) biphenylide in THF at 222°K.

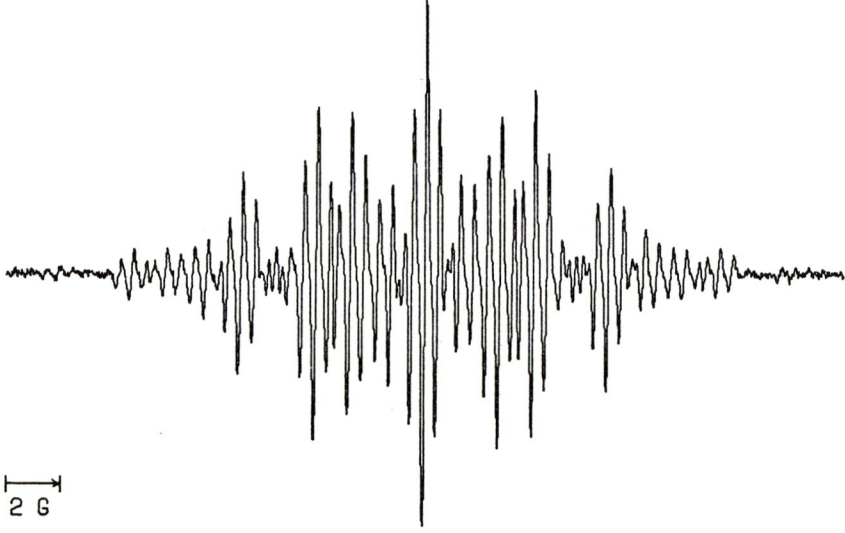

Fig. 4. The ESR spectrum of potassium 4,4'-diethylbiphenylide in THF at 222°K.

1. The Large Computer in Chemical Instrumentation 9

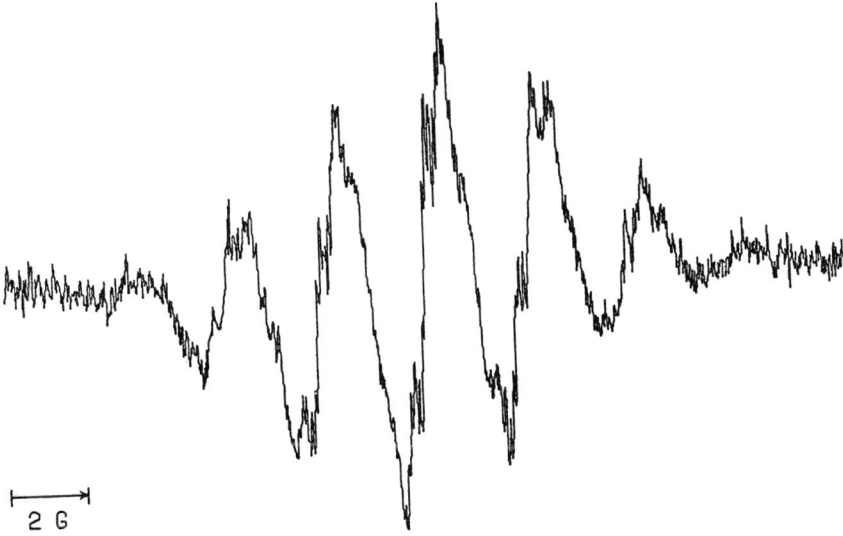

Fig. 5. The ESR spectrum of potassium 4,4'-diisopropylbiphenylide in THF at 222° K.

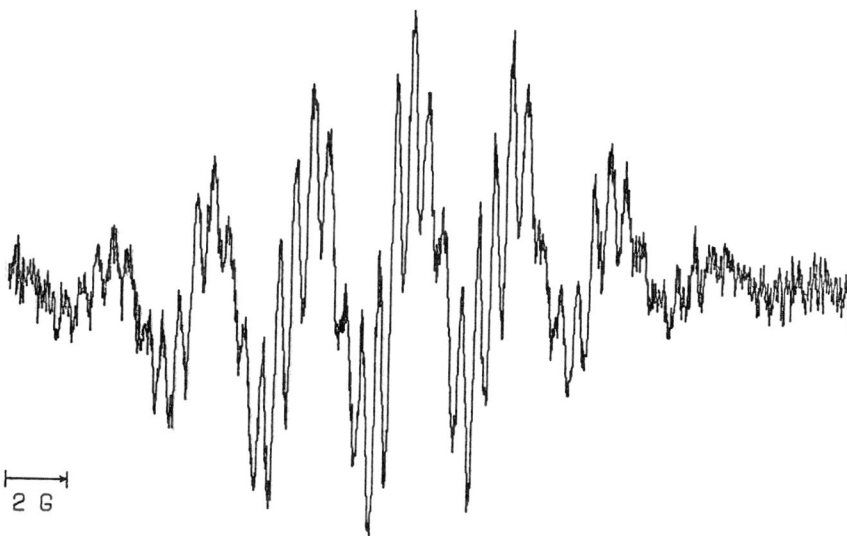

Fig. 6. The ESR spectrum of potassium 4,4'-di-n-butylbiphenylide in THF at 222° K.

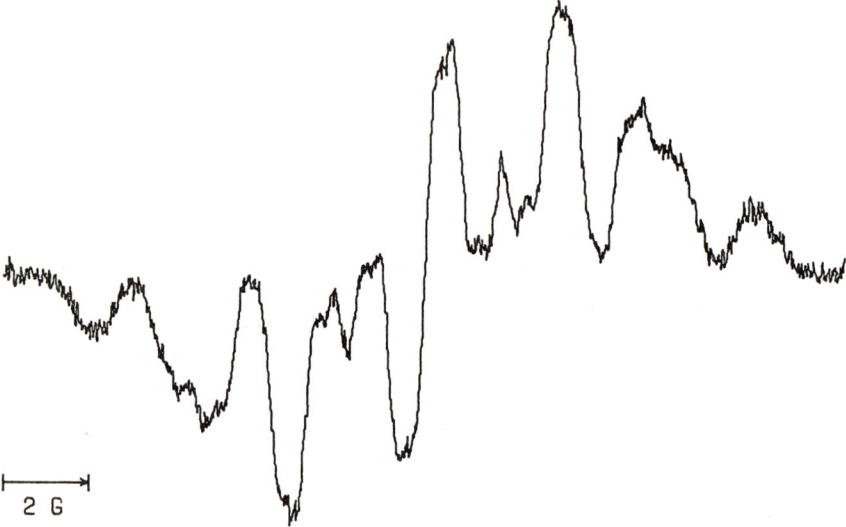

Fig. 7. The ESR spectrum of potassium 4,4'-di-sec-butylbiphenylide in THF at 222°K.

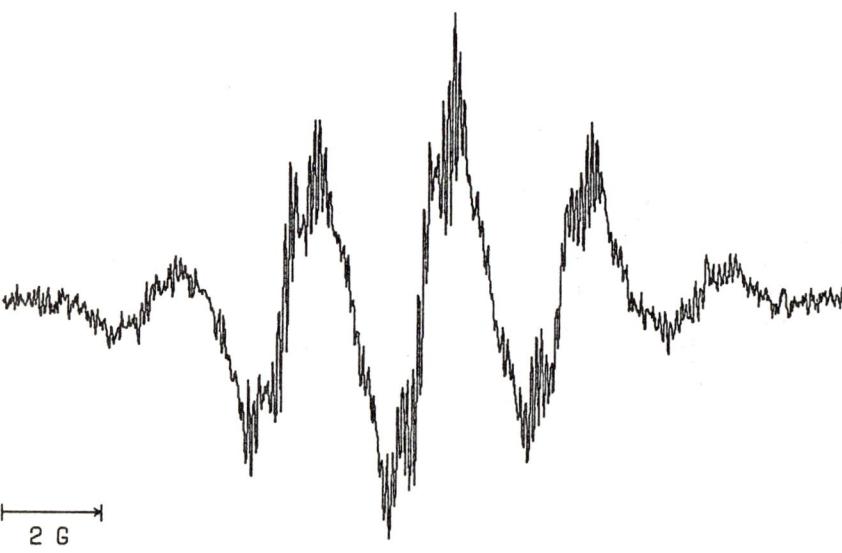

Fig. 8. The ESR spectrum of potassium 4,4'-di-tert-butylbiphenylide in THF at 222°K.

$$C^\delta - C^\gamma - C^\beta - C^\alpha - \text{(ring 1: 4,3,2,1,5,6)} - \text{(ring 2: 1',2',3',4',6',5')} - C^{\alpha'} - C^{\beta'} - C^{\gamma'} - C^{\delta'}$$
with $H^\epsilon, H^\delta, H^\gamma, H^\beta$ on left side and $H^{\beta'}, H^{\gamma'}, H^{\delta'}, H^{\epsilon'}$ on right side; ring protons H on positions 3,2 (top) and 5,6 (bottom) of ring 1, and 2',3' (top) and 6',5' (bottom) of ring 2.

The nomenclature for the substituted biphenyls is given in the diagram. The protons on the ring are numbered according to the carbon to which they are attached. Therefore, the 2-proton is attached to the 2-carbon. The protons on the side chain are named according to the Greek alphabet depending on the number of bonds separating the proton from the ring. Note, however, that the β-proton is not attached to the β-carbon, and so on. For the 4,4'-dialkylbiphenyls, the 2,6,2',6' and the 3,5,3',5' protons comprise equivalent sets and will be referred to as the 2- and the 3-protons.

The spectrum of the 4,4'-dimethylbiphenylide in Fig. 2 has only three coupling constants associated with it. There are two 1:4:6:4:1 quintuplets with different coupling constants from the 2- and the 3-protons and a 1:6:15:20:15:6:1 septet from the six equivalent methyl group protons. The total number of lines is then 5 x 5 x 7 or 175. The intensity of some of these lines is so small that they are indistinguishable from the noise. As the number of magnetic nuclei on the alkyl substituent increases, the spectrum becomes more difficult to analyze for the coupling constants. If the technique of electron nuclear double resonance (ENDOR) is available, it may be used to determine the coupling from each set of magnetic nuclei individually. A large computer may also be used to simulate the spectra from an initial guess and then to refine the couplings by comparing the experimental with the theoretical spectra.

III. DATA ACQUISITION — THE INTERFACE DEVICE

Constructing an interface device to acquire and relay the experimental data to the computer is fairly simple in theory but more complex in practice. Ideally, all relevant information from the experiment would be produced in digital form, and a multiwire connector would transfer the information onto a computer storage device. The large computer with its variety of peripheral equipment could then read, store, and process these data (Fig. 9).

However, the output from most instruments is not suitable for an interface device. The process of recording the data requires a definite

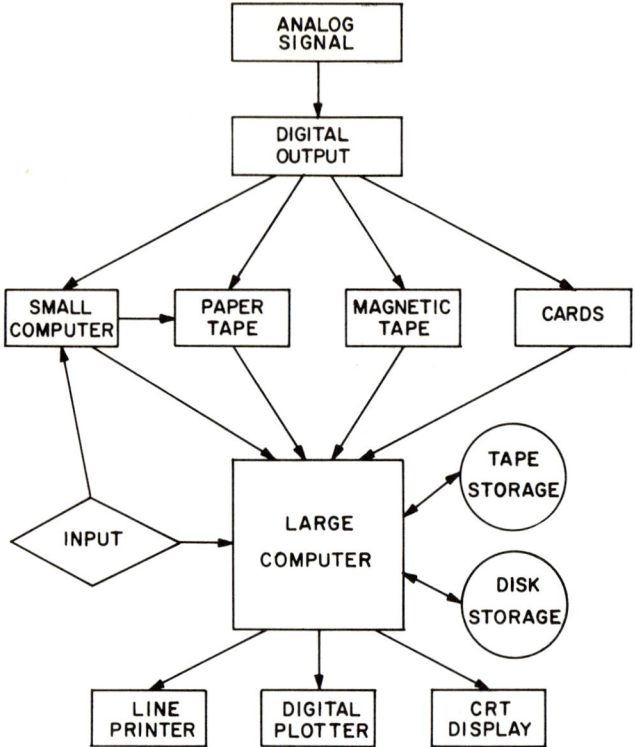

Fig. 9. The experiment interface to the large computer.

time interval and the data must be displayed in digital form. The responses from the instrument must be accumulated over a time which is greater than the time necessary to record the data. The accumulated responses are then converted to digital form and recorded on the interface device. Therefore, during each time interval, data are accumulated while the previous data are being recorded. A continuous experimental response is thereby broken down into a series of points. The number of points recorded is determined by the length of the time interval selected and the duration of the experiment. Each data point consists of the response from each parameter measured in sequence within the recorded point.

All the points in the data set are stored so the experiment can operate independently from the large computer. The time delay involved before the large computer is available for use is dependent on a variety of local

1. The Large Computer in Chemical Instrumentation

conditions. The data set must be transferred to the computer, and the computer must be given instructions (input) for data processing. The data may be transferred by direct cable, phone line, on foot, on a bicycle, or by truck. Input instructions may be sent to the computer via cards or a keyboard terminal. If the computer is operational, the time delay is influenced by the job priority, the length of the queue, and the speed of the operator in providing the necessary tapes or disks. Once the computer has begun the job, time is needed by the central and peripheral processors before the results are printed, plotted, or displayed. The longest delay of all in some installations is the procedure of returning the output from the large central computer to the user. Of course, there is always the possibility that the computer program did not function correctly, thus requiring the entire process to be repeated.

A. Storage Devices

The storage device used is the most important part of the interface system. The capabilities and limitations of this device determine how the experimental data will be accumulated and transferred to the large computer. Four major storage alternatives are cards, magnetic tape, paper tape, and a small computer.

Ordinary 80-column punched cards are acceptable input to all large computers. Some computers may have several terminals where card input can be read; others may have only one. Keypunching cards in order to provide the computer with a small amount of information is not very time-consuming, but attempting to keypunch by hand all the numbers corresponding to a spectrum is both tedious and subject to error. The data may be punched directly by a machine connected to the experimental apparatus, but this method is not desirable. The keypunch machine is expensive to own and operate; the rate of recording is restricted by the speed of the punch; the cards are bulky and difficult to handle; and the mechanical operations of the keypunch are easily disturbed. One advantage is that the data can be modified by hand.

Either seven- or nine-track half-inch magnetic tape can be read by most large computers. Most remote terminals do not have facilities for reading magnetic tape, so the tape must be sent to the main computer installation. Data can be recorded on magnetic tape at a high rate of speed, and a tremendous amount of data may be stored on one magnetic tape. The problem is that the format of the data on the tape may not be acceptable to the computer. Parity errors may occur or special software may be necessary to allow the computer to read the tape.

Paper tape cannot be read by some large computers. Others have a high-speed paper-tape reader or have the facilities available for remote terminals which can handle paper tape. The same terminal may also be used to enter instructions into the computer. Paper tape is inexpensive and convenient to record a small amount of data. Disadvantages are the slow speed of the mechanical punch and the unwieldly, fragile length of paper tape produced by a large amount of data.

A magnetic tape cassette is used like paper tape, but it is able to record data faster and store the data in a more convenient fashion. The cassette is limited by the length of magnetic tape available and the cost of the facilities necessary to use this form of recorder. When using phone lines, a cassette can record data rapidly and then play the data back slowly for information transfer.

A small computer is the most versatile interface device. It can be used both as a terminal for the large computer and as a device to control the experiment and store the resultant data. Unfortunately, many large computers do not have the facilities to handle this form of interface device. The variety of small and minicomputers available makes the provision of the appropriate connection difficult. If the facilities are available to interface a certain kind of minicomputer, then an expensive direct cable connection may have to be used. The input and output operations of the small computer terminal are limited only by the peripheral equipment available.

The small computer is also an excellent device to store data from an experiment. The data can be stored instantaneously and can be processed before the large computer is used. Many operations such as Fourier transforms are more conveniently handled by the small computer. If a large amount of information is produced by an experiment, additional core or peripheral storage may be required. The facilities of the small computer are also available to control the experiment or the data acquisition process.

The greatest drawback to using a small computer is the difficulty in using the machine language for the instructions. Higher level languages are generally available but are not desirable for use on a small computer. The time and effort involved in programming a small computer can be substantially reduced by assembling and storing the machine language programs for the small computer on the large computer. The small machine is then free to handle experiment-related tasks, and it is not occupied with the time-consuming operation of programming. The large computer can do a better job of assembling programs since it operates faster and does not have severe core memory restrictions.

1. The Large Computer in Chemical Instrumentation

The choice of interface method is a function of the hardware, software, and operation of the large computer, the requirements of the experiment, and the amount of money available. A Teletype may serve as an inexpensive interface device, or a small computer terminal with its own digital plotter, CRT display, and mass storage device may be employed. The amount of subsequent application must be balanced against the initial installation labor.

The specific interface device constructed for the ESR experiment is described in detail next as an example of the construction and operation of an interface. A magnetic tape system was chosen for several reasons: The ESR experiment produces a large amount of data; the local computer center is only 50 yards away, and batch processing time is under an hour. The actual electronic equipment used is mentioned, but recent advances in electronics have been so rapid that equipment that is a few years old is already obsolete. Modern devices provide more capability for less money.

B. A Digital Tape System

The digital tape system depicted in Figs. 10 and 11 [2] automatically records analog voltage, frequency, or digital data onto computer-compatible magnetic tape. The system consists of a differential amplifier, a Matrix Corporation Model 1605 Voltage to Frequency Converter, two Systron-Donner Model 114 Counters, a Digi-Data Digital Stepping Recorder (DSR), a Digi-Data Model 1640 DSR Coupler, and the associated electronic interfacing. There are four different modes of data input into the system: a time base, binary coded decimal (BCD) information, a frequency signal, and an analog voltage.

The time base simultaneously controls the rate at which the data are sampled and recorded onto the magnetic tape. The time base used may be either internal or external. The internal time base uses the 60-Hz power line voltage as a basis. There are two switches which divide the 60 Hz frequency; one will divide by 1 or 10 and the other will divide by 2, 4, 8, or 16. Therefore, the possible internal time bases range from 60/2 (30) readings per second to 60/160 (0.375) readings per second. Alternatively, an external time base can be supplied to the system.

Data in binary coded decimal (BCD) form may be recorded directly onto the tape through the coupler. The coupler will handle up to 12 characters of data per sampling. The rate of recording onto the magnetic tape is 200 characters per second or less. Twelve-digit numbers may be recorded directly onto the tape through the coupler by means of the fixed data option. The numbers in the 12 registers on the front of the coupler are set and then written onto the magnetic tape.

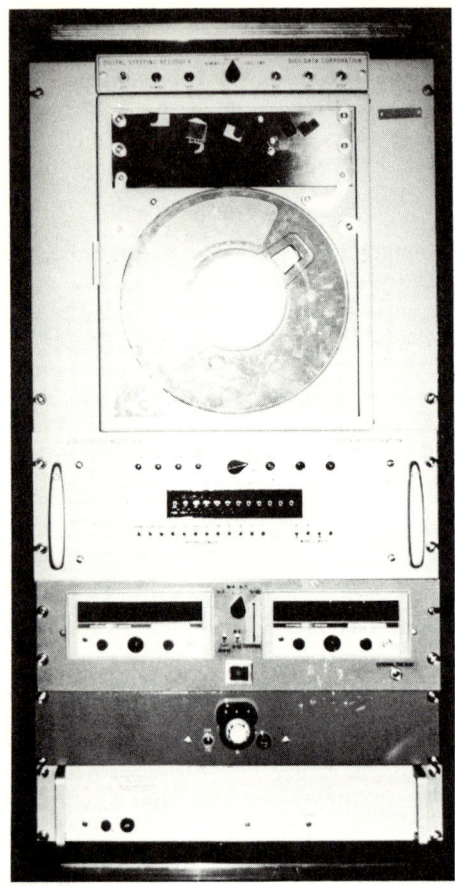

Fig. 10. A front view of the digital tape system.

A frequency signal or a series of pulses is accumulated and then stored by the counters during the interval determined by the time base. While one counter is accumulating data, the other counter displays a BCD signal corresponding to the number of counts previously registered for the coupler. When the time base gating signal is received, the counters exchange functions, and the one that was previously accumulating the frequency now displays the data in BCD form for the coupler while the other counter accumulates data. Therefore, no input data are lost while the data are being recorded onto the magnetic tape. There are four digits in the counters, so if the number accumulated is greater than four digits the information in the fifth and higher digits is lost.

1. The Large Computer in Chemical Instrumentation

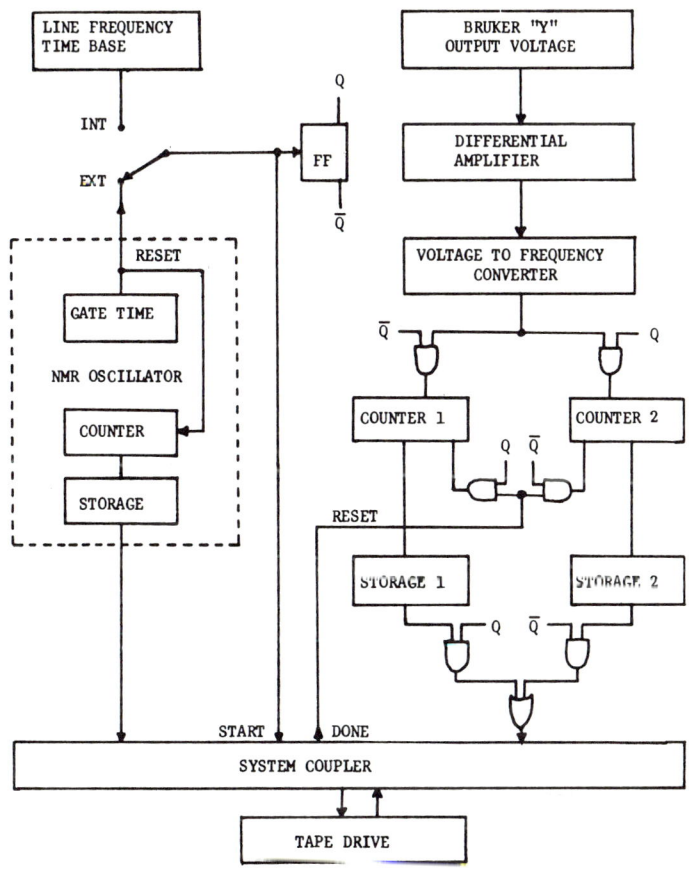

Fig. 11. Digital tape system schematic.

A voltage signal is processed by first passing it through a differential amplifier which first amplifies the signal by an appropriate scaling factor and then adds a sufficient bias voltage so that the signal is never negative. The positive voltage signal then goes into a voltage-to-frequency converter whose output is 1000 Hz for each volt of input up to maximum output of 10,000 Hz. Above this value, the voltage response becomes nonlinear. This voltage-proportional frequency output is then used as input to the counters. The amount of amplification and the correct added voltage are determined by the strength of the input signal, the 0-10 V range of the voltage-to-frequency converter, and the 9999 accumulation limit of the counters.

C. Gaussmeter Readings

A nuclear resonance oscillator is used in the Bruker spectrometer to indicate the magnetic field. The value of the magnetic field is directly proportional to the resonance signal from the standard sample in the probe according to the equation:

$$H = \frac{h\nu}{g\beta} = \frac{6.625 \times 10^{-27} \text{ erg-sec } \nu \text{ (sec}^{-1})}{(5.5855)(5.0504 \times 10^{-24} \text{ erg/gauss})} = \frac{10^{-3} \nu}{4(1.0644)} \text{ gauss.}$$

In order to convert the frequency from the probe into gauss, the value is first divided precisely by 4 and used as the input to a Hewlett-Packard Model 5216A Counter. An external frequency standard of 1.0644 MHz is then provided to the counter in place of the normal 1.0000 MHz internal standard of the counter. Therefore, the numerical value of the counter reading will be equal to the number of gauss times 1000 Hz.

The Model 5216A Counter has a storage register that stores the previous value while the accumulation register is recording the frequency. At the end of the gating period, the value from the accumulation register is transferred to the storage register and the accumulation register is reset to zero. No data can be accumulated during this transfer and reset procedure. The variable delay interval in the counter for the reset procedure was unsatisfactory, and the reset time was rewired to provide a reduced constant delay interval of 0.032 sec.

The pulse that begins the gating period of the counter is also used as the external time base gating signal for the digital tape system. In this procedure, the time interval for the accumulation of intensity data will be equal to the time interval used for the collection of gauss data plus the constant reset interval needed by the one display counter on the gaussmeter. A counter with a more rapid reset interval would be more desirable. The actual times available for gating periods are 0.13, 0.97, and 9.42 sec, so intensity data recorded at the smallest time interval will not accurately represent the actual intensity over the period used to record the magnetic field strength.

Only the output from the right-hand six digits of the counter is recorded on the tape, so the most significant digit must be added to the data by the processing program for the standard interval used in the experiment of 0.97 sec. No significant digits are lost with the 0.13-sec interval, and the two most significant digits are lost with the 9.42-sec interval.

D. Recording of Signal Intensity

The signal used for the digital tape system is taken directly from the "Y" signal of the Bruker spectrometer. The controls in the oscilloscope and the chart recorder are bypassed. This output signal is an analog voltage (± 200 mV maximum) which is proportional to the intensity of the ESR derivative signal. The voltage is used as the input to the differential amplifier of the digital tape system.

One problem in using the digital tape system is that the numerical value for the intensity is arbitrary and is determined by the controls on the differential amplifier and the nature of the voltage signal input. This numerical value may only be an integer number between 0 and 9999. In order to minimize the errors in the rounding off of the spectrum, the difference in the intensity readings should be as high as possible. For recording first-derivative ESR spectra, the optimal situation would be to have the base line at 5000, the minimum in the spectrum just over 0, and the maximum in the spectrum just under 9999. Such an extreme difference in readings is not normally necessary or desirable, since the limits of the counters may be exceeded.

When recording ESR spectra using the 0.1-sec gate time, the limiting component is the voltage-to-frequency converter, which is only capable of 10,000 Hz. The maximum number of counts possible in the actual time interval of 0.13 sec is 1300, and therefore the correct choice for a base line would be 650.

The differential amplifier in the digital tape system first amplifies the -200 to $+200$ mV signal to -5 to $+5$ V and then adds a constant bias voltage of $+5$ V. The amount of amplification and the bias voltage are variable for use on other instruments. The resultant 0-10 V signal is then used as input to the voltage-to-frequency converter. The flip-flop switch (labeled FF in Fig. 11) determines which of the counters monitors the frequency output and which counter is scanned by the system coupler. After the recording process is finished, a signal comes from the coupler which resets the accumulator section of one of the counters to zero in preparation for the next input frequency signal.

E. Magnetic Tape Format

The Digi-Data Digital Stepping Recorder and Coupler writes $\frac{1}{2}$ in., seven-channel, even-parity magnetic tape at 556 characters (bytes) per inch. The format in which the tape is written is described by the following series of terms:

Bit: A tape is divided into seven channels across the tape containing magnetic particles oriented in random fashion. The magnetic tape unit normally orients all the particles in the same direction (0 polarity). A bit is a location on the tape which is oriented in the opposite direction to the rest of the particles (reversed or 1 polarity).

Character: Each character consists of the seven locations in the channels at a given longitudinal position on the magnetic tape. There are six channels (1, 2, 4, 8, A, B) which are used to record the data, and a parity channel (C) which is set (in even parity) such that the sum of reversed polarity (1) bits in the character is even. The A and the B channels are not needed to write numbers on the tape, and thus only four on (1)-off(0) signals are necessary for each number.

Word: A word is composed of a number of characters written sequentially on a magnetic tape. The word length may be adjusted to be 1-12 characters in length. The orientation of each bit in a sample ESR word is given in Table 1.

TABLE 1

An ESR Data Point Recorded on Magnetic Tape

Y reading = 5016   Gauss = 3294.738

Channel	\multicolumn{10}{c}{Character number}
	1	2	3	4	5	6	7	8	9	10
Parity	0	0	1	0	1	0	1	1	0	1
B	0	0	0	0	0	0	0	0	0	0
A	0	0	0	0	0	0	0	0	0	0
8	0	1	0	0	0	1	0	0	0	1
4	1	0	0	1	0	0	1	1	0	0
2	0	1	0	1	1	0	0	1	1	0
1	1	0	1	0	0	1	0	1	1	0
	5	0	1	6	2	9	4	7	3	8

1. The Large Computer in Chemical Instrumentation

Record: A record consists of any number of words up to 4095 written sequentially with no separation on the magnetic tape. Each record is separated from the next record by 3/4 in. of 0 polarity magnetic tape containing a longitudinal check character. The bits in this character are set so that the sum of all the bits in the channel in the previous record is even. The recorder requires 0.3 sec to write the interrecord gap.

File: A file is composed of any number of records written on the magnetic tape. Files are separated from each other by 3-3/4 in. of 0 polarity magnetic tape containing a longitudinal check character and the end-of-file character which is a bit in the 1, 2, 4, and 8 channels. The tape drives on the large computer expect that a file marker will be $7\frac{1}{2}$ in. of 0 polarity tape, so two consecutive end-of-file markers must be used.

Tape: A tape may have any number of files written on it subject to the limitation of the length of the tape. The tape has a reflective marker near the beginning which is used as the indicator to the tape unit to begin writing or reading from the tape. The first datum will be written at a point 3.4 in. from this marker. The tape also has a reflective marker at the end which is used to prevent the tape from running off the reel.

ESR spectra are written on the magnetic tape as ten-character words. When the coupler is given the write command, the 40 wires containing the BCD information are scanned for a negative potential. If a sufficiently negative potential is present, a bit will be written in the corresponding location on the tape. Four characters from the Systron-Donner counters ("Y" intensity) and six digits from the gaussmeter are recorded. When the number of words recorded reaches the preset record length, or the experiment ends, the coupler writes an interrecord gap on the tape. This gap serves as a marker to separate one spectrum from the next. One end-of-file mark written on the tape may be used to separate data. Two end-of-file marks signify the end of information on the tape.

IV. ESR DATA HANDLING AND ANALYSIS

Once the data are in digital form and on a file accessible by the large computer, the data are read and processed according to the instructions written in the form of a computer program. Additional information on how the experimental data are to be processed may be given to the program at the time of execution.

The first step performed by the computer should be to read the data and produce output showing that the data have been recorded correctly. The next step should be to store the data in a computer-readable mass storage device. The information from the experiment is thereby made available for future use should unforeseen eventualities occur. An error may be later found in the program, new data analysis procedures may be developed, or a different form of graphic display may be necessary for publication or slides.

This section describes the program used for the processing of ESR data and the results obtained with that program on the spectra shown in Figs. 1-8. The next section discusses more general procedures for handling data from instrumentation.

A. General Description

Program DIZZY, a FORTRAN IV program [3], processes the data from an ESR spectrum which is recorded on digital magnetic tape or cards. The program is specifically designed for ESR spectra with only one g value, one line width, and hyperfine splitting from any number of nuclei with any spin (such as solution spectra of organic free radicals). A least-squares curve-fitting procedure may be used to analyze the ESR spectrum in order to obtain accurately the relevant spectroscopic constants. The theoretical spectrum may then be compared to the experimental spectrum to verify the results.

The following operations may be performed by program DIZZY:

1. Print a listing of the readings recorded from the gaussmeter and the counters.

2. Store the spectrum numerically in binary format on another magnetic tape.

3. Punch the spectrum onto cards.

4. Plot the experimental spectrum on a CALCOMP plotter.

5. Modify the spectrum for the least-squares procedure by reducing the number of points used or averaging points.

6. Compute the theoretical line spectrum from the coupling constants and nuclear spins, and then compute a simulated first-derivative spectrum by applying a Lorentzian line-shape function to each line.

7. Fit the simulated spectrum and the experimental spectrum using a nonlinear least-squares iterative procedure to modify the theoretical constants.

1. The Large Computer in Chemical Instrumentation

8. Plot the simulated spectrum and the difference between the simulated and experimental spectra at each iteration.

9. Simulate a spectrum over a given number of points by taking a line-shape function for one line, and computing the splitting from each group of nuclei for each point.

B. Theoretical Description

The first-order solution of the ESR spin Hamiltonian for one g value and N nuclei gives $\prod_i (2I_i + 1)$ lines at the positions:

$$H_K = H_o - \sum_{i=1}^{n} a_i M_{I_i}$$

a_i = hyperfine splitting constant for nucleus i,

M_{I_i} = spin quantum number for nucleus i.

The program assigns a Lorentzian line shape to each line in the spectrum. Accidental degeneracy is ignored.

$$F_K(H) = \frac{A\tau}{\pi((H_K - H)^2 + \tau^2)}$$

A = total integrated intensity of absorption line,

τ = half-width at half-height of absorption line.

The spectrum with K lines is recorded as the first derivative of the absorption, so the value F for the intensity at H gauss is given by the equation:

$$F(H) = \frac{2A\tau}{\pi} \sum_K R_K \frac{H_K - H}{((H_K - H)^2 + \tau^2)^2} + \alpha + \beta H$$

R_K = ratio of the intensity of line K to the total intensity A,

α = base-line shift from an arbitrary zero value,

β = base-line drift (assuming a linear drift of the derivative curve with respect to the field).

Therefore, the parameters α, β, A, τ, H_o, and a_i (the hyperfine splitting constant for the ith group of nuclei with spin M_I) describe the theoretical spectrum, and the partial derivatives with respect to each of these parameters at a given number of points in the spectrum are computed.

For each point ℓ in the spectrum, a nonlinear first-order Taylor series with respect to the parameters may be written:

$$F_\ell^{CALC} - F_\ell^{OBS} = \sum_{j=1}^{m} \frac{\partial F_\ell}{\partial X_j} \Delta X_j$$

X_j = parameters describing the value for F_ℓ^{CALC},

m = number of parameters X.

If a unit weighting factor is applied to all the points ℓ in the spectrum, the above equation may be rewritten in matrix notation:

$$\Delta \underline{F} = D \Delta \underline{X}.$$

This equation may be solved directly for $\Delta \underline{X}$:

$$\Delta \underline{X} = (D^T D)^{-1} D^T \Delta \underline{F}.$$

Therefore, if an initial estimate is given for the parameters X, a calculated spectrum and the associated derivatives may be computed, the normal equation solved, and the corrections for the parameters found. Since the function F(H) is not first order in the parameters X, the procedure must be reiterated until the sum of the squares of the difference at each point between the observed and calculated intensity values becomes stable. This procedure may give a local minimum in parameters X instead of the best possible solution; the computed spectrum should be visually compared with the experimental spectrum to verify the results. If the values for the input parameters are changed, a different solution for the parameters may result.

C. Subroutines

The following subroutines are used by program DIZZY. If a major portion of a subroutine was obtained from another source, the source is indicated.

1. The Large Computer in Chemical Instrumentation

REDIN   The input parameters from cards 1-5 are read and parameters are inserted in the case of blank or zero numbers. The parameters are arranged in common storage for use by the program.

CHECK   The data are read from the tape from the digital recording system (TAPE4), the binary storage tape (TAPE5), or cards. The data are processed into internal floating point code in intensity and gauss units.

POST   The binary storage tape (TAPE5) is positioned at the end-of-information on the tape for writing a new spectrum or at the beginning of a given file for reading a previously stored spectrum.

HCALIB   The intensity readings from the spectrum are assigned positions in gauss according to any spectrum previously recorded on the binary storage tape.

MPOLY   A polynomial least-squares curve-fitting procedure is performed on a series of (x, y) points.

TAPRED   The points from the spectrum may be averaged. An initial estimate for the parameters X(M) is made. The gauss and intensity readings are printed and plotted. The number of points in the spectrum is reduced for the least-squares fit.

XVALUE   An initial set of X parameters is read, or the set of X parameters used in the least-squares iteration is punched onto cards.

RATIO   The program calculates the theoretical ratio of the intensity of each hyperfine line in the spectrum to the total intensity [4].

LINFOR   The program calculates the position of each hyperfine line with respect to the center. The theoretical intensity and the partial derivatives with respect to the parameters X(M) at each experimental point used in the least-squares fit are computed and stored [4].

MATINV   The derivative matrix is inverted, and the linear equations are solved for the changes in the parameters X(M) [4].

LSFIT   The least-squares fit is organized, the theoretical spectrum is printed, the error analysis and output parameters are printed, and the decision to execute another iteration is made.

PEGLEG   A matrix is printed.

TME   The central processor time used is printed.

WRIPLO   The spectrum and information describing the spectrum are
          plotted. The difference between the theoretical and experi-
          mental spectra may be plotted.

SESRS   The full spectrum is simulated from the coupling constants and
          the line width provided that no nuclei with spin greater than one
          are present. A Lorentzian line shape is assumed, and the
          intensity contribution from each group of nuclei is calculated at
          a series of equally spaced points [5].

A flow chart (Fig. 12) shows the interrelation of the main subroutines in
the program.

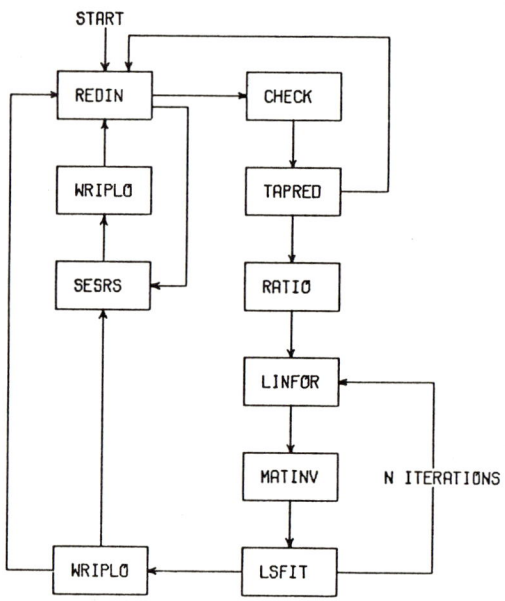

Fig. 12. DIZZY flow chart.

1. The Large Computer in Chemical Instrumentation   27

D. Program Usage

The program uses four tapes:

1. TAPE4 — the input tape from the digital tape system.

2. TAPE5 — the storage tape for all ESR spectra.

3. TAPE6 — the location of the cards punched from the data.

4. TAPE99 — the location of the plot file produced by the program.

Each of these tapes must be requested separately if they are to be used by the program.

The input spectrum is a scan of the ESR spectrum from low to high field. The data may be recorded by the digital tape system on TAPE4 or punched on cards. If the program reads the spectrum from TAPE4, it will be stored on TAPE5 and punched on TAPE6. If the program reads the spectrum from cards, it will be stored on TAPE5, which is a permanent storage tape for the data. Each spectrum on the tape is identified by its file number for convenient access by the program. The plots produced by the program are placed on TAPE99.

If storage on TAPE5, cards, or plots is not wanted, then the appropriate card should be omitted from the job control language sequence.

E. Input Cards

The following input cards are needed to control the execution of the program: Card 1 is a title card. Card 2 dictates the size of the plot. Cards 3-5 contain the initial guesses for the parameters of the spectrum. Card 6 controls the source of the data and the reduction of the data for use in simulation. Cards 7 and 8 are punched by the program after each iteration of the least-squares analysis. They provide a useful record of the parameters used by the least-squares fit and may be used to change the parameters used by the program.

An attempt has been made to minimize the amount of data needed for input. If an input parameter has been left blank, the program assigns a reasonable value based on the nature of the spectrum.

A plot of one experimental spectrum may be obtained with four blank cards.

For a least-squares analysis of the spectrum of biphenyl, a sample set of input data is the following:

Card 1:   BIPHENYL RADICAL ANION

Card 2:   Columns 11-20, 2.0 (gauss per inch)

Card 3:   Column 5, 3 (sets of equivalent nuclei)

Card 4:   Column 1, 4; Columns 2-10, 2.70 (coupling in gauss)
          Column 11, 4; Columns 12-20, 0.45 (coupling in gauss)
          Column 21, 2; Columns 22-30, 5.40 (coupling in gauss)

Card 6:   blank

Card 7:   blank (cards 5 and 8 are not needed)

The program will punch a new card 6 giving the spectrum location on TAPE5 and the parameters used for the data reduction.

The blank columns of the sample set of input cards instruct the program to choose the standard substitution for this input data. The only input needed is the structure of the molecule in terms of number of equivalent sets of nuclei, spin of the nuclei, the number of nuclei in a set, and an approximate guess for the coupling constant for the set. If the ESR spectrum does not contain a large central peak, then a guess for the line width must be supplied. The entire spectrum need not be used for simulation and provisions are available to select any portion of the spectrum and average or skip points if desired.

F. Results from ESR Spectra Analysis

The program alters the spectral parameters by a least-squares process in an attempt to find the minimum difference between the observed and calculated spectra. The analysis is said to have converged or diverged depending on whether or not a minimum was found. The parameter refinement diverged if the spectrum was asymmetric or if the initial estimates for the spectral parameters were insufficiently close to the correct values. As in most least-squares processes, local minima may be occasionally found. They may be avoided by choosing different values for the input parameters.

The spectral parameters used to define the spectrum are the base-line shift, the base-line drift, the total intensity, the absorption line half-width, the magnetic field at the center of the spectrum, and the hyperfine splitting constants for each group of nuclei. The initial approximations were refined for most spectra by the program within six iterations to give parameters for a theoretical spectrum which closely approximated the experimental spectrum. When the two spectra were overlaid and aligned, they were virtually indistinguishable. The error graph of the difference between the two spectra revealed any deviations. Often the error graph had an appearance similar to the inherent noise level of the spectrum. The error graph tended to be much larger than the noise level for those spectra which had a very small line width and therefore had sharply resolved lines.

The inherent resolution of this procedure was checked frequently with a dilute sample of perylene dissolved in concentrated sulfuric acid. The ESR spectrum of the radical cation was recorded at room temperature.

Perylene

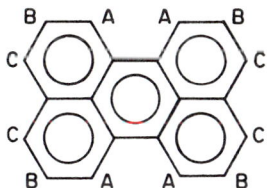

Three sets (labeled A, B, and C) of four equivalent protons couple with the unpaired electron to give a hyperfine splitting pattern of 125 lines. The analysis of this spectrum proceeded to give a calculated line width which varied between 0.04 and 0.05 G and three calculated coupling constants of 4.049, 3.049, and 0.443 G.

The standard deviation calculated from the least-squares process was usually much less than the deviation observed among consecutive spectra of the same sample. Thus, instrumental and sample variations are more important in determining the error than is the random scattering in a particular experiment. For perylene the standard deviation, on the basis of six spectra, for each of the coupling constants was calculated to be about 0.0015 G.

The calculated coupling constants for a series of 4,4'-dialkylbiphenylide radical anions at different temperatures are given in Table 2. The actual spectra at 222° K are shown in Figs 1-8. The table also shows the calculated line width and a column labeled RERR. The relative error

TABLE 2

Coupling Constants for 4,4'-Dialkylbiphenylides

Alkyl	Temperature (°K)	Proton Coupling (G)				Coupling (G) α-^{13}C	Line width	RERR[a]
		2-	3-	β-	γ-			
CD$_3$	192	2.747	0.464	0.871[b]	—	—	0.130	6.27
	222	2.753	0.493	0.857[b]	—	—	0.153	9.00
CH$_3$	192	2.753	0.481	5.764	—	—	0.106	5.32
	222	2.755	0.504	5.724	—	—	0.214	3.05
^{13}CH$_3$ (50% ^{13}C)	192	2.750	0.496	5.758	—	3.581	0.129	4.37
	222	2.753	0.505	5.748	—	3.554	0.129	4.62
C$_2$H$_5$	192	2.730	0.457	3.886	0.052	—	0.067	3.73
	222	2.732	0.475	3.973	0.050	—	0.067	2.81
iso-C$_3$H$_7$	177	2.769	0.452	2.284	0.108	—	0.114	5.38
	192	2.732	0.473	2.393	0.098	—	0.116	4.32
	207	2.739	0.478	2.481	0.092	—	0.095	5.02
	222	2.737	0.480	2.547	0.086	—	0.093	4.70
	237	2.734	0.487	2.611	0.082	—	0.095	3.39
n-C$_4$H$_9$	192	2.719	0.464	3.608	0.135	—	0.260	2.94
	222	2.710	0.471	3.712	0.134	—	0.201	5.40

sec-C_4H_9	177	2.76	0.41	1.20	(0.2)	—	(0.36)	8.04
	192	2.757	0.434	1.403	0.153	—	0.348	4.83
	207	2.747	0.454	1.516	0.144	—	0.245	3.51
	222	2.743	0.460	1.593	0.138	—	0.221	3.76
	237	2.740	0.475	1.665	0.130	—	0.196	8.02
	252	2.737	0.477	1.683	0.127	—	0.197	5.51
tert-C_4H_9	192	2.746	0.459	—	0.107	—	0.110	4.78
	222	2.732	0.481	—	0.102	—	0.091	5.56

^{a} REFR = $\left\{\left[100\left(\sum_{i=1}^{N}(\text{Observed}_i - \text{Calculated}_i)/N \text{ points}\right)^{\frac{1}{2}}\right]/\text{Maximum observed value}\right\}$ for the intensities of points every 0.033 G from the beginning to the end of the spectrum.

^{b} Deuterium coupling constant.

(RERR) is a number which determines the closeness of the fit of the calculated and observed spectra. The relative error is smaller for spectra with less base-line noise and more poorly resolved lines. The coupling constants greater than 0.3 G had calculated standard deviations of 0.0003 to 0.002 G. This is a smaller error than the errors expected from possible variations in temperature of $\pm 2°$K and the variations in the concentration of a sample compound.

This method of analysis can determine the coupling constants accurately from poorly resolved spectra as well as from complex spectra with distinct lines. The computer process can also distinguish the line width from a coupling constant of similar magnitude because the entire line shape is examined and not just the position of the line. Small variations in the coupling constants with a temperature change are accurately measured. The total intensity of the radical present is calculated directly in the procedure rather than calculating a double integral of the derivative spectrum. All the parameters are determined more accurately with the use of the computer.

The coupling constants found were used to hypothesize the mechanism of hyperfine splitting in radicals with planar structures. The alkyl proton coupling constants were explained by using a hindered rotation model for the alkyl carbon-carbon bonds and the alkyl bond to the aromatic ring. With this model, the coupling constants are a function of the orientation of the protons and the barrier to internal rotation of the alkyl bonds.

V. DATA PROCESSING — THE COMPUTER PROGRAM

Once experimental data have been converted to digital form and recorded on a large computer-compatible device, all the advantages of a large computer can be utilized to make the data handling and processing task simpler. Some general characteristics of a large computer are listed in Table 3. The positive features are found on all large computers, whereas the negative aspects of the large computer are very installation dependent. Local conditions are what make the extensive use of a large computer for experimental data processing either desirable or undesirable.

The ability to program large computers in high-level compiler languages reduces the programming effort involved in data processing. The large number of routines (especially in the FORTRAN IV language) that have been previously written can be adapted for many basic data analysis procedures. Programs written in compiler languages are mostly independent of the machine being used so they can be readily transferred from one installation to another.

TABLE 3

Large Computer Characteristics

(+)	Core storage size
(+)	Mass storage devices
(+)	File manipulation
(+)	Floating point arithmetic
(+)	High-level languages
(+)	Knowledge pool
(+)	Fast output devices
(−)	Periodic inoperation
(−)	Batch processing
(−)	Fixed monitor system
(−)	Priority time delay
(−)	Usage cost
(−)	Operator errors
(−)	Output manipulation

A. The Experimenter

The computer simply does the mechanical operations necessary to produce data in a more meaningful form for creative thought. The computer program designed to analyze the experimental data may be directed toward finding specific numerical values as in the ESR example, or it may assist the experimenter to refine and evaluate the data. A great deal of work that was previously involved in modifying the experimental results can be thereby eliminated, and minor corrections to the data that were previously ignored can be easily handled by the computer. The accuracy and versatility of the experiment are therefore increased.

No capability is worthwhile unless it can be utilized by the experimenter. Programs should be written so the computations that they perform are intelligible later to the person who is using the program. The operations performed on the data should be clearly defined so erroneous conclusions are not made. One of the greatest advantages of a digital computer is the flexibility available for data processing. If an experiment is performed that requires a new kind of data processing or examination, the program should be capable of being easily modified to accommodate this situation.

The programs to process data from an experiment will probably be used by persons who have no interest in computers and have no desire to learn. Often, these people are neglected by the computer enthusiasts who design the program. A large number of computer operations can be executed with a minimal number of clearly defined instructions. The inertia of the reluctant experimenter can be overcome by such initially simple computer usage.

The computer processing advantage is lost when the time required to program and use the computer is greater than the time necessary to perform the mechanical operations by other methods. One main goal in using the computer should be to spend as little time as possible actually using the computer.

B. Program Objectives

The facilities available with a large computer are often poorly utilized by data processing programs. The computer is often just used as a "number cruncher" to calculate the desired results. Even the calculation procedures in the program are sometimes inadequate. It is incongruous to see someone with computer output sitting at a desk calculator making further computations. Many errors are made because the output numbers have been inadequately labeled. The other capabilities are ignored or misused by the computer program.

The experimental data are conveniently stored in the computer on tape, disk, or drum. A library of all the experimental data is then available for future calculations or comparisons. Data need not be stored in a formatted code since the numbers can be transferred directly into storage from core memory. A routine may be written that automatically stores and retrieves the data upon command. Other routines may search the stored data and compare various features of a spectrum to similar features on spectra stored in the library.

The graphic display devices on a large computer provide a variety of methods to display the experimental data. A digital plotter produces graphs in two or simulated three dimensions. This output can be designed and labeled so it is suitable for publication or a notebook. Slide preparation effort and expense can be reduced by having the computer display the data and appropriate titles in a form immediately suitable for photography. Different forms of graphical display can often help illuminate the problems involved in data interpretation. The display devices for different computers do not always use the same instructions so a program written for one device may have to be rewritten to conform to the standards of a different device. Programs from other computers often have incomprehensible display statements. The documentation for such statements should be included with the program.

Input requirements are often excessively complicated or unnecessary. Many input numbers can be calculated by the program and need not be given. A desirable compromise is to have places for input parameters and have the computer use standard values if no input numbers are used. A description of the input parameters and the values used by the program should be presented on the output to check for input errors and for reference. Possible data or input errors that terminate the program abnormally should be anticipated and the appropriate diagnostics issued by the program. Nothing is more frustrating to a computer novice than to use a program and have the computer respond with ERROR or DUMP followed by an indecipherable maze of numbers. Coherent criticism of specific input errors is more welcome. Programs which describe every step clearly and outline the available choices are most desirable.

The structure of the program often inhibits modifications and increases computer costs. Most data processing programs are suitable for an overlay form where the data array and the control information are retained in core storage while the subroutines needed are transferred in and out of core storage from disk storage. The amount of core needed is therefore substantially reduced. Data sets may also be temporarily stored on disk. Disk files should be handled internally by the program to avoid complicated job control language. If the job control language is standardized for all jobs with a given program, a great deal of frustration in dealing with control errors can be avoided.

C. Modular Construction

The experimental response from most instruments consists of a series of peaks on some noise. The kinds of data processing necessary to analyze a spectrum of this nature are similar irregardless of the units of the X and Y axes. In order to save programming time and effort, the processes involved in analyzing these spectra can be written as modular routines which are then assembled together to provide the required computer operations for the data. Such modular construction of the program takes advantage of already proven calculation procedures and data handling routines.

An extensive list of computer operations divided into seven basic categories is given in Table 4. The routines on the list are examples of operations that can be performed by modular programs. Of course, not all of these routines will be necessary to process the data from an instrument. The choice of which routines will be used depends on the experiment.

D. A Spectroscopy Program — RAMAN

The RAMAN spectroscopy program [6] is designed to process and analyze Raman spectroscopic data which have been recorded in digital form. The program is written to be used routinely and to conform to the objectives previously mentioned.

The general structure of the program is shown in Fig. 13. The processing sequence of the data is controlled by a series of numbers on the program control card. This control card also contains the location of the spectroscopic data set and a title. The operations and their respective code numbers are listed in Table 5. The data are first read from the location given on the control card which may be either a tape from the Raman spectrometer or a permanent storage tape in the computer. The data are then processed by the modular routines available according to the sequence given on the control card. For example, if the control operations are listed as "1587," the data will be stored on the permanent storage tape, smoothed, the peaks listed, and the spectrum plotted on the digital plotter. Each operation requires an additional control card which allows for various options in the processing routines. For the operations "1587" given, these cards may be blank.

The last ten characters of the title are used by the program to indicate operations that modify the data. For example, if the data are smoothed, a "5" will be added to the title.

Data may be stored temporarily on five files which may be either tape or disk storage. These files are used to transfer more than the one set of data that can remain in core storage. These files can also be used to store data while the experimenter decides on further modifications.

A description of each of the subprograms in RAMAN data processing follows. The choice of which modular routines are implemented in the program depends on the availability of theoretical procedures to interpret the

TABLE 4

Modular Data Processing Routines

Category	Routine
Storage	Data storage Information retrieval Search technique Comparison
Graphics	Digital plotting CRT display Printed plot Printed output
Correction	Axis unit modification Base-line shift Base-line drift Calibration Normalization Scaling of data
Improvement	Point averaging Time averaging Background reduction Lagrangian interpolation Digital filtering Line sharpening Digital smoothing Summation of spectra Difference spectra
Transformation	Derivative spectrum Integral spectrum Convolution Deconvolution Fourier transform Integral transforms
Calculation	Noise level Peak location Peak integration Spectrum integration Line following Curve fitting
Analysis	Least-squares fitting Direct search minimization Simulation Parameter improvement Normal equation solution Band-shape analysis

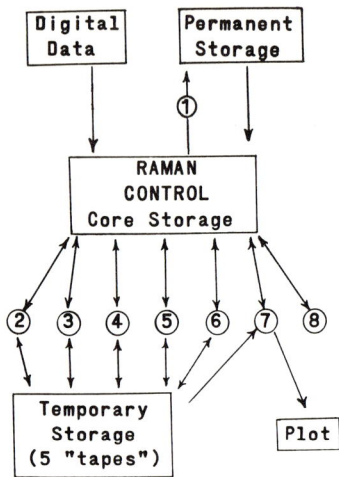

Fig. 13. RAMAN data processing.

TABLE 5

The RAMAN Program

Code	Subprogram	Function
—	READY	Decode digital data
1	—	Store spectral data
2	CORRECT	Base-line correction
3	AVERAGE	Ordinate averaging
4	ADDSUB	Add or subtract spectra
5	SMOOTH	Smoothing of spectra
6	FIT	Curve fitting
7	RMNPLT	Plotting
8	PEAK	Peak finding

data. All of the subprograms can produce listings of the data and other output to describe their operation and results.

READY: The data from the experiment are recorded on the magnetic tape system described in Section III. Numbers describing the parameters of the data are first recorded on the tape by means of the fixed data option, and then the spectrum is recorded from a stepping motor drive on the X axis and a photomultiplier counter on the Y axis. The subprogram reads the data from the magnetic tape and decodes them into internal computer number representation. Vacuum and photomultiplier corrections are made if desired, and the units of the X and Y axes can be changed into more convenient ones for data handling. The usual units chosen are cm^{-1} for the X axis and counts for the Y axis.

CORRECT: This subprogram computes a base line for the spectrum and then computes an approximation to the Rayleigh curve of a Raman spectrum and subtracts it from the spectrum. The Rayleigh curve is assumed to be a product function that is partly pure exponential and partly Gaussian. The parameters for the product function are found by a least-squares fit of the regions of the Rayleigh curve that are free of peaks. The regions of the spectrum to be used in calculating the base line and the Rayleigh curve are selected by the experimenter.

AVERAGE: The spectrum as recorded may contain an excessive number of data points. The number of points can be reduced by taking a weighted average of the ordinate (Y) values on either side of a given point and placing the average in the array for that point. The results can also be normalized. The number of points used in the averaging process and the number of points retained by the program are two separate input parameters, so the program may also be used as a simple smoothing routine.

ADDSUB: This subprogram adds or subtracts any number of spectra to or from each other to produce a composite spectrum. Each ordinate in a particular set of data is multiplied by an input factor and added to the corresponding ordinate of the previous set of data. Only the portions of the spectra where the abscissa values overlap within a given tolerance are calculated. For convenience, this subprogram can also read data into core from the permanent storage tape. Some applications are signal averaging, base-line correction, and the creation of a spectrum arising from only the isotopic polarizability derivative tensor elements.

SMOOTH [7]: The spectrum is smoothed by either a five- or nine-point smoothing function. For each point on the spectrum, that point and the two or four points on either side are used to produce a smoothed point.

The smoothing cycle can be repeated as often as specified. Each cycle will eliminate the end points of the spectrum unless the program is instructed to retain the unsmoothed end points in the data array. The subprogram may also be used to differentiate the spectrum to obtain first- and second-derivative spectra. The smoothing routine is able to remove spurious spikes that occur in the data.

FIT [8]: This subprogram was originally designed to fit a combination of Cauchy and Gaussian functions to the envelope of an infrared spectrum. In order to adapt the program to Raman data, the ordinate values from the Raman data are converted to the equivalence of transmittance for calculations and then converted back to the original units at the conclusion of the computation. The parameters for the composite function are first estimated from the spectrum, and an iteration process is used by the program to minimize the sum of the squared residuals. The output consists of the computed spectrum of the composite fitted curve and the spectrum of each of the component curves. The areas of each of the component curves are also calculated. Intermediate values for the function parameters may be placed in temporary storage for later modification.

RMNPLT: This subprogram provides the graphical output for the data on a CALCOMP plotter. A complete plot of the data can be obtained with no extra input information, or the nature of the plot may be selected by modifying the variables used by the plotting routine. Title information for the axes and the graph may be given to the routine. The plot may be either a dashed or solid line, and the dimensions and axis increments may be altered. An arbitrary number of data sets may be plotted on the same graph. A portion of a long spectrum may be enlarged in size and plotted individually. A great deal of versatility is therefore available to improve the visual presentation of the data.

PEAK [9, 10]: The subprogram locates the positions and intensities of the peaks and valleys in a spectrum. The noise level of the spectrum is calculated, and then the slope of the curve is evaluated and inflection points are noted. A cubic least-squares function is then fit to either nine or fifteen points in the region of each inflection point, and the noise rejection criterion evaluated. The calculated positions and intensities of the inflection points from the cubic functions are then listed.

VI. CONCLUSION

Future improvements in hardware and software will facilitate the use of the large computer as an analysis tool for instrumentation. More instruments are being designed with internal digital circuitry and digital output of results. Computer terminals are becoming more versatile and

1. The Large Computer in Chemical Instrumentation

much lower in cost. The software for the interface of many small terminals to a large computer is improving rapidly. The library of programs to analyze instrumental data is constantly increasing. Once all these advances are put together in a unified system, the large computer will no longer be an isolated entity but will become an integral part of the experiment.

REFERENCES

[1]. L. W. Bush, Ph.D. Thesis, Northwestern University, Evanston, Illinois, 1967.

[2]. Constructed by R. J. Loyd, Chemistry Department, Northwestern University, Evanston, Illinois.

[3]. L. G. Newman, Ph.D. Thesis, Northwestern University, Evanston, Illinois, 1971.

[4]. A. Bauder and R. J. Myers, J. Mol. Spectrosc., 27; 110 (1968).

[5]. E. W. Stone and A. H. Maki, J. Chem. Phys., 38; 1999 (1963).

[6]. W. S. Davis, L. G. Newman, and D. F. Shriver, private communication, Northwestern University, Evanston, Illinois, 1971.

[7]. R. N. Jones, et al., Program PC-132, Bulletin 11, National Research Council of Canada, 1968.

[8]. J. Pitha and R. N. Jones, Program PC-116, Bulletin 11, National Research Council of Canada, 1968.

[9]. A. Savitsky, Anal. Chem., 33; 25A (1961).

[10]. R. N. Jones, et al., Program PC-180, Bulletin 11, National Research Council of Canada, 1968.

Chapter 2

THE CALCULATION OF THE OPTICAL PROPERTIES OF THIN METAL
FILMS FROM INTERNAL REFLECTANCE DATA

Eric N. C. Randall and Harry B. Mark, Jr.

Department of Chemistry
University of Cincinnati
Cincinnati, Ohio 45221

I.	INTRODUCTION	44
II.	THE PRINCIPLES AND TERMINOLOGY OF INTERNAL REFLECTION SPECTROSCOPY	47
	A. Perpendicular Polarization	56
	B. Parallel Polarization	56
III.	THIN FILMS AND THIN-FILM OPTICS	57
IV.	EXPERIMENTAL APPARATUS AND TECHNIQUES	60
V.	METHOD FOR DETERMINATION OF FILM THICKNESS AND OPTICAL CONSTANTS OF THIN FILMS	64
	A. Introduction	64
	B. Data Acquisition	66
	C. Data Reduction	67
	D. Error Analysis	78
	E. Internal Self-Consistency of Results	81
VI.	DISCUSSION AND SUMMARY	82
	APPENDIX I. SYMBOLS AND ABBREVIATIONS	86
	APPENDIX II. THE CALCULATION OF THE REFLECTION ABSORBANCES OF AN n-PHASE SYSTEM	88

Copyright © 1973 by Marcel Dekker, Inc. All Rights Reserved. Neither this work nor any part may be reproduced or transmitted in any form or by any means, electronic or mechanical, including photocopying, microfilming, and recording, or by any information storage and retrieval system, without permission in writing from the publisher.

APPENDIX III. A FORTRAN IV LANGUAGE COM-
PUTER PROGRAM FOR THE CALCU-
LATION OF REFLECTION ABSORB-
ANCES OF AN n-PHASE SYSTEM . . .   90

APPENDIX IV. A FORTRAN IV LANGUAGE COM-
PUTER PROGRAM FOR THE CAL-
CULATION OF THE OPTICAL PROP-
ERTIES OF THIN METAL FILMS
FROM IRS MEASUREMENTS   95

REFERENCES   101

I. INTRODUCTION

The interest in obtaining the optical constants of thin films in general has been a long-standing problem owing to the extensive optical uses of thin films in other applications, such as the beam splitters of interferometers [1], protective lens coatings [2], near infrared photoconductors [3], multilayer filters [4], cathode-ray tube screen coatings [5], protective coatings for hygroscopic optical components of infrared spectrometers [6], optically transparent electrodes [7-15], and so on. Numerous methods, using mechanical, electrical, and optical techniques, have been employed to determine the optical constants (i.e., the refractive index n and extinction coefficient k) and the thickness of thin films h. However, the restrictions and disadvantages inherent in these methods make them unsuitable for general application.

The methods employing mechanical techniques include gravimetry and molecular ray calculations, which are used for the determination of film thickness [16]. Gravimetry, or direct weighing, has several major disadvantages. Foremost among these is that it is a destructive technique. Another objection is that even when accurate measurements have been made, bulk density of the film material must be assumed to pertain when calculating the film thickness. This assumption is unwarranted, especially for films of thickness less than 150 Å [17]. The molecular ray method is also based on knowledge of the mass of the film, but in this case the measurements are made during the evaporation process. When the film is prepared under molecular streaming conditions, knowledge of experimental conditions such as the film material vapor pressure, the source temperature, the time of evaporation, and so on, can be used to calculate the mass of the film if it is assumed that all incident atoms condense at the substrate. Bulk density must be assumed again in this case, increasing the

2. Optical Properties of Thin Metal Films

uncertainty of the results. Furthermore, the availability of quality, commercially available films has in many laboratories eliminated the occasion of preparing evaporated films.

Electrical methods, such as resistance measurements and coulometric measurements, have been employed to determine film thickness [18]. The primary objection to the use of resistance measurements is the disagreement between bulk and thin-film resistivities. The resistivity of a thin film is highly dependent on its microstructure, the microstructure, in turn, being greatly affected by posttreatment and environment. Prostak [19], for example, noted that the resistance of a thin gold film decreased from 24 to 18 Ω per square when heated to 300°C for 15 min. If the film had exhibited bulk gold resistivity, its resistance would have been 5 Ω per square. These results clearly demonstrate that resistance measurements are unsuitable for determination of film thickness. Coulometric methods entail current-time measurements during the course of an electrolytic deposition of a thin film. The method lacks general applicability (a conducting substrate being required) and uncertainty is introduced by the necessity of assuming that bulk densities apply. Electrolytically deposited films are also typically of poor quality in that they lack mechanical durability [20].

Optical methods may also be used to measure film thickness. Interference effects permit the accurate determination of film thicknesses over a broad range, but the film and substrate must both be transparent (i.e., k = 0) [21]. The refractive index of relatively thick films has also been measured optically by means of an Abbe refractometer [22]. The method is suitable only for films of thickness greater than about 10,000 Å. All of the above methods are objectionable because separate measurements are required for each property being determined. The ideal technique would yield all three properties with but one set of measurements.

This has been approached by more recent optical methods. Perhaps the most useful of the photometric methods is that reported by Male' [23]. By this method n, k, and h are simultaneously determined, but the measurements are experimentally difficult, the data treatment is time-consuming, and the results lack precision. The transmittance of the film and the reflectance at both the substrate and air sides of the film are measured at near normal incidence. The results are then used to prepare curves of n versus k for each reflectance measurement from a series of nomograms. The intersection of the curves represents the solution of n and k. The film thickness is then found by graphical interpolation from a series of plots of phase change versus refractive index for various values of the extinction coefficient. The film thickness is directly related to the phase change.

Polarimetric methods have also been employed for the simultaneous determination of optical properties [24]. In most of these methods plane-polarized light is reflected from the film surface at an angle of 45°. The reflected light is generally elliptically polarized and by observing the azimuth and phase change of the reflected light, one is able to calculate n, k, and h. Explicit equations for these properties cannot be obtained in terms of observables, thus curve-fitting or approximation methods must be employed [25]. In addition, the computations are time-consuming and laborious and the necessary instrumentation is elaborate and expensive.

It is seen that each of the above methods suffers one or more of the following disadvantages: The method is specific to a particular type of film; it is destructive to the film; it requires separate measurements for each unknown; it lacks accuracy; it requires special instrumentation; it is too time-consuming to be used on a day-to-day basis; or it lacks general applicability with respect to the range of film thickness or optical properties which might be encountered.

This chapter presents a method of determining all three variables (n, k, and h) simultaneously and rapidly from internal reflection data [26-30] without being handicapped by any of the above restrictions. The method is generally applicable so that a wide range of values of optical properties can be determined.

Among the advantages of this method, in addition to those cited above, are that it is exact, to the extent that no approximations were required for the calculations, and it lends itself to computerized data gathering and data reduction without significant modification. On the other hand, there are minor inherent difficulties associated with the method which should be noted. The main difficulty is that the precision of the optical data is limited by the spectrophotometer employed in the measurements, and furthermore, reflectance measurements of high accuracy are required.

The initial portions of this chapter provide background information, the definitions of terms, and some of the experimental details necessary for an understanding of the nature of the problem. With this background in mind, it then describes in detail the method used for the simultaneous determination of the optical constants and film thickness of thin films with sample calculations given for a platinum film of intermediate film thickness (450 Å thick). A comparison is given of the calculated and observed spectra obtained under various conditions to illustrate the internal self-consistency of the calculated constants. The last section summarizes the results of other films studied and suggests new areas of research opened by the availability of this technique.

2. Optical Properties of Thin Metal Films

II. THE PRINCIPLES AND TERMINOLOGY OF INTERNAL REFLECTION SPECTROSCOPY

The fundamental optical phenomenon associated with attenuated total reflectance spectroscopy (ATR) is that known as total internal reflection (TIR). The phenomenon of TIR is a common occurrence and nearly everyone has observed it at one time or another; for example, when looking into a glass of water at some oblique angle it is noticed that objects close to (but not in intimate contact with) the glass cannot be seen. The light has been totally reflected at the glass-air interface, and objects beyond the interface will not be seen.

Total internal reflection occurs when light passing through an optically dense medium (i.e., a medium with a relatively high refractive index) impinges upon an interface of an optically rarer medium at an oblique angle. External reflection, in contrast, arises in the case where light passes through an optically rare medium prior to reflection from an interface of a substance which is optically more dense. The physical optics of TIR for a simple two-phase system will be initially described below, deriving TIR expressions from Snell's law of refraction, as it is instructive to describe a simple system before going to the more complex systems employed in this study.

It is assumed for the simple case (see Fig. 1) that two homogeneous transparent phases, separated by a planar interface, have refractive indices n_1 and n_2, respectively, where $n_1 > n_2$. Snell's equation [see Eq. (1)] shows

$$n_1 \sin \theta_1 = n_2 \sin \theta_2 \tag{1}$$

that light passing through phase 1 and incident on the interface at an angle θ_1 will be refracted with respect to the normal to a larger angle θ_2 (i.e., $\theta_2 > \theta_1$). It is evident that as θ_1 increases, θ_2 will approach $90°$. The angle of θ_1 at which θ_2 just equals $90°$ can be determined from Eq. (1). When $\theta_2 = 90°$,

$$\theta_1 = \sin^{-1}(n_{21}) \equiv \theta_c, \tag{2}$$

where $n_{21} = n_2/n_1$ and the angle of incidence (θ_1) is referred to as the critical angle (θ_c). When θ_1 is greater than θ_c, θ_2 becomes a complex quantity and loses its simple geometric interpretation of an angle. This is the situation in the case of TIR, and Maxwell's equations can be used to show that the light is totally reflected.

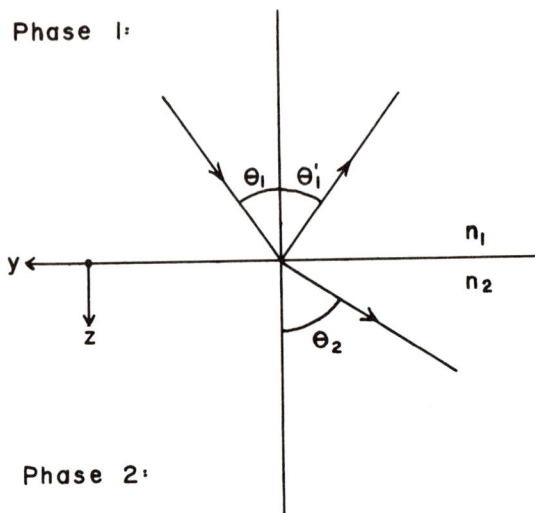

Fig. 1. Representation of reflection at the boundary of a two-phase system.

The derivations of the following equations will not be shown here because they are cumbersome, but they are well documented [28, 31, 32]. A particularly good treatment is given in Chapter 33 of Ref. [31]. It should be pointed out that the sign conventions and nomenclature employed are different from those used by Hansen [28] and others in the recent literature. The convention and nomenclature of Hansen are used throughout this dissertation.

Assuming the incident beam of Fig. 1 to be a plane electromagnetic wave, the complex electric field (E-field) is expressed as

$$\vec{E} = \vec{E}_o \exp[-\hat{i}(\omega t - \vec{\varkappa} \cdot \vec{r})] \qquad (3)$$

where $\vec{E}$ is the amplitude at point $\vec{r}$ from the origin at time t and the propagation vector $\vec{\varkappa}$ points in the direction the wave is traveling and whose magnitude is $|\vec{\varkappa}| = \varkappa = 2\pi n_1/\lambda_o$. Equation (3) is the mathematical representation of a sinusoidally varying plane wave. For the case where θ_1 is greater than θ_c, the E-field in phase 2 takes the form

$$\vec{E} = \vec{E}_o \exp(-\hat{\varkappa} z) \exp[-\hat{i}(\omega t - \varkappa_y y)]. \qquad (4)$$

2. Optical Properties of Thin Metal Films

The wave amplitude of Eq. (4) is seen to decrease exponentially with increasing distance into phase 2 (along the z axis) and the wave is called the evanescent wave. Because $\varkappa$ is of the order ω/c (i.e., λ_0), the light, even though totally reflected at the interface, has E-fields penetrating the surface to depths of the order of the wavelength of the light. The depth of penetration, d_p, of the E-fields is usually defined as the distance required for the time-averaged E-field to decay to e^{-1} of its value at the interface [26]. Thus, for a transparent phase 2, the penetration depth is given by

$$d_p = \frac{\theta}{4\pi n_1 (\sin^2 \theta_1 - n_{21}^2)^{1/2}} \qquad (5)$$

Depending on the values of the various parameters of Eq. (5), the penetration depth is ordinarily in the range $\lambda_0/10 < d_p < \lambda_0$.* For example, to minimize the depth of penetration at a given wavelength, an internal reflection element of high refractive index relative to that of phase 2 and large angles of incidence would be employed. It is seen that a sinusoidal standing wave is formed by the incident and reflected waves in phase 1 and an exponentially decaying evanescent wave is formed in phase 2 in the case of TIR.

Attenuated total reflection, the basis of ATR, or IRS,[†] occurs when phase 2 is light absorbing. In this case the ordinary refractive index is ordinarily replaced by a term known as the complex refractive index, $\hat{n} = n + ik$, where n is the real refractive index and k is the extinction coefficient,[‡] a measure of the light absorbing capability of the medium. It should be noted that some authors will arbitrarily define the complex retractive index as $\hat{n} = n - \hat{i}k$, $\hat{n} = n(1 + \hat{i}n\varkappa)$, or $\hat{n} = n(1 - \hat{i}n\varkappa)$, where $\varkappa$ is known as the attenuation index, and when using the literature it is important to know how the term has been defined [19, 26, 28]. Also, it should be noted that the complex refractive index is merely a notational convenience and bears no relationship to the relative velocity of light in the medium, nor does it conform to Snell's equation. When the complex refractive index replaces the real refractive index and other appropriate substitutions are made in the general E-field equation for a plane wave in

---

*Actually the penetration depth becomes indefinitely large near the critical angle.

[†] Technically IRS is the preferred nomenclature, although historically, ATR and IRS are used synonymously.

[‡] Using the nomenclature of Hansen [28].

phase 1, it becomes

$$E = E_o \exp\left[-\hat{i}2\pi\left(\frac{ct}{\lambda_o} - \frac{\hat{n}z}{\lambda_o}\right)\right],\quad (6)$$

and substituting $\hat{n} = n + \hat{i}k$, one obtains

$$E = E_o \exp\left[-\hat{i}2\pi\left(\frac{ct}{\lambda_o} - \frac{nz}{\lambda_o}\right)\right]\exp\left(\frac{-2\pi kz}{\lambda_o}\right).\quad (7)$$

The presence of the multiplicative exponentially decaying term accounts for the absorbance of light as a function of distance in the medium. On the other hand, when phase 2 is light absorbing and θ_1 is greater than θ_c, the electric field equation is of the form

$$\vec{E} = \vec{E}_o \exp[-\hat{i}(\omega t - \hat{\varkappa}_y y - \hat{\varkappa}_z z)]\quad (8)$$

where $\hat{\varkappa}_y$ and $\hat{\varkappa}_z$ are complex and related to $\hat{n}_2$ [31].

The result is a loss of energy to phase 2; that is, there is an attenuation of the radiant energy — hence, the name attenuated total reflection for this "lossy" coupling mechanism [29].

It should be pointed out that the presence of additional phases between the first (optically dense) phase and the last (optically rare) phase has little effect on the above discussion, regardless of the optical properties (n and k values) of these intermediate phases (providing they are sufficiently thin and/or are not opaque). This can be seen by recalling that Snell's equation [Eq. (1)] can be represented as

$$n_1 \sin \theta_1 = n_{\text{last phase}} \sin \theta_{\text{last phase}}.\quad (9)$$

It follows directly that TIR or ATR, depending on the value of $k_{\text{last phase}}$ will still occur when θ_1 is greater than θ_c and energy reaches the last phase.

It is obvious that the above principles can be employed to monitor the spectrum of the last phase. The necessary conditions are a first phase of relatively high refractive index, intermediate phases of sufficient transparency that energy reaches the last phase, and a means of regulating the angle of incidence of the light beam.

2. Optical Properties of Thin Metal Films

In conventional transmission spectroscopy, the transmittance T is given by

$$T = I/I_0 \tag{10}$$

where I_0 is the incident intensity of the light and I is the resultant intensity after passing through the sample. Alternatively, rather than measuring the relative light transmitted, the relative light absorbed (A) is measured. The relationship between T and A is given by [28]

$$A = -\log_{10}(T) = \log_{10}(I_0/I). \tag{11}$$

The light intensity I at any point in space is directly proportional to the time average of the E-field vector magnitude squared, $<E^2>$, which is given by [56]

$$<E^2> = \tfrac{1}{2} E \cdot E^* = \tfrac{1}{2} E_{max}^2. \tag{12}$$

Combining Eq. (6) with Eq. (12), one obtains

$$T = e^{-4\pi k z/\lambda_0} = e^{-\alpha z} \tag{13}$$

where α is the absorption coefficient given by

$$\alpha = 4\pi k/\lambda_0. \tag{14}$$

Additional equations in common usage are

$$I/I_0 = 10^{-abc} = 10^{-\epsilon bc} \tag{15}$$

or

$$A = abc \tag{16}$$

where a is the absorptivity, b is the sample thickness, c is the concentration (in appropriate units), and ϵ is the molar absorptivity. The above discussion of transmission formulas is introduced for comparison with their counterparts in IRS.

In IRS the relative intensities before and after a reflection are monitored and the ratio I/I_0 is known as the reflectivity R, that is,

$$R = I/I_0. \tag{17}$$

When a system has a high reflectivity (i.e., as $R \to 1$), a convenient method of enhancing reflectivity measurements is by the use of multiple reflections [29]. If, for example, n reflections are used, the reflectivity is given by

$$R = R_1^n \tag{18}$$

where R_1 is the reflectivity for a single reflection.

There is also an absorbance relationship corresponding to the conventional normal-incidence absorbance given by

$$A_r = -\log_{10}(R) = \log_{10}(I_0/I) \tag{19}$$

where A_r is called the reflection absorbance. The reflection absorbance can also be expressed as

$$A_r = ab_{eff}c \tag{20}$$

which corresponds with the same form of the absorbance expression used in transmission spectroscopy as given by Eq. (16). Here b_{eff}, the effective "cell" thickness, is defined as the thickness the sample would have if the absorbance measurement had been made by conventional transmission methods [7]. In this equation a and c have their usual meanings.

In IRS, the intensities are also related to $<E^2>$, just as in the case of transmission spectroscopy, but it is obvious that even in the two-phase system the situation is more complicated by the fact that $<E^2>$ varies in a nonlinear way according to distance from the interface and the extinction coefficient of the medium. Naturally, as a system becomes more complicated by the presence of additional phases, the calculation of R or A_r becomes more and more complex, even when the optical properties of all phases are known.

Because of the small depth of penetration of the evanescent wave, the availability of multiple reflection techniques, and other factors, IRS is clearly seen to be a valuable tool for studies of a wide variety of materials where other methods would be nearly useless. Among these are studies of opaque (or nearly opaque) substances [33], studies of nearly transparent substances [33], and studies of very thin films (thicknesses less than $\lambda/100$) [34]. However, in addition to the information available from the above, there is one very valuable source of information still unmentioned. This is through the use of polarized incident light.

2. Optical Properties of Thin Metal Films

The definition of the two modes of polarization used may be examined by referring to Fig. 1 (where a right-handed coordinate system is employed, such that the positive x axis rises from the plane of the figure and is normal to it). One of these modes entails the plane polarization of the E-field perpendicular to the plane formed by the incident and reflected rays (the plane of incidence) and is known as a transverse electric field and is designated as TE (transverse electric), s (senkrecht*), or $\perp$ (perpendicular) polarization. It is seen that $<E^2>$ will only have one component for $\perp$-polarization, the $<E_x^2>$ component. The other mode employs linear polarization of the E-field parallel to the plane of incidence and is called transverse magnetic polarization (TM) or is designated as p (parallel †) or $\parallel$ (parallel) polarization. In this case $<E^2>$ has two components — an $<E_y^2>$ component and an $<E_z^2>$ component. In general, $<E_x^2>$ does not equal the sum of $<E_y^2>$ and $<E_z^2>$. The significance of this phenomenon is that the reflection absorbances using $\perp$- and $\parallel$-polarizations will ordinarily differ. Of course, at normal incidence (as is the case for transmission spectroscopy) and at grazing angles ($\theta_1 \simeq 90°$) degeneracies arise such that physically there can be no distinction made between $\perp$- and $\parallel$-polarization and the corresponding absorbances ($A\perp$ and $A\parallel$) are identical. Figure 2, a plot of reflectivity versus angle of incidence, illustrates the differences in reflectivity for $\perp$- and $\parallel$-polarization for the two systems, a two-phase glass-air system and a three-phase glass-gold film-air system. The very rapid changes in R which occur in the vicinity of the critical angle should be noted. This phenomenon is always evidenced for TIR and reference is made to it in Section IV of this chapter in the calibration of the angle setting of a variable angle cell. The two systems are included for comparison. Figure 3 contains the reflection absorbance curves as a function of the angle of incidence and illustrates the differences between R plots and A_r plots, although the same information is contained in both forms.

The presence of the metal film in Figs. 2 and 3 has two effects on the measurements. The first of these effects is especially pronounced for metal films (as opposed to other solid films) owing to the generally high extinction coefficients of metals. There is a marked absorbance of light by the metal film, with a resultant effect on the E-fields in the third phase. The other effect is related to reflections at the 1-2 and 2-3 interfaces. In this case, depending on the optical properties of the phases, constructive or destructive interference of the incident and reflected waves occurs [19]. These absorbance and resonance effects are taken into account in the equations of Appendix II

---

*German: perpendicular.

†German: parallel.

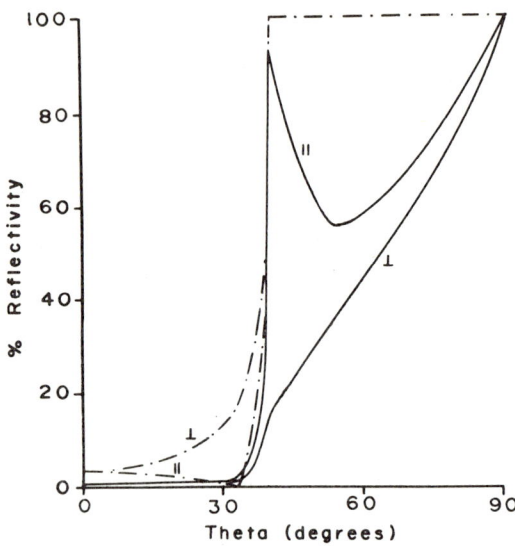

Fig. 2. Reflectivity as a function of angle of incidence for representative two- and three-phase systems: —·—·—·, two-phase system (glass-air); and ———, three-phase system (glass-gold film-air).

These equations were derived from Fresnel's reflection coefficients and Maxwell's equations by Hansen [28] and permit the exact calculation of both the E-fields and the reflectivities (or reflection absorbances) for an n-phase system. The equations assume coherence of the light and isotropic, homogeneous, plane-parallel-sided phases. The solutions of the equations are exact only to the extent that these conditions are met. Although the generalized equations for an n-phase system are to be found in Appendix II, the specific equations for a three-phase system are given here because they are significantly less complex than the generalized equations and are particularly useful for the calculation of the optical properties of thin films. (A FORTRAN IV language computer program for the solution of the generalized equations of the reflectivities of n-phase systems appears in Appendix III.)

For a three-phase system, phases 1 and 3 are semiinfinite. For the second phase it is convenient to formulate a reduced thickness (d), given by

$$d = h/\lambda_0 \tag{21}$$

2. Optical Properties of Thin Metal Films

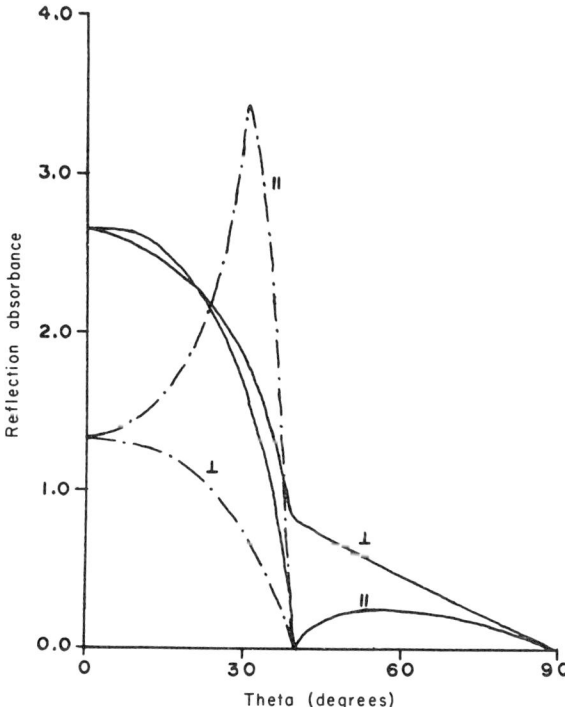

Fig. 3. Absorbance as a function of the angle of incidence for representative two- and three-phase systems: —·—·—·, two-phase system (glass-air); ———, three-phase system (glass-gold film-air).

where h is the film thickness and λ_0 is the vacuum wavelength expressed in the same units as h. In order to simplify the form of later expressions the terms ξ and β are defined. The former term is given by

$$\xi_j \equiv \hat{n}_j \cos \theta_j = (\hat{n}_j^2 - n_1^2 \sin^2 \theta_1)^{1/2}$$
$$= |\operatorname{Re} \xi_j| + \hat{i} |\operatorname{Im} \xi_j| \tag{22}$$

where the complex quantity ξ consists of a real term ($\operatorname{Re} \xi$) and an imaginary term ($\operatorname{Im} \xi$), both of which must be positive [28]. The latter term is given by

$$\beta_2 = 2\pi d \xi_2. \tag{23}$$

A. Perpendicular Polarization

The reflection coefficient for the boundary between phases j and k is given by

$$r_{\perp jk} = \frac{\xi_j - \xi_k}{\xi_j + \xi_k} \qquad (24)$$

in which the magnetic permeability μ (a term appearing in Hansen's equations) [28] has been set equal to 1. This approximation is valid at optical frequencies. For a three-phase system the coefficient representing light reflected in the first phase is given by

$$r_\perp = \frac{r_{\perp 12} + r_{\perp 23}\, e^{2i\hat{\beta}}}{1 + r_{\perp 12}\, r_{\perp 23}\, e^{2i\hat{\beta}}} \qquad (25)$$

The reflectivity for the three-phase system is then given by

$$R_\perp = |r_\perp|^2 = r_\perp \cdot r_\perp^* \qquad (26)$$

where $r_\perp^*$ is the complex conjugate of $r_\perp$. The reflection absorbance is related to the reflectivity as mentioned above and is given by

$$A_{r_\perp} = -\log_{10}(R_\perp). \qquad (27)$$

B. Parallel Polarization

The corresponding equations when the incident light is parallel polarized are given by

$$r_{\| jk} = \frac{\hat{n}_k^2\, \xi_j - \hat{n}_j^2\, \xi_k}{\hat{n}_k^2\, \xi_j + \hat{n}_j^2\, \xi_k}, \qquad (28)$$

$$r_\| = \frac{r_{\|12} + r_{\|23}\, e^{2i\hat{\beta}}}{1 + r_{\|12}\, r_{\|23}\, e^{2i\hat{\beta}}}, \qquad (29)$$

$$R_\| = |r_\||^2 = r_\| \cdot r_\|^*, \qquad (30)$$

and

$$A_{r\|} = -\log_{10}(R_\|). \tag{31}$$

It is seen by inspection of these equations that even for the relatively simple case of a three-phase system it is virtually impossible to invert the equations mathematically to solve directly for n_2, k_2, or h. It might be possible to simplify the equations to some extent by the introduction of power series expansions and other approximations, however it is desirable to retain the exact equations wherever possible. The importance of this observation will become more obvious in Section V.

III. THIN FILMS AND THIN-FILM OPTICS

The importance of being able to measure quantitatively the optical constants of thin metal films in relation to spectroelectrochemistry has already been described; however, this scientific area is only one small and relatively new application and example of a need for this information. The use of thin films for purposes requiring electrooptical information has mushroomed in recent years due to advances in thin-film technology, particularly with respect to high vacuum techniques and electron microscopy.

The availability of economical pumps capable of producing and, in some cases, studying thin films has made it possible for virtually any laboratory with an interest in thin films to undertake research of this type. Innovations in high vacuum technology have also facilitated the commercial production of thin films by permitting thin films with very large surface areas (several square feet) to be prepared under high vacuum conditions [35]. Electron microscopy has made possible the direct observation of structural faults, pores, channels, and other details which could only be inferred from indirect measurements in the past.

Some examples of these uses, as mentioned earlier, include reflecting films for interferometry, lens coatings, protective coatings for hygroscopic components of infrared spectrophotometers, and coatings of cathode-ray tubes to minimize electrostatic charging. This increased use of thin films led to increased attention to the determination of their optical properties. This attention has been reflected by the large number of recent publications devoted to this topic [36-39].

Three features of the recently published attempts to characterize the optical properties of thin films are immediately apparent. First, specialized equipment is frequently required which is ordinarily expensive and unavailable to the chemist. Such equipment would include reflectometers and ellipsometers. This is objectionable in that a chemist would have a hard time justifying the expense of such an instrument whose sole purpose is the characterization of the optical properties of thin-film electrodes. A second feature is the time and effort required to gather the data and reduce them to optical constants. When every electrochemical experiment must be expected to alter the optical properties, the task of keeping track of these properties from experiment to experiment would be overwhelming. Consequently, a much more rapid method has been needed. Third, the wide disparity in the published results is almost bewildering until the reasons for these disparities are made clear.

Table 1 compares recently published constants for platinum (both bulk and thin-film data are included for comparison). Some of the reasons for the disagreement found in these literature values are listed below, but it is interesting to note that very large uncertainties would be incorporated by reliance on these data. In most of these cases the data were obtained under conditions of high vacuum as soon as possible after the films had been deposited. This discounts their use by the spectroelectrochemist because aging and environmental factors will not have become evident.

Heavens and Smith [40] have compiled an impressive list of reasons for the extreme variations found in the reported optical constants of evaporated thin films. These reasons are based on the influence of the structure of the thin film on thin-film optical properties, and include the following factors: (1) pressure in the evaporation chamber; (2) rate of deposition; (3) thickness of the film; (4) nature and temperature of the substrate; (5) state of the evaporate in the vapor phase; (6) crystal form and growth habit; (7) crystal binding energy; (8) crystal-substrate binding; (9) mobilities of evaporate molecules on substrate and growing film; (10) purity and physical state of the evaporate; (11) evaporation temperature and velocity of the impinging molecules; and (12) the angle of incidence of the impinging molecules at the surface. The film thickness, the third factor mentioned above, also has a pronounced direct effect on the optical constants n and k. This results in the relative enhancement of small errors in measured film thickness when calculating n and k. Of course, the optical constants are also wavelength dependent.

Electron microscopy reveals that thin films are somewhat disordered and that the first few molecular layers tend to conform to the surface structure of the substrate [41]. As additional layers are deposited, the tendency shifts to the more ordered crystal structure of the evaporate. Other studies [12] showed that in some cases surface discontinuities, such as islands of the evaporate and microscopic pores and channels, may

2. Optical Properties of Thin Metal Films

TABLE 1

Literature Values for the Optical Constants of Platinum

λ (nm)	Bulk[a]		Film[b]		Film[c]		Film[d]		Bulk[d]	
	n	k	n	k	n	k	n	k	n	k
439	1.63	2.08	—	—	—	—	—	—	—	—
466	1.71	2.11	—	—	—	—	—	—	—	—
477	1.72	2.14	—	—	—	—	—	—	—	—
480	—	—	0.91	2.50	—	—	—	—	—	—
488	1.76	2.11	—	—	—	—	—	—	—	—
500	—	—	0.91	2.60	1.1	4.1	—	—	—	—
503	1.79	2.15	—	—	—	—	—	—	—	—
517	1.85	2.12	—	—	—	—	—	—	—	—
520	—	—	1.01	2.75	—	—	—	—	—	—
540	—	—	1.05	2.80	—	—	—	—	—	—
560	—	—	1.11	3.05	—	—	—	—	—	—
580	—	—	1.12	3.20	—	—	—	—	—	—
589	2.06	2.06	—	—	—	—	2.63	3.54	2.06	4.26
600	—	—	1.25	3.40	1.2	4.5	—	—	—	—
630	2.16	2.06	—	—	—	—	—	—	—	—
640	—	—	1.27	3.50	—	—	—	—	—	—
660	—	—	1.38	3.80	—	—	—	—	—	—
665	2.34	2.08	—	—	—	—	—	—	—	—
668	—	—	—	—	—	—	2.91	3.66	—	—
700	—	—	1.53	3.95	1.3	5.1	—	—	—	—

[a] International Critical Tables, Vol. V, 1st ed., McGraw-Hill, New York, 1929, p. 250.
[b] V. L. Rideout and S. H. Wemple, J. Opt. Soc. Am., 56:749 (1966).
[c] B. T. Barnes, J. Opt. Soc. Am., 56:1546 (1966).
[d] G. Hass and L. Hadley, American Institute of Physics Handbook, (D. Gray, ed.), 2nd ed., McGraw-Hill, New York, 1963, 6:114.

develop depending on the conditions at the time of the deposition. Inhomogeneities of this sort cannot be treated adequately by assuming the model of the ideal thin metal film proposed by Hansen [28] (i.e., phases which are isotropic, homogeneous, and bound by plane parallel surfaces). These factors do, however, help to account for the differences noted between bulk and thin-film optical constants.

IV. EXPERIMENTAL APPARATUS AND TECHNIQUES

The need for quantitative spectroscopic measurements of the highest possible accuracy required the use of the best equipment available and special precautions to eliminate or minimize systematic sources of error.

The IRS measurements reported here were obtained by use of a variable-angle cell constructed for this study by N. J. Harrick (Harrick Scientific Corporation, Ossining, N.Y.).* The cell used was the RMVA (retromirror variable-angle attachment), an attachment offering many advantages essential to the attainment of quantitative data (see Fig. 4). The use of high-quality, front-surfaced mirrors (suitable for measurements in both the UV and visible portions of the spectrum) and a fused silica hemicylinder in this attachment made it possible to make IRS measurements in which defocusing of the light beam and absorbance due to the presence of the cell are minimized, while the definition of the angle of incidence is optimized. A detailed description of the cell is given in Ref. [42].

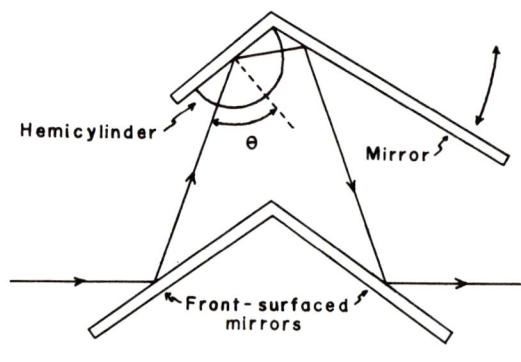

θ = 60°

Fig. 4. Schematic representation of the retromirror variable-angle internal reflection attachment [29, 42].

---

*This unit is now commercially available.

2. Optical Properties of Thin Metal Films

The angle of incidence of the RMVA was calibrated by use of the very rapid change in reflectivity at the critical angle when parallel-polarized light is utilized (see Fig. 5). For this example, where the IRE (internal reflection element) is fused silica and the reflectivity versus air is observed as a function of the angle of incidence at a wavelength of 486.1 nm, the reflectivity of parallel-polarized light is seen to drop from 1.0 at the critical angle (43.1°) to almost 0.0 in just a 3° change in the angle of incidence. It should be noted that the calibration curve does not have exactly the same shape as the theoretical curve for this system. This is largely due to the fact that the specular reflectance of the mirrors of the cell varies as a function of angle of incidence and that the light beam of the Cary 14 is not truly parallel; that is, there will be a small but finite range of angles of incidence at any given cell setting. This phenomenon provides a very sensitive method of calibrating the angle of incidence. Any error in the angle is assumed to be linear throughout the range of angles available (approximately 35 to 75°).

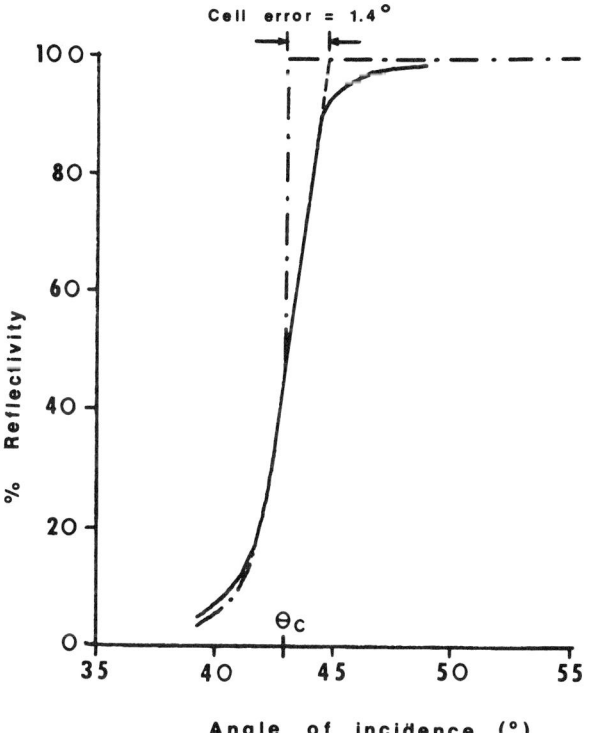

Fig. 5. Reflectivity as a function of angle of incidence for the calibration of a variable-angle cell: -·-·-, theoretical curve; ———— experimental results.

The thin films used for this part of the study were platinum films deposited directly by evaporation on a clean, optically polished, 1-mm thick, fused silica substrate. Chemical analyses were not performed on these films, but it may be assumed that tungsten is present in trace amounts due to the evaporation of the filament used to heat the platinum. The films were assumed, for the purposes of calculations, to be isotropic, homogeneous, and consistent of plane parallel surfaces. This latter assumption can be checked by measuring the transmittance of the film, rotating the film, and remeasuring the transmittance. Due to the high absorptivity of platinum in the visible portion of the spectrum, any significant wedge-like structure of the film can be detected in this manner [43]. Any samples exhibiting a pronounced wedge structure were not used for the determination of optical constants. Techniques used to minimize anisotropy are dealt with at a later point in this chapter; the assumption of homogeneity seemed to have no adverse effects. Recent theoretical models are available to deal with inhomogeneities [44] but were unnecessary, at least for the films used in these studies.

Two types of polarizers were used for different aspects of these studies. For ATR measurements schlieren quality Glan-Thompson prisms were employed. Special sheet Polaroid polarizers were used for the instrument calibrations as recommended by Bennett [45].

All measurements were made on a Cary Model 14 recording spectrophotometer. Some modification of the instrument was required to accommodate the RMVA cell described above and to enhance the quality of the measurements (see Fig. 6). One modification consisted of drilling a hole in the sample compartment wall so that a control arm could be used to facilitate cell alignment. Another modification consisted of the placement of a 1 x 5 mm slit in the chopper compartment of the spectrophotometer [46]. This change is beneficial in two ways: The spectral slit width is reduced and the beam is collimated to a 30-min spread. This condition is mandatory for the accuracy desired in these measurements. As there is sufficient energy available in the visible region of the spectrum for this instrument, this practice does not adversely affect the measurements.

In order to make accurate optical measurements, certain precautions must be taken to ensure the minimization of systematic sources of error. Some of these sources of error are inherent in the instrument being used and others are caused by the introduction of the RMVA cell.

The manufacturers of the Cary Model 14, and many other manufacturers of quality spectrophotometers, will provide calibration tables for both wavelength readout and recorder slidewire linearity for the correction of raw data. If these services have not been provided, it is

2. Optical Properties of Thin Metal Films 63

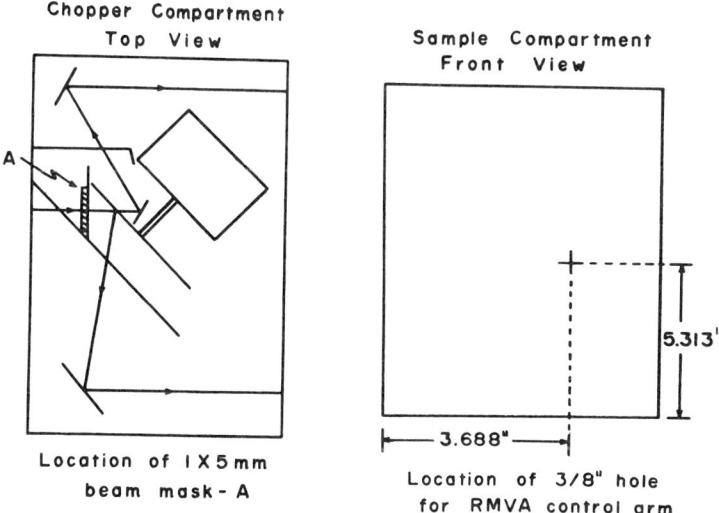

Fig. 6. Schematic diagram of the modifications to the Cary 14 spectrophotometer (not to scale).

necessary to compose these correction tables by methods described in the literature [43].

In general, the presence of an IRS cell in the sample compartment has the capability of introducing several sources of systematic error. Among these are defocusing of the light beam, depolarization of a polarized light beam, and light scattering due to surface irregularities of the optical components of the cell. The design of the Harrick RMVA is such that horizontal defocusing is negligible at the narrow slit widths employed in these studies and vertical defocusing is minimized by using a minimum slit height and the above-mentioned beam mask. Depolarization caused by the cell was measured by a modified three-polarizer technique introduced by Bennett [45]. The high optical densities encountered in these measurements lead to some unreliability; however the order of the depolarization was found to be only about 0.01%. Light scattered by the optical components of the cell, while not measured directly, was assumed to be insignificant because of the optical flatness of the substrates and the apparent lack of surface flaws. Some care is necessary, though, to ensure that the surfaces of the components (especially the front surface mirrors) are clean and free of surface damage.

V. METHOD FOR DETERMINATION OF FILM THICKNESS AND OPTICAL CONSTANTS OF THIN FILMS

A. Introduction

Ordinarily the use of thin-film electrodes for simultaneous IRS spectroelectrochemistry involves a three-phase system; the same type three-phase system is employed when IRS techniques are used to determine n, k, and h of a thin film (see Fig. 7). The first phase is the internal reflection element (IRE) itself, and for visible IRS studies the IRE is usually quartz, glass, or fused silica in order that it be essentially transparent throughout the visible spectrum. Fused silica is somewhat preferred because its refractive index as a function of wavelength is well known, and it can be optically polished well. The second phase is the thin metal film (Au, Pt, Pd, etc.) to be used as an electrode and the third phase is the environment, which for spectroelectrochemistry is the electrolysis solution, but which may vary for other studies. The restrictions placed on the third phase are that it must have a refractive index less than that of phase 1 and that, for the purpose of calculating n, k, and h of phase 2, its optical properties be known. These conditions, though, are not excessively restrictive.

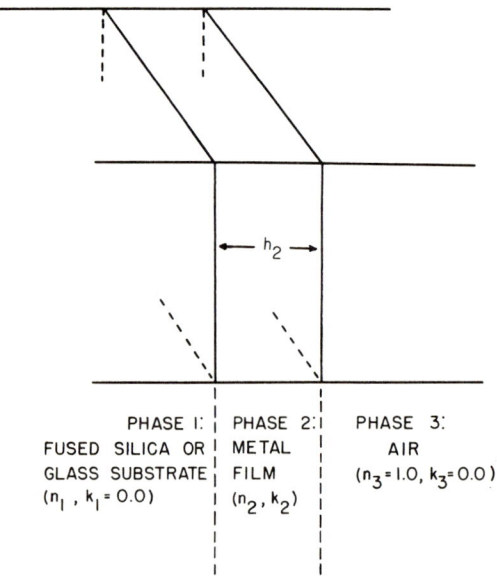

Fig. 7. Schematic representation of the three-phase system used for the determination of the optical properties of thin films.

2. Optical Properties of Thin Metal Films

In order to examine the feasibility of determining the optical constants of a thin metal film electrode under the conditions of an electrochemical experiment, a study was made of a highly light-absorbing aqueous solution of potassium permanganate. The capability of making the determinations under these conditions would allow examination of changes in the properties of an electrode before and after an electrolysis. In particular, the thickness of the electrode, which may be suspected to change during an electrode reaction, may be monitored. Additionally, the capability of determining the optical properties of the electrode as a function of applied potential would be made possible. Marked changes have been observed in reflection absorbance (hence, optical properties) as a function of applied potential [47-49]. This effect, known as the electroreflectance effect, is currently attributed to perturbations of the free electron plasma which tunnels outward from the electrode surface into the solution phase [49]. In order to characterize quantitatively the optical properties of an electrode for studies of electrode reactions, this capability is seen to be a necessity.

The conditions of a typical electrochemical study were simulated by examination of a dilute ($\sim$ 10 mM) aqueous solution of potassium permanganate, which visually is an intense red-purple solution. Single reflection absorbance measurements of this solution were identical (within experimental error) to those obtained for triply distilled water. At the 525-nm transmission absorbance peak, this peak having the greatest optical density, the molar absorptivity was found to be approximately 2450 M^{-1} cm^{-1}. This value is in good agreement with the value 2440 M^{-1} cm^{-1} obtained at a wavelength of 526.5 as reported by Den Boef, et al. [50]. The extinction coefficient k of the 10 mM solution was calculated for the conditions of λ_{max} to be about 2.4×10^{-4}. This value is considerably below the range of values of k representing the lower limits of detection for a single reflection. It is, therefore, not surprising that the IRS spectrum of the permanganate solution is essentially identical to that obtained for pure water. Because water meets the requirements of materials suitable to be used as the third phase, dilute aqueous solutions (even if highly colored) may also be used.

Although measurements have been made under a variety of conditions (i.e., various three-phase systems), the method of determining film constants described here was evaluated with air constituting the third phase, as it introduces the least uncertainty in the calculation of the optical constants, and for the sake of simplicity of measurements. The results of studies with a number of different films which have also been studied in the course of this investigation are reported in a subsequent chapter. However, in this chapter, the method of determining optical properties of thin films is applied and tested using a particular platinum film. The platinum film used for this study was obtained from Harrick Scientific Corporation and was vacuum evaporated using a tungsten filament on a plasma-cleaned fused silica substrate 1 mm thick. The platinum film data and results are given so that the calculation process may be followed more conveniently and with greater understanding.

The primary purpose of this section is to demonstrate a method by which optical properties of thin films may be rapidly and easily obtained and to show the validity of the method. There will not be any attempt to catalog the optical properties of thin metal films, especially in view of the great differences observed for films due to the effects of prior conditions and usage. Moreover, not all of the films examined were suitable for evaluation studies. Some very thick films showed "anomalous" behavior with respect to the IRS spectra obtained (see Section VI), some films seemed to have a significant wedge-like structure as determined by methods reported in Section IV, and some films had obvious surface defects.

B. Data Acquisition

For reasons which are discussed in detail later, a minimum of four spectrophotometric measurements must be made at each wavelength to characterize uniquely the optical properties n, k, and h at that wavelength. It was found that a semiglobal approach to the problem of three equations in three unknowns typically gave more than one physically valid solution set for n, k, and h. A fourth spectrophotometric measurement is then required as an additional constraint to limit the results to one solution set.

One of the measurements used was the transmission absorbance obtained using unpolarized light. This type of measurement was used because of relative ease and high accuracy. And, too, this measurement is required to help determine the suitability of a film for these studies as described above. Besides, this type of measurement tends to minimize the effects of film anisotropies [43]. The remaining three measurements used were all IRS measurements.

A generalized expression for the reflection absorbance is given by

$$A_r = f(n_1, n_2, k_2, h_2, n_3, \theta, \lambda, P) \tag{32}$$

where P represents the plane of polarization of the light and the other symbols are as defined in Section II and Appendix I. It can be seen from this expression that both θ and P can be varied without affecting the other parameters. With two modes of polarization and a large number of discrete angles of incidence available, one is faced with a large selection of

2. Optical Properties of Thin Metal Films

potential experimental conditions from which to choose. Somewhat arbitrarily it was decided to make measurements at two angles of incidence (both well above the critical angle) using both $\perp$- and $\|$- polarization. Prostak [51] has recommended the use of IRS measurements obtained just above and below the critical angle for these calculations; however, attempts at the use of such measurements in this study proved to be unsuccessful. In some cases no solutions for n, k, and h could be obtained, or in other cases the solutions obtained were self-inconsistent (e.g., neighboring points of a refractive index spectrum or extinction coefficient spectrum would be vastly different). Because A_r changes very rapidly in the vicinity of the critical angle (see Fig. 3), a small error in the measured angle of incidence can be expected to have a pronounced effect on the calculated optical constants. There is further justification for this choice in that the position of the polarizer in the spectrophotometer could be reproduced more accurately than the position of the RMVA cell, which requires translations in the horizontal plane at each angle setting for alignment purposes. The alignment procedure is described in detail in the instruction pamphlet provided by the manufacturer (Harrick Scientific Corporation, Box 824, Ossining, N. Y.), and is performed in order that the light beam strike the IRE at normal incidence. By restricting the number of angles of incidence as much as possible, one minimizes any effects of anisotropies which might occur if different portions of the film were monitored at each angle setting.

Both the transmission measurements and IRS measurements were made for the film-substrate system versus the substrate only. Hence, the transmission measurement represents the absorbance of the film alone; while the IRS measurements still represent a three-phase system. The effect of making the IRS measurements versus the substrate is that the absorbances of the IRS cell components (especially the front-surfaced mirrors) do not appear in the final result.

Figure 8 shows five spectra of a platinum film as described above, the transmission absorbance spectrum and the perpendicular and parallel IRS absorbance spectra obtained at angles of incidence of 50 and 60°. These angles are both several degrees above the critical angle, which averages about 43° in the visible region of the spectrum. Because more spectra have been obtained than are needed for the calculation of n_2, k_2, and h_2, the extra spectrum is used to check for internal self-consistency of the calculated optical constants.

C. Data Reduction

In order to derive a refractive index spectrum or extinction coefficient spectrum from an essentially featureless IRS spectrum (as is the case for

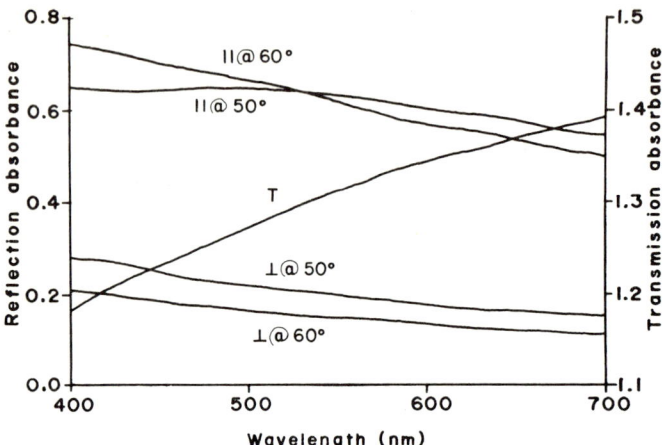

Fig. 8. Transmission and internal reflection spectra of a thin platinum film.

platinum films), it is only necessary to analyze a relatively small number of points of the IRS spectrum. Sixteen points were analyzed for this platinum film at 20-nm increments over the wavelength region from 400 to 700 nm. Of course, if the IRS spectrum of a film has more structure, a proportionately larger number of points would have to be analyzed. Fortunately, the spectra of metal films typically display little structure.

The use of a digital computer is necessary for the calculations which are described below. Otherwise, the data processing, which relies heavily on iterative techniques, would be virtually impossible on the basis of the calculation time required. Figure 9 is a flow chart of the data treatment process in the form encountered in the computer software literature. (Some of the terms found in Fig. 9 are computer related, but the text should make these terms clear.) Frequent reference to Fig. 9 should help one to follow more easily the description of the process of treating the data. The "step numbers" found below refer to the numbers found in Fig. 9 and are present to facilitate cross-referencing. The corresponding FORTRAN IV language computer program is found in Appendix IV.

Step 1: Assuming, arbitrarily, that the analysis of the IRS data is to begin at the wavelength 400 nm and recalling that four measurements are required, the following information must be furnished to the computer in the proper format:

2. Optical Properties of Thin Metal Films

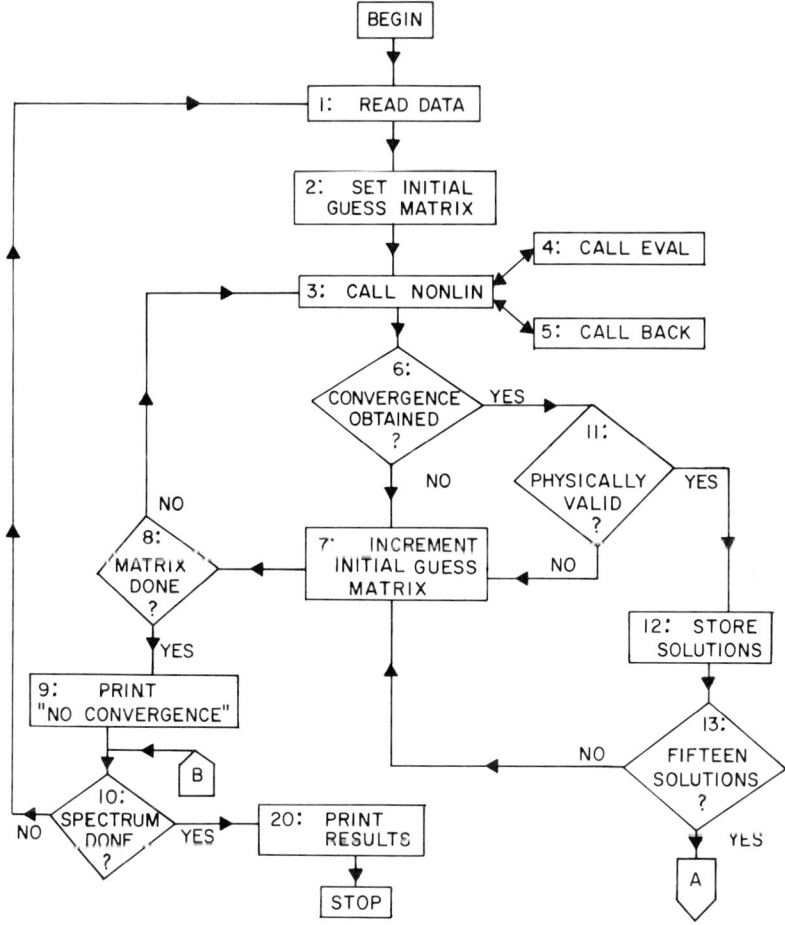

Fig. 9. Flow chart of the data processing sequence employed for the calculation of the optical properties of thin metal films.

a. The three reflection absorbance measurements and one transmission absorbance measurement to be processed.

b. The optical properties of phases 1 and 2.

c. The experimental conditions, i.e., the wavelength and angles of incidence.

This process is referred to as "reading in" the data. Figure 10 contains the data provided to the computer for the derivation of the first set of optical constants of the platinum film. The refractive index of phase 1, the fused silica IRE, was obtained by numerical interpolation from refractive versus wavelength data provided by a specification sheet made available by the manufacturer (Dynasil Corporation of America, Box D, Berlin, N.J.). The difference between the refractive indices of air and a perfect vacuum (exactly 1.0 by definition) is small enough that it may be considered negligible throughout the visible region of the spectrum.

Step 2: It was pointed out in Section II, B that the equations for reflection absorbance cannot be inverted for direct solution of n_2, k_2, and h_2 without making use of approximations. A number of methods have been developed for the solution of equations of this sort, including graphical methods [52], direct search techniques [53], regression analysis [53], and iterative processes based on the differential calculus, such as Newton's method [54].

For these studies, Newton's method, and more recent modifications of Newton's method [55] were most successful. The essential features of the technique are shown by Fig. 11, which demonstrates Newton's method for a function of a single variable [$y = f(x)$]. An approximate root (x_0) of the equation is chosen and the first derivative [$f'(x_0)$] is evaluated for $f(x)$ at this point. As the graphical interpretation of a first derivative is

A_T	$A_{r\perp}@\theta_1$	$A_{r\parallel}@\theta_1$	$A_{r\parallel}@\theta_2$
1.185	0.215	0.750	0.653

$\lambda(\text{Å})$	θ_1	θ_2
4000.0	60.0	50.0

n_1	n_3
1.470	1.000

Fig. 10. Sample input data.

2. Optical Properties of Thin Metal Films

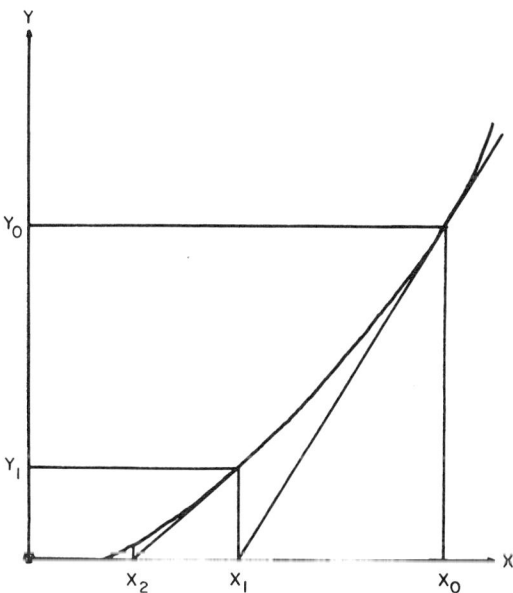

Fig. 11. Graphical representation of Newton's method for the solution of a function of a single variable.

that of the slope of the line tangent to the curve at the point of evaluation, thus the first derivative is given by

$$f'(x_0) = \frac{y_0 - 0}{x_0 - x_1} = \frac{f(x_0)}{x_0 - x_1} . \tag{33}$$

Rearranging Eq. (33) gives

$$x_1 = x_0 - \frac{f(x_0)}{f'(x_0)} . \tag{34}$$

By extension, the general relationship is

$$x_{n+1} = x_n - \frac{f(x_n)}{f'(x_n)} \tag{35}$$

where each successive iteration approaches the exact solution of the equation, as indicated by Fig. 11. This method is successful whenever no singular point [i.e., a point at which f'(x) = 0 such as a maximum, minimum, or inflection point] exists between the original approximation and the actual solution. The method as modified for n-dimensional systems is described later.

At this stage of the process an approximation, or initial guess, is made for each variable n_2, k_2, and h_2. These values are completely arbitrary and were in this instance set as follows: $n_{2(init)} = 0.5$, $k_{2(init)} = 0.5$, and $h_{2(init)} = 50.0$ (Å).

Step 3: Because there are three unknowns (n_2, k_2, and h_2), it was assumed that three equations would adequately define these three unknowns. An algorithm recently published by Brown [56] for the solution of simultaneous nonlinear equations was found to be successful and to offer advantages that result in relatively small computer processing times when compared with other iterative techniques.

Early attempts to solve for these three unknowns using both direct search techniques and a secant method for the solution of simultaneous nonlinear equations were found to be highly dependent on the initial guess employed and/or the magnitude of the incrementation. Among the phenomena observed was a semiperiodicity of the solutions for the thickness of the films. It was found that the calculations described below avoided this difficulty.

The name "NONLIN" was given to the computer subroutine which performs these calculations. The technique consists of numerically expanding the first equation in a Taylor series about the initial guess. Only linear terms are retained and the series is then set equal to zero. The variable having the partial derivative of largest absolute magnitude is then solved for as a linear combination of the remaining two variables. This linear combination is substituted in the second equation so that the first variable is eliminated. The second equation is expanded in a Taylor series in a like manner and a second term is eliminated such that for the third equation, a single equation in one unknown results. This equation is solved by Newton's method, which was described earlier. The solution obtained by the Newton step is then substituted back into the linear equations obtained above.

Among the advantages offered by Brown's method is that fewer function evaluations are required per iteration than by Newton's method. When the calculation of first partial derivation is included, the number of function evaluations required by Brown's method is given by

2. Optical Properties of Thin Metal Films

$$N = \frac{n^2}{2} + \frac{3n}{2} \tag{36}$$

where N is the number of function evaluations required and n is the dimensionality of the system [55]. The number of function evaluations required by Newton's method is given by

$$N = n^2 + n. \tag{37}$$

Thus, for a system of three simultaneous equations Brown's method requires nine function evaluations, as opposed to twelve function evaluations required by Newton's method. Reducing the number of function evaluations by 25% has the effect of decreasing the total calculation time significantly, although not by 25%. In addition, the partial derivatives are evaluated numerically so that only the equations to be solved need be furnished. The need to determine analytical expressions of the partial derivatives is therefore eliminated. Moreover, by eliminating variables in the manner described above, round-off errors are minimized. At each stage a variable is selected to be eliminated such that the coefficients of the linearized Taylor series expansion are all less than one. This has a partial pivoting effect which tends to prevent the growth of round-off errors [55].

Provisions are made in NONLIN to control the number of significant figures desired in the solution, the maximum number of iterations to be allowed, and tests to determine if internal matrices are singular. Singularity indicates that the process is "blowing up"; that is, the process is diverging rather than converging.

Steps 4 and 5: Subroutine "EVAL" contains the three functions being evaluated. The three functions contained in this subroutine are the transmission absorbance of the film (A), and the reflection absorbances for both $\perp$- and $\parallel$-polarizations ($A_{r\perp}$ and $A_{r\parallel}$, respectively). Provision is made so that only one function is evaluated at each call of EVAL. That is, when evaluation is desired of the transmission absorbance, only the transmission absorbance is calculated. This capability represents a considerable savings of time for a highly iterative process.

Subroutine "BACK" performs the backsubstitution of improved solutions in the linear equations as described in Step 3. This routine is also described in the algorithm described by Brown [56].

Step 6: When the calculations of Steps 3-5 have been completed, a check is made to determine whether or not the calculations ceased because the process was "blowing up." If such was the case and convergence was not obtained, control is passed to Step 7; if the process should converge, the sequence progresses with Step 11.

Step 7: Initial attempts to use this approach for the determination of the optical properties of thin films under controlled conditions indicated possible adverse effects dependent on the values selected for initial guesses. Under some circumstances a failure to converge would be observed. Varying regions of convergence yielding multiple solution sets were observed when initial guesses were varied. It was found that direct search techniques applied to an initial guess matrix circumvented this dependence on the initial values selected. This means that the values used as the initial guesses were incremented one by one after successive calls upon subroutine NONLIN. In this way convergence is ultimately guaranteed if a solution actually exists.

Trial and error led to increments of the following magnitude: $\Delta n_{2 \, (init)} = 0.5$, $\Delta k_{2 \, (init)} = 0.5$, and $\Delta h_{2 \, (init)} = 10.0$ (Å). The ranges over which the initial guesses were allowed to vary are given by $0.5 \leq n_{2 \, (init)} \leq 4.0$, $0.5 \leq k_{2 \, (init)} \leq 4.0$, and $50.0 \leq h_{2 \, (init)} \leq 800.0$. This was not intended, nor did it in fact limit the solutions to these ranges. The actual incrementation sequence proceeded by way of $h_{2 \, (init)}$ being varied over its allowable range. Then $k_{2 \, (init)}$ is incremented and $h_{2 \, (init)}$ is again varied over its allowable range. This process is repeated until all allowable pairs of $k_{2 \, (init)}$ and $h_{2 \, (init)}$ have been used in combination with the first value of $n_{2 \, (init)}$. The value of $n_{2 \, (init)}$ is then incremented and combined with all allowable pairs of $k_{2 \, (init)}$ and $h_{2 \, (init)}$. This sequence is continued so that all possible combinations of the permitted initial values might eventually be used.

Step 8: At each successive incrementation of the initial values a check is made to determine if all of the permitted combinations of the initial values have already been used. Provision was made to terminate the process (Step 9) if this were the case. Otherwise another call was made upon subroutine NONLIN (Step 3).

Step 9: If all possible combinations of the initial guesses are exhausted without success, instructions are programmed into the computer to print a message to this effect. This eventuality has not occurred for the many sets of data thus far analyzed, but it is conceivable that spectrophotometric data lacking sufficient accuracy might lead to this conclusion.

2. Optical Properties of Thin Metal Films

Step 10: When there is a failure to obtain convergence, or, alternatively, when solutions have been obtained for the data of a given wavelength, a check is made to ascertain whether the entire spectrum has been analyzed. If so, the data treatment process is terminated (Step 20); and if not, data for the next wavelength are read for analysis (Step 1).

Step 11: Experience has shown that even when convergence is obtained, the solutions do not always have physical significance. For instance, negative refractive indices or extinction coefficients, while they might be found in numerically valid solutions, are not physically valid. Each set of solutions of n_2, k_2, and h_2 is checked for physical validity. When the solutions lack physical significance, the situation is treated as though there had been a failure to converge and the initial guesses are incremented (Step 7) prior to another attempt to obtain valid solutions.

Steps 12 and 13: If the solutions appear physically valid, the solutions are kept for further analysis. Because the range of potential solutions is quite broad, the possibility of several regions of convergence (i.e., sets of solutions) must be considered. This possibility was explored while developing this technique by keeping all physically valid solutions [generated by trying all allowable combinations of $n_{2\,(init)}$, $k_{2\,(init)}$, and $h_{2\,(init)}$]. Examination of these results (obtained for several sets of data) revealed that although there might be multiple sets of solutions, generally all discrete sets of solutions would be found among the first five to ten sets. For this reason the first 15 sets of physically valid solutions generated are felt to be a number sufficient in most cases to include all discrete sets of solutions.

Table 2 gives the 15 sets of solutions generated for the platinum film data of Fig. 10. It is seen that the solutions fall into distinct groups, but that no unique solution exists.

Figure 12 contains the flow chart of the process for restricting the available sets of solutions to a unique best solution for the data used and is called "PASS2." It is actually a continuance of the flow chart given in Fig. 9, and is connected by the entry and exit points labeled A and B.

Steps 14-16: Each solution set is selected in turn. The reflection absorbance is calculated using the solution set at hand for the values of n_2, k_2, and h_2 under the condition of the fourth constraining data measurement. For the platinum film at hand, $A_{r\|}$ at an angle of incidence of 50° was used as the constraining measurement. The absolute value of the difference between the calculated reflection absorbance and that obtained

TABLE 2

Sample Computer Output in the Calculation of the Optical Properties of a Thin Platinum Film[a]

n	k	h (Å)
4.366	0.702	718.2
4.367	0.703	717.7
3.848	1.624	358.4
4.366	0.702	718.2
4.366	0.702	718.3
1.185	2.028	439.6
1.185	2.026	440.0
1.182	2.021	441.1
1.182	2.020	441.3
1.183	2.022	440.9
1.184	2.025	440.3
1.181	2.017	441.8
1.184[b]	2.023[b]	440.5[b]
1.182	2.018	441.6
1.186	2.029	439.5

[a] Conditions: $\lambda = 400.0$ nm, $\theta_1 = 60.0°$, $\theta_2 = 50°$, $n_1 = 1.47$, $n_3 = 1.00$, $A = 1.185$, $A_{r\perp}$ @ $\theta_1 = 0.215$, $A_{r\|}$ @ $\theta_1 = 0.750$, $A_{r\|}$ @ $\theta_2 = 0.653$.

[b] Best fit solutions; $\Delta A = |A_{r\,(\text{calc.})} - A_{r\,(\text{exp.})}|_{50°} = 0.003$.

experimentally (ΔA) is then obtained in order to determine which set of solutions gives the best fit to the fourth constraint.

Steps 17-19: The value of ΔA and the first solution set are saved during the first pass through this subroutine. During each subsequent

2. Optical Properties of Thin Metal Films

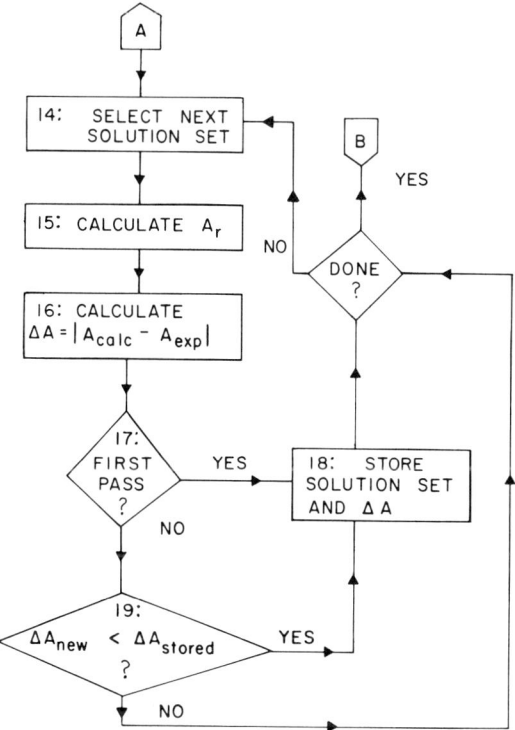

Fig. 12. Flow chart of the data processing sequence employed in a curve-fitting routine.

pass the new value of ΔA is compared with the value previously saved; and, if a better fit is found, the new value of ΔA replaces the old and the corresponding solution set replaces that previously being stored. The result is that when all 15 solution sets have been examined, only the solution set giving rise to the minimum absolute difference between absorbances experimentally observed and those calculated (i.e., ΔA) and ΔA itself, as a measure of the quality of the solutions, have been retained.

Step 20: The data treatment process is terminated and the desired results are printed.

Table 2 contains the results of the above analysis for the solution sets obtained in the analysis of the platinum film data obtained at 400 nm. It should be noted that only solution sets belonging to the same general group gave even reasonably good agreement between experimentally observed and calculated reflection absorbances. Table 3 summarizes those values of n_2, k_2, and h_2 which gave the best fit for each of the 16 points analyzed in the visible spectrum.

At least two criteria regarding the quality of the results can be seen to be fulfilled. First, the calculated thickness of the film is roughly constant throughout the spectrum, which must, of course, be the case for the results to have physical significance. The second criterion satisfied is that the optical constants n_2 and k_2 seem to vary continuously with wavelength. The results would be suspect if marked discontinuities were observed or "sawtooth" spectra of n_2 and k_2 were obtained.

D. Error Analysis

Error coefficients were used in a general error analysis to ascertain the effect of expected errors in measured parameters on the calculated optical properties. The optical properties of the first and third phases and the incident beam wavelength were neglected as sources of error. The maximum errors which might be anticipated for these parameters are insignificant (parts per hundred thousand) relative to those anticipated for the other measurable parameters (parts per thousand). The measured parameters which are expected to have significant errors which will affect the calculated optical constants are A, $A_{r\perp}$, $A_{r\parallel}$, and θ. The three expressions for the estimated errors introduced in each of the optical constants by errors in measurement are given by

$$\Delta n_2 \quad \{\alpha_1^2 (\Delta A)^2 + \beta_1^2 (\Delta A_{r\perp})^2 + \gamma_1^2 (\Delta A_{r\parallel})^2 + \delta_1^2 (\Delta\theta)^2 \}^{1/2} \qquad (38)$$

$$\Delta k_2 \quad \{\alpha_2^2 (\Delta A)^2 + \beta_2^2 (\Delta A_{r\perp})^2 + \gamma_2^2 (\Delta A_{r\parallel})^2 + \delta_2^2 (\Delta\theta)^2 \}^{1/2} \qquad (39)$$

$$\Delta h_2 \quad \{\alpha_3^2 (\Delta A)^2 + \beta_3^2 (\Delta A_{r\perp})^2 + \gamma_3^2 (\Delta A_{r\parallel})^2 + \delta_3^2 (\Delta\theta)^2 \}^{1/2}. \qquad (40)$$

In these expressions α_i, β_i, γ_i, and δ_i are the error coefficients of the measured variables, and Δ is "the estimated error in" the appropriately measured variable. The error coefficients in these expressions are defined as follows:

$$\alpha_1 = \left(\frac{\partial n_2}{\partial A}\right)_{A_{r\perp}, A_{r\parallel}, \theta} \qquad (41)$$

$$\beta_1 = \left(\frac{\partial n_2}{\partial A_{r\perp}}\right)_{A, A_{r\parallel}, \theta} \qquad (42)$$

TABLE 3

The Calculated Optical Constants and Film Thickness of a Thin Platinum Film in the 400-700 nm Region of the Spectrum

λ (nm)	n	k	h (Å)
400	1.18	2.02	440.5
420	1.19	2.16	435.7
440	1.19	2.28	436.2
460	1.20	2.36	442.7
480	1.23	2.52	432.2
500	1.18	2.53	450.2
520	1.17	2.63	453.0
540	1.21	2.74	451.5
560	1.24	2.90	444.8
580	1.20	2.94	455.1
600	1.21	3.08	445.0
620	1.25	3.23	434.1
640	1.25	3.29	445.5
660	1.25	3.37	448.0
680	1.25	3.45	453.3
700	1.30	3.68	426.9

$$\bar{h} = 443.4$$

$$\gamma_1 = \left(\frac{\partial n_2}{\partial A_{r\|}} \right)_{A, A_{r\perp}, \theta} \quad (43)$$

$$\delta_1 = \left(\frac{\partial n_2}{\partial \theta} \right)_{A, A_{r\perp}, A_{r\|}} \quad (44)$$

$$\alpha_2 = \left(\frac{\partial k_2}{\partial A} \right)_{A_{r\perp}, A_{r\|}, \theta} \quad (45)$$

$$\beta_2 = \left(\frac{\partial k_2}{\partial A_{r\perp}}\right)_{A, A_{r\|}, \theta} \tag{46}$$

$$\gamma_2 = \left(\frac{\partial k_2}{\partial A_{r\|}}\right)_{A, A_{r\perp}, \theta} \tag{47}$$

$$\delta_2 = \left(\frac{\partial k_2}{\partial \theta}\right)_{A, A_{r\perp}, A_{r\|}} \tag{48}$$

$$\alpha_3 = \left(\frac{\partial h_2}{\partial A}\right)_{A_{r\perp}, A_{r\|}, \theta} \tag{49}$$

$$\beta_3 = \left(\frac{\partial h_2}{\partial A_{r\perp}}\right)_{A, A_{r\|}, \theta} \tag{50}$$

$$\gamma_3 = \left(\frac{\partial h_2}{\partial A_{r\|}}\right)_{A, A_{r\perp}, \theta} \tag{51}$$

$$\delta_3 = \left(\frac{\partial h_2}{\partial \theta}\right)_{A, A_{r\perp}, A_{r\|}}. \tag{52}$$

Each of the error coefficients was evaluated numerically for the platinum film data at a 400-nm wavelength. Numerical evaluation by first differences was used because obtaining analytical expressions for these first partial derivatives would be difficult. The estimated errors in the optical properties were calculated using Eqs. (38) - (40). For these calculations it was assumed that the errors in the measurable parameters would not exceed the following values: $\Delta A = 0.005$, $\Delta A_{r\perp} = 0.002$, $\Delta A_{r\|} = 0.002$, and $\Delta \theta = 0.1$. The 0-1, 1-2 A slidewire used for the spectrometric measurements had a manufacturer's stated maximum nonlinearity of 0.002A on the 0-1 scale and 0.005A on the 1-2 scale. The RMVA cell has a vernier for angle measurement which can be read to the nearest tenth of a degree. After calibration, the reproducibility of the cell setting is limited to $\pm 0.1°$ accuracy.

The estimated errors in the calculated optical properties of the thin platinum film at $\lambda = 400$ nm were found to be $\Delta n = 0.009$, $\Delta k = 0.025$, and

2. Optical Properties of Thin Metal Films

$\Delta h = 6.24$ Å. These values represent estimated errors of less than 2% in n, k, and h.

E. Internal Self-Consistency of Results

The validity of the calculated optical constants was verified by comparing experimentally observed spectra with a calculated spectra obtained by substituting the calculated optical constants in Eqs. (21) - (31), along with the known experimental conditions. The first test used for the platinum film was the comparison of the calculated IRS spectrum for ⊥-polarization at an angle of incidence of 50° with that obtained experimentally as reported on page 68. Figure 13 shows both the calculated and experimental spectra, which are in excellent agreement.

An additional test, devised to give further confidence in the calculated constants and to test the general applicability of the results, involved the use of water as the third phase. As described earlier this system simulates most dilute aqueous solutions. The IRS spectrum was obtained for this system at an angle of incidence of 70° using both ⊥- and ∥-polarized light. The water phase was assumed transparent (k = 0.0) and to have a refractive index of 1.33 throughout the visible spectrum. The experimental and calculated spectra are given in Fig. 14. The quality of the agreement

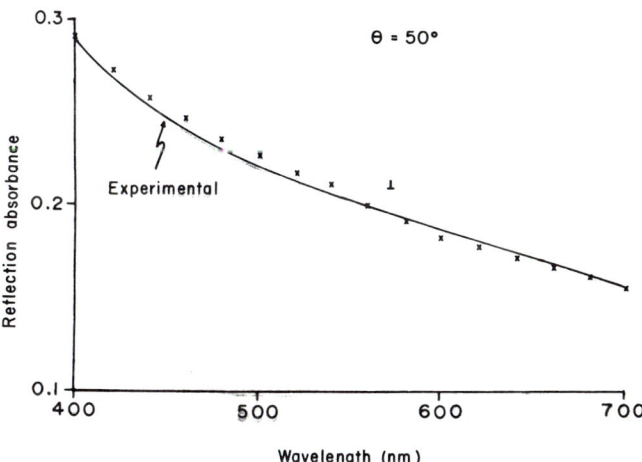

Fig. 13. Calculated and experimental internal reflection spectra of a thin platinum film.

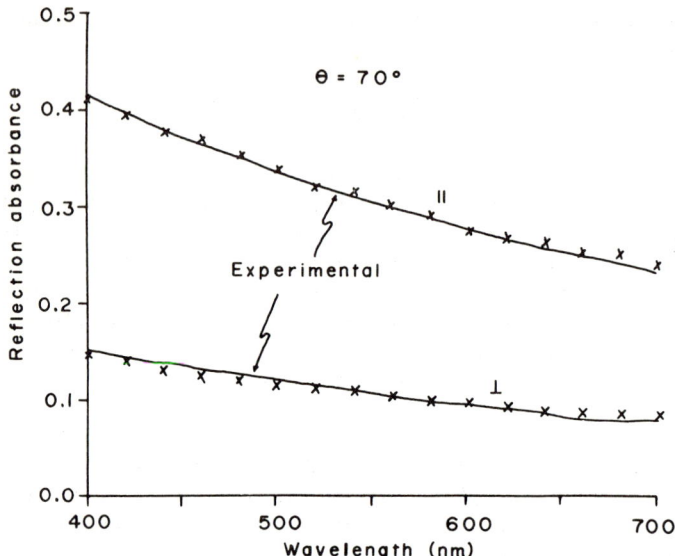

Fig. 14. Calculated and experimental internal reflection spectra of a thin platinum film with an aqueous third phase.

clearly demonstrates the validity of the calculated optical constants. Moreover, it indicates that the optical constants of a metal film electrode can probably be calculated from data obtained during (or immediately before or after) an electrochemical experiment.

VI. DISCUSSION AND SUMMARY

The method of calculating optical properties of thin metal films described in Section V has been used in the study of several films with generally good success. The results of some of these studies are described below in order to illustrate some of the problems which can be encountered and to illustrate further the general utility of the method with respect to the variety of films which can be studied.

The results obtained for each of three different gold films serve to illustrate several important points. The dependence of the refractive index and extinction coefficient on the film thickness is illustrated by the first pair of films. The possibility of anomalous behavior for very thick films is illustrated in the third case. The IRS spectra of one such film were markedly different from those obtained for other gold films and

2. Optical Properties of Thin Metal Films

exhibited an absorbance maximum at about 480 nm. This maximum was most pronounced for parallel polarization and larger angles of incidence.

The three gold films appeared visually to be of markedly different thicknesses. In the following discussion the film numbers are in order of increasing apparent thickness. Gold film #1, the thinnest and most transparent film, yielded the optical properties found in the first column of Table 4. A comparison of experimental absorbance measurements and the absorbances calculated using these optical properties and the experimental conditions is found in Fig. 15. The calculated optical constants of gold film #2 are given for a portion of the spectrum in the second column of Table 4. A comparison of the refractive indices and extinction coefficients of gold films #1 and #2 indicates the marked differences which may exist for films of different thicknesses. These differences are due to the effects of structural changes, the structure of a very thin film being significantly different from that of a bulk metal.

Calculations based on spectral data obtained for gold film #3, the thickest of the gold-film samples, gave anomalous results. The calculated constants had vastly varying thicknesses, discontinuous trends in n or k (i.e., a plot of these properties vs. wavelength consisted of randomly scattered points), and failed to yield calculated IRS spectra in agreement with experimental results. It is felt that this phenomenon is in some way linked to the thickness of the film, this film apparently being much thicker

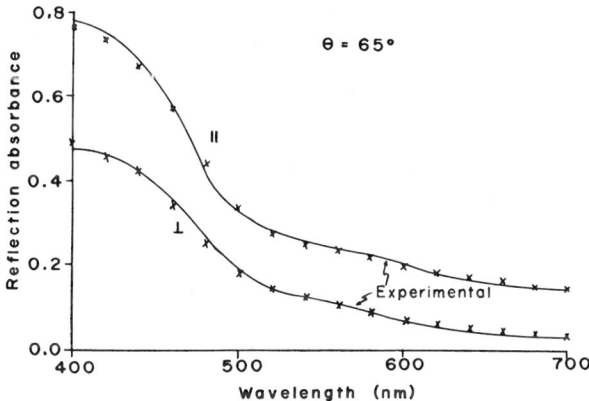

Fig. 15. Calculated and experimental internal reflection spectra of a thin gold film.

TABLE 4

The Calculated Optical Constants and Film Thicknesses of Two Thin Gold Films in the Visible Region of the Spectrum

	Film 1			Film 2		
λ (nm)	n	k	h (Å)	n	k	h (Å)
400	1.82	1.73	128.6	1.57	0.86	309.2
420	1.80	1.80	124.6	1.55	0.86	307.6
440	1.69	1.78	127.9	1.51	0.84	308.1
460	1.47	1.89	124.0	1.46	0.80	308.8
480	1.12	1.91	126.5	1.36	0.72	313.9
500	0.85	2.05	123.3	—	—	—
520	0.70	2.06	126.8	—	—	—
540	0.63	2.29	125.1	—	—	—
560	0.57	2.52	126.7	—	—	—
580	0.49	2.59	138.9	—	—	—
600	0.47	3.16	117.3	—	—	—
620	0.41	3.27	128.0	—	—	—
640	0.39	3.63	120.3	—	—	—
660	0.36	3.76	125.8	—	—	—
680	0.18[a]	1.00[a]	917.5[a]	—	—	—
700	0.32	3.99	135.8	—	—	—

[a] These results must be discounted due to the lack of agreement with the remainder of the spectrum and the lack of fit with the fourth absorbance measurement ($\Delta A = 0.05$).

than film #1 or #2. The IRS spectrum of this film (see Fig. 16) is quite different in character from spectra obtained for other gold films. The IRS spectra of Fig. 15 are representative of those typically obtained for very thin gold films. It seems as though film #3 is becoming an

2. Optical Properties of Thin Metal Films 85

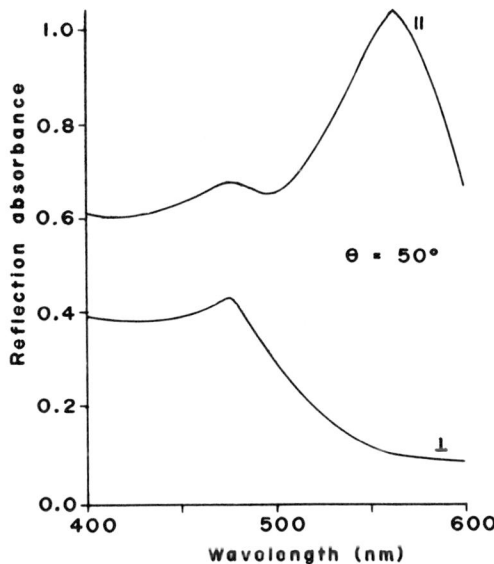

Fig. 16. Internal reflection spectra of a thick gold film.

increasingly better reflector toward the high energy end of the spectrum, possibly due to an approach to bulk-metal properties. Although this behavior is not totally understood, the results are cited to accentuate the fact that the IRS approach to the calculation of the optical properties of thin films is not universally applicable. This approach is, however, useful as a means of rapidly and accurately determining the optical properties of thin films under a wide variety of conditions as described below.

Insofar as the film itself is concerned, the only restrictions are, essentially, that the film be sufficiently thin and nonabsorbing so that energy enters the third phase and that it satisfy the model assumed in making the calculations. This last requirement was tested by attempting to calculate the optical constants and thickness of an ultrathin gold film which was obviously inhomogeneous and did not have plane parallel surfaces. The film was black in color and the resistance was greater than could be measured with a voltohmmeter (VOM). As might be expected, solutions were returned by the first pass at a 400-nm wavelength, but the "best fit" solution obtained by imposing a fourth constraining measurement, however, was found to be invalid due to a very large ΔA ($\Delta A = [A_{calc} - A_{exp}]$ for the constraining measurement). The value of ΔA in this instance was 0.26, compared with values typically less than 0.01 for homogeneous films. Furthermore, when the "best values" of the optical properties (at an angle of incidence of 70°) were used to calculate the anticipated absorbances when water constituted the third phase, $\Delta A_\perp$ was found to be 0.09 and $\Delta A_\parallel$ to be 0.29. These results accentuate the need to employ systems which satisfy the theoretical model as closely as possible.

The only restriction on the third phase, in essence, is that the refractive index be less than that of the thin-film substrate. When the third phase is a solution suitable for electrochemical studies, it may, in general, be considered transparent and to have a refractive index identical to that of the solvent. These approximations are valid whenever the solution is sufficiently dilute and single-reflection IRS spectra are being employed. The concentration limit of the solution is dependent, of course, on the molar absorptivity of the solute, but may be several millimolar even for highly light-absorbing species such as potassium permanganate.

The capability of obtaining the optical constants during the course of an electrolysis may be useful in the quantitative study of electromodulation effects, specific adsorbtion, and optical effects of the double layer.

APPENDIX I: SYMBOLS AND ABBREVIATIONS

Abbreviations:

ATR	Attenuated total reflection
IR	Infrared
IRE	Internal reflection element
IRS	Internal reflection spectroscopy
RMVA	Retromirror variable angle (attachment)
TE	Transverse electric
TIR	Total internal reflection
TM	Transverse magnetic
UV	Ultraviolet

Symbols:

a	Absorptivity
A (A_T)	Transmission absorbance

2. Optical Properties of Thin Metal Films

A_r	Reflection absorbance
Å	Ångstrom
b	Cell thickness
b_{eff}	Effective cell thickness
c	Concentration or speed of light (in vacuo)
°C	Degrees centigrade
d	Reduced film thickness $(= h/\lambda_0)$
d_p	Penetration depth
e	Base of natural logarithms $(= 2.71828\ldots)$
E	Electric field vector
$<E^2>$	Time average of electric field
E_0	Electric field amplitude
h	Film thickness (or phase thickness)
$\hat{i}$	Imaginary unit $(= \sqrt{-1})$
I	Light intensity
I_0	Incident light intensity
$k\ (k_i)$	Extinction coefficient (in ith phase)
$n\ (n_i)$	Refractive index (in ith phase)
$\hat{n}\ (\hat{n}_i)$	Complex refractive index $(= n + \hat{i}k)$
n_{ij}	Ratio of refractive indices of ith and jth phases
p	Parallel
P	Polarized
$\vec{r}$	Position vector
r_{ij}	Reflection coefficient at boundary of ith and jth phases
R	Reflectivity
s	Perpendicular (senkrecht)
t	Time (in seconds)
T	Transmittance
$|\vec{\nu}|$	Wave number $(= 2\pi/\lambda)$
x, y, z	Cartesian coordinates

α	Absorption coefficient
$\alpha_i, \beta_i, \gamma_i, \delta_i$	Error coefficients (of $\underline{i}$th variable)
β_i	$= 2\pi d \xi_i$
Δ	Difference delta
ϵ	Molar absorptivity
ξ_i	$= \hat{n}_i \cos \theta_i$
$\varkappa, \vec{\varkappa}$	Attenuation index, propagation vector
$\lambda \, (\lambda_o)$	Wavelength (in vacuo)
Ω	Resistance (ohms)
π	Pi ($= 3.14159\ldots$)
$\theta \, (\theta_i)$	Angle of incidence (in $\underline{i}$th phase)
θ_c	Critical angle
ω	Angular frequency
$\perp$	Perpendicular
$\parallel$	Parallel

APPENDIX II: THE CALCULATION OF THE REFLECTION ABSORBANCES OF AN n-PHASE SYSTEM

A matrix method for the calculation of reflection absorbances of n-phase systems is given in this Appendix. The equations given below were published in a paper by Hansen [28]. For additional details related to these equations and the equations for the related calculations of phase changes, transmittances, and field intensities, Hansen's paper should be consulted.

It is convenient to define the following terms for the jth layer of an n-phase system to simplify further equations:

$$\xi_j = \hat{n}_j \cos \theta_j = (n_j^2 - n_1^2 \sin^2 \theta_1)^{1/2} \tag{A1}$$

$$\beta_j = 2\pi (h_j / \lambda) \xi_j \tag{A2}$$

$$p_j = \xi_j = \hat{n}^2 q_j \tag{A3}$$

2. Optical Properties of Thin Metal Films

where $\hat{n}_j = n_j + ik_j$ and n and k have their usual meanings. For each phase between the first phase and the last there exists a characteristic matrix which is polarization dependent.

A. Perpendicular Polarization

For the jth layer the characteristic matrix is given by

$$M_j = \begin{bmatrix} \cos(\beta_j) & -\hat{i}\sin(\beta_j)/p_j \\ -\hat{i}p_j\sin(\beta_j) & \cos(\beta_j) \end{bmatrix}. \tag{A4}$$

A characteristic matrix for the group of intermediate phases is defined by

$$M = M_2 M_3 \cdots M_{n-1} = \begin{bmatrix} m_{11} & m_{12} \\ m_{21} & m_{22} \end{bmatrix}. \tag{A5}$$

The reflection coefficient for the entire system is then expressed in terms of the matrix elements of Eq. (A5) as

$$r_\perp = \frac{(m_{11} + m_{12}p_n)p_1 - (m_{21} + m_{22}p_n)}{(m_{11} + m_{12}p_n)p_1 + (m_{21} + m_{22}p_n)}. \tag{A6}$$

The reflectivity is

$$R_\perp = |r_\perp|^2 = r_\perp \cdot r_\perp^* \tag{A7}$$

and the reflection absorbance is

$$A_{r\perp} = -\log_{10}(R_\perp). \tag{A8}$$

B. Parallel Polarization

The characteristic matrix of the jth layer in this case is given by

$$M_j = \begin{bmatrix} \cos(\beta_j) & -\hat{i} \sin(\beta_j)/q_j \\ -\hat{i}q_j \sin(\beta_j) & \cos(\beta_j) \end{bmatrix} \quad . \tag{A9}$$

Using the matrix elements of the group matrix given by Eq. (A5), the reflection coefficient is

$$r_{\|} = \frac{(m_{11} + m_{12}q_n)q_1 - (m_{21} + m_{22}q_n)}{(m_{11} + m_{12}q_n)q_1 + (m_{21} + m_{22}q_n)} \quad . \tag{A10}$$

The reflectivity is expressed in the ordinary way by

$$R_{\|} = |r_{\|}|^2 = r_{\|} \cdot r_{\|}^* \tag{A11}$$

and the reflection absorbance by

$$A_{r\|} = -\log_{10}(R_{\|}) \quad . \tag{A12}$$

APPENDIX III: A FORTRAN IV LANGUAGE COMPUTER PROGRAM FOR THE CALCULATION OF REFLECTION ABSORBANCES OF AN n-PHASE SYSTEM

In this Appendix a listing is given of a FORTRAN IV computer program for the calculation of reflection absorbances of an n-phase system with sample results. Because it is frequently desired to calculate absorbances as a function of the angle of incidence, provision is made to calculate up to 100 pairs of absorbances over any range of angles of incidence. The number of pairs of absorbances may be varied to suit the user. The statements preceded by a "C" are comments intended for clarification and may ordinarily be retained without change. The program, as it is written, may be used for no more than a ten-phase system, which will ordinarily be more than sufficient.

2. Optical Properties of Thin Metal Films

Program Listing:

```
      COMPLEX CI/(0.0,1.0)/, CN(10), CXI(10), CR, CQ(10), CBETA(10), CM
     $(10,2,2), CMN(2,2), CS1, CS2, CS3, CS4, CP1, CP2
C
C     A DOLLAR SIGN IN COLUMN 6 INDICATES A CONTINUATION OF THE
C     PREVIOUS FORTRAN STATEMENT
C
      REAL N(10), K(10), H(10), LAMB
      DIMENSION THETA(100), RS(100), AS(100), RP(100), AP(100)
      DATA RADIAN/57.29577951/, TWOPI/6.283185307/
    1 FORMAT (I2)
    2 FORMAT (8F10.5)
    3 FORMAT (2F10.5, I4)
    4 FORMAT (5F10.5)
    5 FORMAT ('2',T50,'LAMBDA = ',F6.1,2X,'ANGSTROMS')
    7 FORMAT ('-',T20,'PHASE:',T40,'N:',T60,'K:',T80,'H:')
    8 FORMAT ('0',T23,'1',T37,F6.4,T55,'TRANSPARENT',T74,'SEMI-INFINITE
     $')
    9 FORMAT (' ',T22,I2,T37,F6.4,T57,F6.4,T77,F6.1)
   10 FORMAT (' ',T22,I2,T37,F6.4,T57,F6.4,T74,'SEMI-INFINITE',///)
   11 FORMAT (' THETA',5X,'R(PERP)%',4X,'A(PERP)',5X,'R(PAR)%',4X,'A(PA
     $R)')
   12 FORMAT (' ',F4.1,7X,F5.1,7X,F5.3,6X,F5.1,7X,F5.3)
C
C     THE FOLLOWING SECTION IS FOR DATA INPUT...
C     IN AN INTERACTIVE MODE IT IS DESIRABLE TO PRECEDE EACH READ
C     STATEMENT TO PROMPT THE DESIRED INPUT...E.G.- THE FIRST
C     STATEMENTS MIGHT BE:
C20   WRITE (6,15)
C15   FORMAT (' ENTER THE NUMBER OF PHASES')
C     READ (5,1) NOP
C
   20 READ (5,1,END=100) NOP
      READ (5,2) LAMB
      READ (5,3) THETA(1), DTHETA, NI
      READ (5,2) N(1)
C
C     A 2 PHASE SYSTEM IS TREATED AS A SPECIAL CASE
C
      IF (NOP .EQ. 2) GO TO 22
      M = NOP - 1
      DO 21 I=2,M
   21 READ (5,2) N(I), K(I), H(I)
      READ (5,2) N(NOP), K(NOP)
      GO TO 24
   22 READ (5,2) N(2), K(2)
      CN(2) = CMPLX(N(2),K(2))
      DO 23 I=1,NI
      THETA(I) = THETA(I)/RADIAN
      STH = SIN(THETA(I))
      CTH = COS(THETA(I))
      TEM = N(1) * N(1) * STH * STH
      XI1 = N(1) * CTH
      CXI(2) = CSQRT((CN(2) * CN(2) - TEM)
      CR = (XI1 - CXI(2))/(XI1 + CXI(2))
```

```
      RS(I) = REAL(CR * CONJG(CR))
      AS(I) = -ALOG10(RS(I))
      RS(I) = RS(I) * 100.
      CR = ((CN(2) * CN(2) * XI1) - (N(1) * N(1) * CXI(2)))/((CN(2) * C
     $N(2) * XI1) + (N(1) * N(1) * CXI(2)))
      RP(I) = REAL(CR * CONJG(CR))
      AP(I) = -ALOG10(RP(I))
      RP(I) = RP(I) * 100.
      THETA(I) = THETA(I) * RADIAN
      J = I + 1
   23 THETA(J) = THETA(I) + DTHETA
      GO TO 90
C
C     THIS SECTION TREATS MULTIPHASE SYSTEMS
C
   24 CN(1) = CMPLX(N(1),0.0)
      DO 25 I=2,NOP
   25 CN(I) = CMPLX(N(I),K(I))
      DO 65 I=1,NI
      THETA(I) = THETA(I)/RADIAN
      STH = SIN(THETA(I))
      CTH = COS(THETA(I))
      TEM = N(1) * N(1) * STH * STH
      DO 26 J=1,NOP
      CXI(J) = CSQRT((CN(J) * CN(J)) - TEM)
   26 CQ(J) = CXI(J)/(CN(J) * CN(J))
      DO 27 J=2,M
      H(J) = H(J)/LAMB
      CBETA(J) = TWOPI * CXI(J) * H(J)
      H(J) = H(J) * LAMB
      CM(J,1,1) = CCOS(CBETA(J))
      CM(J,2,2) = CM(J,1,1)
      CM(J,1,2) = -(CI * CSIN(CBETA(J)))/CXI(J)
   27 CM(J,2,1) = -(CI * CXI(J) * CSIN(CBETA(J)))
      IF (NOP .GT. 3) GO TO 30
      CMN(1,1) = CM(2,1,1)
      CMN(1,2) = CM(2,1,2)
      CMN(2,1) = CM(2,2,1)
      CMN(2,2) = CM(2,2,2)
      GO TO 40
   30 CMN(1,1) = (CM(2,1,1) * CM(3,1,1)) + (CM(2,1,2) * CM(3,2,1))
      CMN(1,2) = (CM(2,1,1) * CM(3,1,2)) + (CM(2,1,2) * CM(3,2,2))
      CMN(2,1) = (CM(2,2,1) * CM(3,1,1)) + (CM(2,2,2) * CM(3,2,1))
      CMN(2,2) = (CM(2,2,1) * CM(3,1,2)) + (CM(2,2,2) * CM(3,2,2))
      IF (NOP .LT. 5) GO TO 40
      DO 35 J=4,M
      CS1 = CMN(1,1)
      CS2 = CMN(1,2)
      CS3 = CMN(2,1)
      CS4 = CMN(2,2)
      CMN(1,1) = (CM(J,1,1) * CS1) + (CM(J,1,2) * CS3)
      CMN(1,2) = (CM(J,1,1) * CS2) + (CM(J,1,2) * CS4)
      CMN(2,1) = (CM(J,2,1) * CS1) + (CM(J,2,2) * CS3)
   35 CMN(2,2) = (CM(J,2,1) * CS2) + (CM(J,2,2) * CS4)
```

2. Optical Properties of Thin Metal Films

```
      40  CP1 = (CMN(1,1) + (CMN(1,2) * CXI(NOP))) * CXI(1)
          CP2 = CMN(2,1) + (CMN(2,2) * CXI(NOP))
          CR = (CP1 - CP2)/(CP1 + CP2)
          RS(I) = REAL(CR * CONJG(CR))
          AS(I) = -ALOG10(RS(I))
          RS(I) = RS(I) * 100.
          DO 45 J=2,M
          CM(J,1,1) = CCOS(CBETA(J))
          CM(J,2,2) = CM(J,1,1)
          CM(J,1,2) = -(CI * CSIN(CBETA(J)))/CQ(J)
      45  CM(J,2,1) = -(CI * CSIN(CBETA(J)) * CQ(J))
          IF (NOP .GT. 3) GO TO 50
          CMN(1,1) = CM(2,1,1)
          CMN(1,2) = CM(2,1,2)
          CMN(2,1) = CM(2,2,1)
          CMN(2,2) = CM(2,2,2)
          GO TO 60
      50  CMN(1,1) = (CM(2,1,1) * CM(3,1,1)) + (CM(2,1,2) * CM(3,2,1))
          CMN(1,2) = (CM(2,1,1) * CM(3,1,2)) + (CM(2,1,2) * CM(3,2,2))
          CMN(2,1) = (CM(2,2,1) * CM(3,1,1)) + (CM(2,2,2) * CM(3,2,1))
          CMN(2,2) = (CM(2,2,1) * CM(3,1,2)) + (CM(2,2,2) * CM(3,2,2))
          IF (NOP .LT. 5) GO TO 60
          DO 55 J=4,M
          CS1 = CMN(1,1)
          CS2 = CMN(1,2)
          CS3 = CMN(2,1)
          CS4 = CMN(2,2)
          CMN(1,1) = (CM(J,1,1) * CS1) + (CM(J,1,2) * CS3)
          CMN(1,2) = (CM(J,1,1) * CS2) + (CM(J,1,2) * CS4)
          CMN(2,1) = (CM(J,2,1) * CS1) + (CM(J,2,2) * CS3)
      55  CMN(2,2) = (CM(J,2,1) * CS2) + (CM(J,2,2) * CS4)
      60  CP1 = (CMN(1,1) + (CMN(1,2) * CQ(NOP))) * CS(1)
          CP2 = CMN(2,1) + (CMN(2,2) * CQ(NOP))
          CR = (CP1 - CP2)/(CP1 + CP2)
          RP(I) = REAL(CR * CONJG(CR))
          AP(I) = -ALOG10(RP(I))
          RP(I) = RP(I) * 100.
          THETA(T) = THETA(I) * RADAIN
          J = I + 1
      65  THETA(J) = THETA(I) + DTHETA
      C
      C   THIS SECTION PRINTS RESULTS OF CALCULATIONS
      C
      90  WRITE (6,5) LAMB
          WRITE (6,7)
          WRITE (6,8) N(1)
          IF (NOP .EQ. 2) GO TO 91
          WRITE (6,9) (I, N(I), K(I), H(I), I=2,M)
      91  WRITE (6,10) NOP, N(NOP), K(NOP)
          WRITE (6,11)
          WRITE (6,12) (THETA(I), RS(I), AS(I), RP(I), AP(I), I=1,NI)
          GO TO 20
     100  STOP
          END
```

Sample Data Cards:

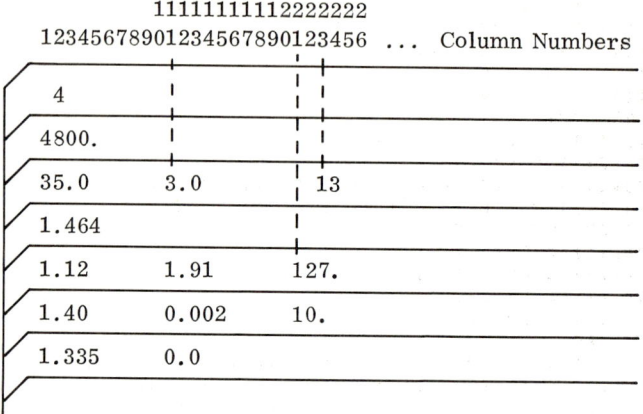

Sample Output:

LAMBDA = 4800.0 ANGSTROMS

PHASE:	N:	K:	H:
1	1.4640	TRANSPARENT	SEMIINFINITE
2	1.1200	1.9100	127.0
3	1.4000	0.0020	10.0
4	1.3350	0.0000	SEMIINFINITE

THETA	R(PERP)%	A(PERP)	R(PAR)%	A(PAR)
35.0	9.0	1.044	6.4	1.196
38.0	9.5	1.020	6.4	1.193
41.0	10.1	0.994	6.5	1.187
44.0	10.8	0.966	6.7	1.175
47.0	11.6	0.935	7.0	1.158
50.0	12.5	0.901	7.4	1.131
53.0	13.7	0.865	8.1	1.092
56.0	15.0	0.825	9.2	1.037
59.0	16.5	0.782	11.1	0.955
62.0	18.5	0.733	15.0	0.824
65.0	21.2	0.673	28.7	0.542
68.0	37.5	0.426	42.3	0.374
71.0	47.0	0.328	28.2	0.550

2. Optical Properties of Thin Metal Films

APPENDIX IV: A FORTRAN IV LANGUAGE COMPUTER PROGRAM FOR THE CALCULATION OF THE OPTICAL PROPERTIES OF THIN METAL FILMS FROM IRS MEASUREMENTS

Program Listing:

```
       DIMENSION X(3), RI(15), EC(15), TH(15)
       COMMON A1, A2, A3, A4, R1, R3, W, T1, T2
       DATA N/3/, NUMSIG/2/, GN/.5/, GK/.5/, GH/50./
1      FORMAT (8F10.5)
2      FORMAT (3(F10.5,5X))
10     READ (5,1,END=100) A1, A2, A3, A4
       NCNT = 1
       READ (5,1) W, T1, T2
       READ (5,1) R1, R3
       SGN = GN
       SGK = GK
       SGH = GH
15     X(1) = SGN
       X(2) = SGK
       X(3) = SGH
16     MAXIT = 20
       CALL NONLIN(N,MAXIT,NUMSIG,ISING,X)
       IF (ISING .EQ. 0) GO TO 20
       IF (X(1) .LT. 0.) GO TO 20
       IF (X(1) .GT. 20.) GO TO 20
       IF (X(2) .LT. 0.) GO TO 20
       IF (X(2) .GT. 20.) GO TO 20
       IF (X(3) .LT. 0.) GO TO 20
       IF (X(3) .GT. 1000.) GO TO 20
       WRITE (6,2) (X(I), I=1,3)
       RI(NCNT) = X(1)
       EC(NCNT) = X(2)
       TH(NCNT) = X(3)
       NCNT = NCNT + 1
       IF (NCNT .GT. 15) GO TO 99
20     CONTINUE
       SGH = SGH + 20.
       IF (SGH .GT. 800.) GO TO 31
       GO TO 15
31     SGH = GH
       SGK = SGK + 0.5
       IF (SGK .GT. 6.) GO TO 32
       GO TO 15
32     SGK = GK
       SGN = SGN + 0.5
       IF (SGN .GT. 6.) GO TO 99
       GO TO 15
99     CALL PASS2(RI,EC,TH)
100    STOP
       END
       SUBROUTINE EVAL(X,Y,K)
       DIMENSION X(3)
       COMMON A1, A2, A3, A4, R1, R3, W, T1, T2
       COMPLEX C2, C3, XI2, XI3, BETA, CI/(0.0,1.0)/, RS12, RS23, RCXP,
      $RFS
       DATA TWOPI/6.283185307/, RADIAN/57.29577951/
```

```
            TH = T1
            IR = 0
            R = R3/R1
            THC = ARSIN(R)
            THC = THC * RADIAN
            S = 0.0
            C2 = CMPLX(X(1),X(2))
            IF (K .GT. 1) GO TO 10
            BETA = (TWOPI * X(3) * C2)/W
            RS12 = (1. + C2) * (C2 + 1.)
            RS23 = (C2 - 1.) * (1. - C2)
            C3 = CI * BETA
            X1 = ABS(REAL(C3))
            IF (X1 .GT. 174.) GO TO 50
            X2 = ABS(AIMAG(C3))
            X3 = TWOPI * 2**17
            IF (X2 .GE. X3) GO TO 50
            REXP = CEXP(C3)
            REXP = 4. * C2 * REXP
            C3 = 2. * C3
            X1 = ABS(REAL(C3))
            IF (X1 .GT. 174.) GO TO 50
            X2 = ABS(AIMAG(C3))
            IF (X2 .GE. X3) GO TO 50
            RFS = CEXP(C3)
            RS23 = RS23 * RFS
            RS23 = RS12 + RS23
            REXP = REXP/RS23
            D = REXP * CONJG(REXP)
            IF (D .EQ. 0) GO TO 50
            P = -ALOG10(D)
            Y = P - A1
            RETURN
     10     T = TH/RADIAN
            IF (TH .LT. THC) IR = 1
            ST = SIN(T)
            CT = COS(T)
            IF (IR .EQ. 0) GO TO 21
            H = 0.0
            GO TO 22
     21     H = X(3)/W
     22     C3 = CMPLX(R3,0.0)
            B = (R1**2) * (ST**2)
            XI1 = CT * R1
            XI2 = CSQRT(C2**2 - B)
            XI2 = CMPLX(ABS(REAL(XI2)),ABS(AIMAG(XI2)))
            XI3 = CSQRT(C3**2 - B)
            XI3 = CMPLX(ABS(REAL(XI3)),ABS(AIMAG(XI3)))
            BETA = TWOPI * H * XI2
            BETA = BETA * 2. * CI
            X1 = ABS(REAL(BETA))
```

2. Optical Properties of Thin Metal Films

```
            IF (X1 .GT. 174.) GO TO 50
            X2 = ABS(AIMAG(BETA))
            IF (X2 .GE. X3) GO TO 50
            REXP = CEXP(BETA)
            IF (K .GT. 2) GO TO 40
            RS12 = (XI1-XI2)/(XI1+XI2)
            RS23 = (XI2-XI3)/(XI2+XI3)
            RFS = (RS12 + RS23 * REXP)/(1. + RS12 * RS23 * REXP)
            RS = RFS * CONJG(RFS)
            P = -ALOG10(RS)
            IF (IR .EQ. 0) GO TO 29
            S = P
            IR = 0
            GO TO 21
   29       Y = P - A2 - S
            RETURN
   40       RS12 = (CT/R1-XI2/C2**2)/(CT/R1+XI2/C2**2)
            RS23 = (XI2/C2**2-XI3/C3**2)/(XI2/C2**2+XI3/C3**2)
            RFS = (RS12+RS23*REXP)/(1.+RS12*RS23*REXP)
            RS = RFS * CONJG(RFS)
            P = -ALOG10(RS)
            IF (IR .EQ. 0) GO TO 39
            S = P
            IR = 0
            GO TO 21
   39       Y = P - A3 - S
            RETURN
   50       Y = 100.
            RETURN
            END
            SUBROUTINE NONLIN(N,MAXIT,NUMSIG,ISING,X)
            INTEGER POINT(3,3)
            DIMENSION X(3), TEMP(3), PART(3), COE(3,4), ISUB(2)
            ICONV = 1
            ISING = 1
            NPLUS = N + 1
            RELCON = 10.**(-NUMSIG)
            DO 90 M=1,MAXIT
            DO 21 J=1,N
   21       POINT(1,J) = J
            DO 70 K=1,N
            IF (K .GT. 1) GO TO 22
            GO TO 23
   22       CALL BACK(K,N,X,ISUB,COE,POINT)
   23       CALL EVAL(X,F,K)
            IF (F .GT. 5.) GO TO 92
            FACTOR = 0.001
   24       ITALY = 0
            DO 41 I=K,N
            ITEMP = POINT(K,I)
            HOLD = X(ITEMP)
```

```
              H = FACTOR * HOLD
              IF (H .EQ. 0.) H = 0.001
              X(ITEMP) = HOLD + H
              IF (K .GT. 1) GO TO 25
              GO TO 26
        25    CALL BACK(K,N,X,ISUB,COE,POINT)
        26    CALL EVAL(X,FPLUS,K)
              IF (FPLUS .GT. 5.) GO TO 92
              PART(ITEMP) = (FPLUS - F)/H
              X(ITEMP) = HOLD
              IF (ABS(PART(ITEMP)) .EQ. 0.) GO TO 35
              IF (ABS(F/PART(ITEMP)) .GT. 1E20) GO TO 35
              GO TO 40
        35    ITALY = ITALY + 1
        40    CONTINUE
        41    CONTINUE
              IF (ITALY .LE. (N-K)) GO TO 42
              FACTOR = FACTOR * 10.
              IF (FACTOR .GT. 0.5) GO TO 92
              GO TO 24
        42    IF (K .LT. N) GO TO 43
              IF (ABS(PART(ITEMP)) .EQ. 0.) GO TO 92
              COE(K,NPLUS) = 0.0
              KMAX = ITEMP
              GO TO 71
        43    KMAX = POINT(K,K)
              DERMAX = ABS(PART(KMAX))
              KPLUS = K + 1
              DO 52 I=KPLUS,N
              JSUB = POINT(K,I)
              TEST = ABS(PART(JSUB))
              IF (TEST .LT. DERMAX) GO TO 51
              DERMAX = TEST
              POINT(KPLUS,I) = KMAX
              KMAX = JSUB
              GO TO 52
        51    POINT(KPLUS,I) = JSUB
        52    CONTINUE
              IF (ABS(PART(KMAX)) .EQ. 0.) GO TO 92
              ISUB(K) = KMAX
              COE(K,NPLUS) = 0.
              DO 53 J=KPLUS,N
              JSUB = POINT(KPLUS,J)
              COE(K,JSUB) = -PART(JSUB)/PART(KMAX)
              COE(K,NPLUS) = COE(K,NPLUS) + (PART(JSUB) * X(JSUB))
        53    CONTINUE
        71    COE(K,NPLUS) = (COE(K,NPLUS) - F)/PART(KMAX) + X(KMAX)
        70    CONTINUE
              X(KMAX) = COE(N,NPLUS)
              IF (N .GT. 1) GO TO 72
              GO TO 73
```

2. Optical Properties of Thin Metal Films

```
72      CALL BACK(N,N,X,ISUB,COE,POINT)
73      IF (M .EQ. 1) GO TO 81
        DO 74 I=1,N
        TRY = ABS((TEMP(I) - X(I))/X(I))
        IF (TRY .GT. RELCON) GO TO 75
        ICONV = ICONV + 1
        IF (ICONV .GE. 3) GO TO 91
74      CONTINUE
        GO TO 81
75      ICONV = 1
81      DO 82 I=1,N
82      TEMP(I) = X(I)
90      CONTINUE
        GO TO 92
91      MAXIT = M
        GO TO 99
92      ISING = 0
99      CONTINUE
        RETURN
        END
        SUBROUTINE BACK(K,N,X,ISUB,COE,POINT)
        INTEGER POINT(3,3)
        DIMENSION ISUB(2), X(3), COE(3,4)
        NPLUS = N + 1
        KMIN = K - 1
        KM = K
1       IF (KM .LT. 2) GO TO 3
        KMAX = ISUB(KMIN)
        X(KMAX) = 0.
        DO 2 J=KM,N
        JSUB = POINT(KM,J)
2       X(KMAX) = X(KMAX) + (COE(KMIN,JSUB) * X(JSUB))
        X(KMAX) = X(KMAX) + COE(KMIN,NPLUS)
        KM = KM - 1
        GO TO 1
3       CONTINUE
        RETURN
        END
        SUBROUTINE PASS2(RI,EC,TH)
        DIMENSION RI(15), EC(15), TH(15), RES(2)
        COMMON A1, A2, A3, A4, R1, R3, W, T1, T2
        COMPLEX C2, C3, XI2, XI3, BETA, CI/(0.0,1.0)/, RS12, RS23, REXP,
       $RFS
        DATA TWOPI/6.283185307/, RADIAN/57.29577951/
        IR = 0
        S = 0.0
        RES(2) = 10.0
        R = R3/R1
        THC = ARSIN(R)
        THC = THC * RADIAN
        IF (T2 .LT. THC) IR = 1
```

```
      T = T2/RADIAN
      ST = SIN(T)
      CT = COS(T)
      C3 = CMPLX(R3,0.0)
      X3 = TWOPI * 2**17
      DO 30 N=1,15
      C2 = CMPLX(RI(N),EC(N))
      IF (IR .EQ. 0) GO TO 15
      H = 0.0
      GO TO 16
   15 H = TH(N)/W
   16 B = (R1**2) * (ST**2)
      XI1 = CT * R1
      XI2 = CSQRT(C2**2 - B)
      XI2 = CMPLX(ABS(REAL(XI2)),ABS(AIMAG(XI2)))
      XI3 = CSQRT(C3**2 - B)
      XI3 = CMPLX(ABS(REAL(XI3)),ABS(AIMAG(XI3)))
      BETA = TWOPI * H * XI2
      BETA = BETA * 2. * CI
      X1 = ABS(REAL(BETA))
      IF (X1 .GT. 174.) GO TO 29
      X2 = ABS(AIMAG(BETA))
      IF (X2 .GE. X3) GO TO 29
      REXP = CEXP(BETA)
      RS12 = (CT/R1-XI2/C2**2)/(CT/R1+XI2/C2**2)
      RS23 = (XI2/C2**2-XI3/C3**2)/(XI2/C2**2+XI3/C3**2)
      RFS = (RS12 + RS23 * REXP)/(1. + RS12 * RS23 * REXP)
      RS = RFS * CONJG(RFS)
      P = -ALOG10(RS)
      IF (IR .EQ. 0) GO TO 20
      S = P
      IR = 0
      GO TO 15
   20 Y = P - S
      DA = ABS(Y - A4)
      IF (DA .GT. RES(2)) GO TO 29
      RI(1) = RI(N)
      EC(1) = EC(N)
      TH(1) = TH(N)
      RES(1) = Y
      RES(2) = DA
   29 CONTINUE
   30 CONTINUE
      WRITE (6,1) RI(1), EC(1), TH(1)
    1 FORMAT (' BEST FIT SOLUTIONS:',/,2(F10.5,5X),F10.5)
      WRITE (6,2) RES(1), RES(2)
    2 FORMAT (///,' A(CALC) = ',F5.3,5X,' DELTA A = ',F5.3)
      RETURN
      END
```

2. Optical Properties of Thin Metal Films

Sample Input:

```
                   Column Numbers:
                   1111111111222222222233333
          12345678901234567890123456789012345

Card 1:   1.212       0.344       0.158       0.512
          (A_T)       (A_{r⊥}@θ_1) (A_{r∥}@θ_1) (A_{r∥}@θ_2)

Card 2:   4000.0      65.0        55.0
          (λ)         (θ_1)       (θ_2)

Card 3:   1.470       1.000
          (n_1)       (n_3)
```

N.B. It is assumed for the purposes of these calculations that the first and last phases are transparent.

REFERENCES

[1]. R. W. Ditchburn, Proc. Roy. Irish Acad., 44:123 (1938).

[2]. L. G. Schulz, J. Opt. Soc. Am., 38:432 (1948).

[3]. O. S. Heavens, Optical Properties of Thin Solid Films, Dover, New York, 1965, p. 254.

[4]. L. G. Schulz, J. Opt. Soc. Am., 37:349 (1947).

[5]. O. S. Heavens, Optical Properties of Thin Solid Films, Dover, New York, 1965, p. 5.

[6]. G. Hass and N. W. Scott, J. Phys., 11:394 (1950).

[7]. W. N. Hansen, T. Kuwana, and R. A. Osteryoung, Anal. Chem., 38:1810 (1966).

[8]. A. Prostak, H. B. Mark, Jr., and W. N. Hansen, J. Phys. Chem., 72:2576 (1968).

[9]. N. Winograd, H. N. Blount, and T. Kuwana, J. Phys. Chem., 73:3456 (1969).

[10]. N. Winograd and T. Kuwana, J. Electroanal. Chem., 23:333 (1969).

[11]. N. Winograd, H. N. Blount, and T. Kuwana, J. Phys. Chem., 73:3456 (1969)

[12]. B. S. Pons, J. S. Mattson, L. O. Winstrom, and H. B. Mark, Jr., Anal. Chem., 39:685 (1967).

[13]. H. B. Mark, Jr., and B. S. Pons, Anal. Chem., 38:119 (1966).

[14]. B. S. Pons, J. S. Mattson, L. O. Winstrom, and H. B. Mark, Jr., Anal. Chem., 39:685 (1967).

[15]. A. Prostak, H. B. Mark, Jr., and W. N. Hansen, J. Phys. Chem., 72:2576 (1968).

[16]. P. L. Clegg and A. W. Crook, J. Sci. Instrum., 29:201 (1952).

[17]. R. S. Sennett and G. D. Scott, J. Opt. Soc. Am., 40:203 (1959).

[18]. I. Langmuir and K. H. Kingdon, Proc. Roy. Soc., A, 107:61 (1925).

[19]. A. Prostak, Ph.D. Dissertation, University of Michigan, Ann Arbor, Michigan, 1969.

[20]. C. Statescu, Ann. Phys. (4), 33:1032 (1910).

[21]. P. D. Fochs, J. Opt. Soc. Am., 40:623 (1959).

[22]. L. G. Schulz, J. Chem. Phys., 17:1153 (1949).

[23]. D. Malé, C. R. Acad. Sci., Paris, 230, 1349 (1950).

[24]. J. F. Archard, P. L. Clegg, and A. M. Taylor, Proc. Phys. Soc., 65:758 (1952).

[25]. O. S. Heavens, Optical Properties of Thin Solid Films, Dover, New York, 1965, p. 140.

[26]. J. Fahrenfort and W. M. Visser, Spectrochim. Acta, 18:1103 (1962).

[27]. J. Fahrenfort, Spectrochim. Acta, 17:698 (1961).

[28]. W. N. Hansen, J. Opt. Soc. Am., 58:380 (1968).

[29]. N. J. Harrick, Internal Reflection Spectroscopy, Wiley-Interscience, New York, 1967.

[30]. W. N. Hansen and J. A. Horton, Anal. Chem., 36:783 (1964).

[31]. R. P. Feynman, R. B. Leighton, and M. Sands, The Feynman Lectures on Physics, Vol. II, Addison-Wesley, Reading, Mass., 1964, p. 33-1 ff.

[32]. M. Born and E. Wolf, Principles of Optics, Pergamon, New York, 1959, pp. 1-103.

[33]. J. S. Mattson, H. B. Mark, Jr., and W. J. Weber, Jr., Anal. Chem., 41:355 (1969).

[34]. L. H. Sharpe, Proc. Chem. Soc., 1961:461 (1961).

[35]. N. J. Harrick, private communication.

[36]. S. Yoshida, T. Yamaguchi, and A. Kinbara, J. Opt. Soc. Am., 61:62 (1971).

[37]. G. W. Rubloff, Phys. Rev. B, 3:285 (1971).

[38]. G. D. Scott and R. Shirahata, J. Opt. Soc. Am., 61:131 (1971).

[39]. F. C. Weinstein, J. D. Dow, and B. Y. Lao, Phys. Stat. Sol.(b), 43:K105 (1971).

[40]. O. S. Heavens and S. D. Smith, J. Opt. Soc. Am., 47:469 (1957).

[41]. H. Levinstein, J. Appl. Phys., 20:306 (1949).

[42]. N. J. Harrick, U.S. Pat. 3,491,366 (1970).

[43]. H. E. Bennett and J. M. Bennett, in Physics of Thin Films (G. Hass and R. E. Thun, eds.), Vol. IV, Academic Press, New York, 1967.

[44]. A. R. Stokes, The Theory of the Optical Properties of Inhomogeneous Materials, E. and F. N. Spon, London, 1963, p. 21 ff.

[45]. H. E. Bennett, Appl. Opt., 5:1265 (1966).

[46]. J. A. Horton and W. N. Hansen, Anal. Chem., 39:1097 (1967).

[47]. A. Prostak and W. N. Hansen, Phys. Rev., 160:600 (1967).

[48]. B. D. Cahan, J. Horkans, and E. Yeager, Discuss. Faraday Soc., in press.

[49]. J. D. E. McIntyre and D. M. Kolb, Discuss. Faraday Soc., in press.

[50]. G. Den Boef, H. J. Van Der Beek, and Th. Braaf, Rec. Trav. Chim. Pays-Bas, 77:1064 (1958).

[51]. A. Prostak, Ph.D. Dissertation, University of Michigan, Ann Arbor, Michigan, 1969, p. 85. Citation of a private communication from W. N. Hansen.

[52]. A. B. Winterbottom, Trans. Faraday Soc., 42:487 (1946).

[53]. G. S. G. Beveridge and R. S. Schechter, Optimization: Theory and Practice, McGraw-Hill, New York, 1970.

[54]. F. T. Gucker and R. L. Seifert, Physical Chemistry, Norton, New York, 1966, p. 16.

[55]. K. M. Brown and S. D. Conte, Proceedings A.C.M. National Meeting (1967), p. 111.

[56]. K. M. Brown, Comm. of the A.C.M., 10:728 (1967).

Part II

KINETICS

Chapter 3

APPLICATIONS OF COMPUTER CIRCUITRY AND TECHNIQUES TO
KINETIC METHODS OF ANALYSIS

S. R. Crouch

Department of Chemistry
Michigan State University
East Lansing, Michigan

I.	INTRODUCTION		108
II.	ANALYTICAL KINETIC TECHNIQUES		108
	A.	General Considerations	108
	B.	Mathematical Basis of Kinetic Methods	110
	C.	Single-Component Techniques	116
	D.	Techniques for Simultaneous Determination of Mixtures	132
III.	INSTRUMENTATION FOR KINETIC METHODS		138
	A.	General Considerations	138
	B.	Reaction Monitor Systems	141
	C.	Signal Modifier Systems	151
	D.	Automated Sampling Systems	152
	E.	Computation Systems for Single-Component Rate Methods	154
	F.	Computation Systems for Simultaneous Analysis of Mixtures	188
IV.	SMALL DIGITAL COMPUTERS IN KINETIC ANALYSIS		190
	A.	Applications in Single-Component Kinetic Analysis	191
	B.	Applications in Simultaneous Kinetic Analysis of Mixtures	196
	C.	Applications in Fundamental Studies	198
V.	FUTURE PERSPECTIVES		203
	REFERENCES		204

Copyright © 1973 by Marcel Dekker, Inc. All Rights Reserved. Neither this work nor any part may be reproduced or transmitted in any form or by any means, electronic or mechanical, including photocopying, microfilming, and recording, or by any information storage and retrieval system, without permission in writing from the publisher.

I. INTRODUCTION

The recent widespread availability of low-cost linear and digital integrated circuits is revolutionizing instrumentation in science. Measurements, experimental controls, and data manipulations, which were impossible or involved extremely complex and often unreliable instrumentation merely a few years ago, are today being done on a routine basis. The trend toward completely automated measurement, control, and computation systems is now well established, and fully automated research laboratories are on the horizon. This trend is clearly visible in the youthful field of kinetic methods of analysis. Seemingly a few days ago, the most sophisticated kinetic measurements were made by reading a meter at times indicated by a laboratory clock or mechanical stopwatch. A measurement precision and accuracy of 10% was considered quite acceptable because of the analytical difficulties involved in obtaining concentrations at various times during a reaction. The introduction of continuous recording devices, such as the oscilloscope, the potentiometric recorder, and others, improved the situation somewhat. The tedium of point by point manual data acquisition was eliminated and operator bias reduced. Measurement precision and accuracy also improved. Still necessary, however, was the manual reduction of the recorded data to a form useful for determining rate laws, rate constants, or reactant concentrations. Today, however, systems are available commercially for completely automated kinetic analyses and with the widespread introduction of the small laboratory digital computer, the age of automated rate measurement, control, and data processing is at hand. New developments in these areas should soon make kinetic methods competitive with or superior to conventional analytical techniques for selected determinations.

In the sections of this review which follow, kinetic methods of analysis are examined from an automated measurement viewpoint. Those measurement parameters which influence signal-to-noise ratio, accuracy, and precision are considered in some detail. The applications of analog and digital computing circuitry to the solution of the problems involved in making automated kinetic measurements constitute the major part of this review. In the final section, consideration is given to the use of small digital computers in kinetic analyses.

II. ANALYTICAL KINETIC TECHNIQUES

A. General Considerations

The kinetic approach to analytical chemistry has gained in importance in recent years. Attesting to the popularity of kinetic techniques are several recent review articles dealing with analytical applications of

3. Kinetic Methods of Analysis

kinetics [1-4] and enzymatic analyses [4-6]. Two recent books have also appeared on the subject [7,8]. The former is a comprehensive treatment of kinetics in analytical chemistry, whereas the latter deals primarily with catalytic reactions.

These are several basic reasons why kinetic-based measurements are advantageous over equilibrium-based techniques in certain situations. Many chemical reactions which might be useful for analyses exhibit nonideal behavior in practice, which can lead to large errors in equilibrium-based methods. For example, many reactions do not produce a quantitative yield of product. Often, however, the rates of these reactions, especially during the initial stages of reaction, are quantitatively related to the concentration of the sought-for species and can be used for chemical analyses. Many other reactions are too slow for equilibrium methods. A kinetic measurement, however, can be made in a small fraction of the time that may be required for the reaction to reach equilibrium. Thus kinetic techniques in conjunction with stoichiometric methods make available for analysis a wide variety of chemical reactions. Another advantage of kinetic techniques is the increased selectivity which can be introduced by making use of kinetic differences. In some cases complete separation of the reactions of two or more species with a common reagent can be obtained. By making a kinetic measurement at the time when only the sought-for species contributes to the reaction, high selectivity can be obtained. In cases where complete kinetic separations are impossible, differential kinetic methods can be employed for the simultaneous in situ analysis of two or more species in mixtures, again offering an increase in selectivity over the equilibrium approach.

Kinetic-based methods have several general limitations which must be carefully considered in practical applications. A most important limitation is imposed by the rate of the reaction being measured. For analytical purposes, reactions whose half-lives are longer than a few hours are clearly undesirable. Instrumental and physical factors determine the upper limit of rates which are analytically useful. In most analytical applications half-lives shorter than a few seconds have not been used. However, in principle, many of the advantages of kinetic methods hold true for fast reactions as well as for slow reactions. A second limitation of kinetic methods when compared to equilibrium methods is that the signal-to-noise ratio of the measurement is inherently lower when measurements are made on a dynamic system. Since kinetic methods do not generally utilize the entire reaction, the measured signal change is inherently of lower magnitude than in equilibrium methods. Thus noise can be quite bothersome in kinetic techniques, particularly low-frequency instrumental drifts, which are difficult to average without distorting the measured signal. A third limitation of kinetic methods is the very careful control of reaction conditions which is necessary to obtain accurate and precise results. Any factor which influences the reaction rate must be carefully controlled or monitored during the reaction. Temperature, pH, reagent concentrations, and ionic strengths are a few of the factors which affect the rate of most reactions. Many of these factors need not be as carefully controlled in equilibrium methods. For example, many reactions are forced to completion by addition of a large excess of reagent, which need not be precisely controlled from reaction to reaction. In a kinetic method, however, reagent concentrations must be precisely controlled.

Kinetic techniques for analysis have been classified by several authors [7, 9, 10]. In this discussion, since the major emphasis is to be placed on computer circuitry for automation of kinetic methods, two types of rate methods are considered. The first type of kinetic method considered involves the use of rate information for the determination of a single species in solution. This type of kinetic method is referred to here as a single-component technique. The second class of methods, differential kinetic methods, involves the use of rate data to determine two or more species in mixtures. Each of these two classes may be implemented with similar rate measurement circuitry. However, multicomponent methods involve additional computational steps in order to calculate the concentrations of the individual species.

This discussion first focuses on the principles of those kinetic methods which have been utilized for automated quantitative analysis. Since not nearly as much research has gone into automation of multicomponent rate methods, the major part of this review deals with single-component techniques.

B. Mathematical Basis of Kinetic Methods

In this section the general equations governing kinetic methods are presented to demonstrate measurement principles. Since the type of chemical reaction often governs the particular technique employed, this discussion is divided into three major categories; first- or pseudo-first-order reactions, enzyme-catalyzed reactions, and other catalytic reactions.

1. First-Order or Pseudo-First-Order Reactions

A first-order or pseudo-first-order irreversible reaction of reactant A to form product P may be written as

$$A \xrightarrow{k_A} xP$$

3. Kinetic Methods of Analysis

where k_A is the first- or pseudo-first-order rate constant. The rate of disappearance of A with time is

$$\frac{-d[A]}{dt} = k_A[A]. \tag{1}$$

Integration of Eq. (1) gives the relationship between the concentration of A at any time, $[A]_t$, and the initial concentration, $[A]_o$, which is the desired result of a kinetic measurement

$$[A]_t = [A]_o \exp(-k_A t). \tag{2}$$

Substitution of Eq. (2) into Eq. (1) gives the reaction rate in terms of $[A]_o$

$$\frac{-d[A]_t}{dt} = k_A[A]_o \exp(-k_A t). \tag{3}$$

Equations (2) and (3) are the proportional equations upon which many kinetic methods are based. Equation (2) is not often applied directly. Instead the change in the concentration of A, $\Delta[A]$, over a time interval, $\Delta t = t_2 - t_1$, is related to $[A]_o$ by Eq. (4)

$$\Delta[A] = [A]_o \{\exp(-k_A t_2) - \exp(-k_A t_1)\}. \tag{4}$$

Equation (4) is the basis for the integral approaches to automated rate analysis, the fixed-time and the variable-time methods [7, 10, 11]. Later sections will show that measurement of the amount of A which has reacted in a fixed-time interval can be related to the initial concentration of A (fixed-time method). Likewise, the time required for a fixed amount of A to react can also be related to $[A]_o$ under certain circumstances (variable-time method). The method of proportional equations for differential kinetic analysis can also be based on the integral form, Eq. (2) [7,9]. If two species A and B react with a common reagent under first- or pseudo-first-order conditions, equations similar to (2) can be written for both species. If the rate constants k_A and k_B are sufficiently different, two fixed-time measurements can be made to determine the amounts of A and B reacted and these can be related by simultaneous equations to the initial concentrations.

Equation (3) is the basis for the derivative approach to rate analysis [10]. Analysis for a single species can be carried out by measuring the rate of the reaction at a specific time and relating this to $[A]_o$.

Alternatively, Eq. (3) is considerably simplified if the rate measurement is made early in the reaction before significant concentration changes have occurred. Under these conditions the reaction exhibits pseudo-zero-order kinetics and the exponential term in Eq. (3) is unity. Equation (3) can also be the basis for the simultaneous determination of two or more species in mixtures [12].

It will be noted in later sections that the distinction between those methods based on integral forms of the rate laws [Eqs. (2) and (4)] and those based on the differential form [Eq. (3)] is often small. If the fixed-time or the variable-time method is applied early in the reaction, the measurement approaches a true initial rate measurement.

Often it is the product concentration which is measured instead of the reactant. In that case, since $\Delta[P] = -x \Delta[A]$, Eqs. (3) and (4) can be rewritten

$$\Delta[P] = x[A]_o \{\exp(-k_A t_1) - \exp(-k_A t_2)\} \qquad (5)$$

$$\frac{d[P]_t}{dt} = xk_A[A]_o \exp(-k_A t). \qquad (6)$$

Equations (5) and (6) relate the amount of product formed during a time interval $(t_2 - t_1)$ or the rate of formation of product at any time to the initial concentration of reactant, $[A]_o$.

Most often, product or reactant concentrations are measured instrumentally, in which case an electrical signal $\mathscr{I}$ is produced which is related to $[A]$ or $[P]$. In such cases the instrumental system's transfer function must be taken into consideration. For an instrumental method in which $\mathscr{I}$ is linearly related to the concentration of product, $\Delta \mathscr{I} = \nu \Delta[P]$ and $d\mathscr{I} = \nu d[P]$, where ν is the transfer function in electrical units per concentration unit. Substituting for $\Delta[P]$ and $d[P]$ in Eqs. (5) and (6) and replacing t_2 by $\Delta t + t_1$ yields

$$\Delta \mathscr{I} = \nu x[A]_o \exp(-k_A t_1) \{1 - \exp(-k_A \Delta t)\} \qquad (7)$$

$$\frac{d\mathscr{I}_t}{dt} = \nu xk_A[A]_o \exp(-k_A t). \qquad (8)$$

Thus $[A]_o$ is related to the rate of change of the signal with time by Eq. (8) and to the change in $\mathscr{I}$ by Eq. (7).

3. Kinetic Methods of Analysis

For an instrumental method in which $\mathscr{S}$ is not a linear function of $[P]$, the situation is more complicated. In many cases electronic circuits can be used to linearize the response before the measurement is made. For example, the output of a photomultiplier tube used in absorption spectrophotometry is proportional to transmittance, which is nonlinearly related to concentration. A logarithmic amplifier is often used to provide a signal proportional to absorbance and thus linearly related to concentration.

In many cases, however, such electrical transformations are not convenient with ordinary circuitry and nonlinear response characteristics must be utilized. For these situations, general equations can be used to relate the measured signal to $[P]$. The instrument response can be written in general as

$$\mathscr{S} = F([P]) \tag{9}$$

where $F([P])$ is an arbitrary function. From Eq. (9)

$$d\mathscr{S} = \left(\frac{\partial F([P])}{\partial [P]}\right)_c d[P] \tag{10}$$

where the subscript c indicates that all other parameters are held constant. Similarly, the change in product concentration, $\Delta [P]$, can be expressed as

$$\Delta [P] = f(\mathscr{S}_2) - f(\mathscr{S}_1) \tag{11}$$

where again $f(\mathscr{S})$ is an arbitrary function.

Equations (11) and (5) may be combined to relate the measured signal to $[A]_o$

$$f(\mathscr{S}_2) - f(\mathscr{S}_1) = x[A]_o \exp(-k_A t_1)\{1 - \exp(-k_A \Delta t)\}. \tag{12}$$

Similarly, the rate of change of $\mathscr{S}$ with time can be related to $[A]_o$ by the following equation:

$$\frac{d\mathscr{S}_t}{dt} = \left(\frac{\partial F([P])}{\partial [P]}\right)_c xk_A[A]_o \exp(-k_A t). \tag{13}$$

Since the partial derivative in Eq. (13) is a constant at a particular value of $[P]$, the rate of change of $\mathscr{S}$ is directly related to $[A]_o$, if measurements are always made at the same value of [P] (equivalent to the same value of $\mathscr{S}$). Later sections show that the variable-time method is often advantageous with nonlinear response characteristics.

2. Enzyme-Catalyzed Reactions

Enzyme-catalyzed reactions are used analytically for the determination of enzyme activities and the determination of substrates. These reactions have become increasingly useful because of the importance of enzyme activity measurements in the diagnosis of disease and the extreme specificity which may be obtained in the determination of substrates. Single-component techniques are most often used with enzyme reactions, although one report has appeared in which an enzymic reaction was used for the simultaneous determination of two species in a mixture [12].

The usual Michaelis-Menten mechanism for enzymatic reactions is

$$E + S \underset{k_{-1}}{\overset{k_1}{\rightleftarrows}} E \cdot S \overset{k_2}{\longrightarrow} P + E$$

where E is the enzyme, S the substrate, and P the product. Steady-state treatment of the above mechanism yields the following rate law:

$$\frac{-d[S]}{dt} = \frac{d[P]}{dt} = \frac{k_2[E]_0[S]}{K_m + [S]} \qquad (14)$$

where

$$K_m = \text{Michaelis constant} = \frac{k_{-1} + k_2}{k_1}.$$

Equation (14) is the basis for determination of enzyme activities or substrate concentrations by the derivative technique. If the product of the reaction is monitored by a linear reaction monitor, the rate of change of the resulting electrical signal at a given time t is given by

$$\frac{d\mathscr{S}_t}{dt} = \frac{\nu k_2[E]_0[S]_t}{K_m + [S]_t}. \qquad (15)$$

Thus the rate of change of the signal with time is directly proportional to the enzyme activity, $[E]_0$. Also, when $[S]_t \ll K_m$, the rate is directly

3. Kinetic Methods of Analysis

proportional to the substrate concentration. If initial rates are measured, the initial substrate concentration can be determined.

If the instrumental techniques used to follow the product formation are nonlinear, Eq. (16) results:

$$\frac{d\mathscr{A}_t}{dt} = \left(\frac{\delta F([P])}{\delta [P]}\right)_c \left(\frac{k_2[E]_0[S]_t}{K_m + [S]_t}\right). \qquad (16)$$

As before, the partial derivative in Eq. (16) is a constant if measurements are always made at a specific value of [P], and the measured rate can be used for determining either enzyme activities or substrate concentrations.

The fixed-time or variable-time methods are based on the integral form of Eq. (14). Integration of Eq. (14) between two substrate concentrations $[S]_1$ and $[S]_2$, which are the concentrations at times t_1 and t_2, respectively, yields

$$-K_m \ln([S]_2/[S]_1) - \Delta[S] = k_2[E]_0(t_2 - t_1). \qquad (17)$$

It is shown later how this equation can be utilized for analysis by the fixed-time and variable-time methods.

3. Other Catalyzed Reactions

Homogeneous catalyzed reactions, other than enzyme reactions, are quite important for the determination of trace amounts of catalysts. Usually these reactions are used for the determination of single components, although in principle two or more catalysts could be determined by differential rate techniques. It is difficult to give a generalized treatment for all catalyzed reactions because of the great variety of mechanisms which are observed. However, the rate expression in terms of the rate of formation of product usually has the form

$$\frac{d[P]}{dt} = K\, F([X_1], [X_2], \ldots, [X_i])[C]_0 \qquad (18)$$

where $[X_i]$ are all the reactant and product concentrations, F is an arbitrary function, K is a proportionality constant, and $[C]_0$ is the initial catalyst concentration.

Equation (18) indicates that rate measurements can be used for determinations of catalyst concentrations.

The mathematical relationships for the integral methods of kinetic analysis can be determined if Eq. (18) is rearranged and integrated as follows:

$$\int_{[P]_1}^{[P]_2} \frac{d[P]}{F([X_1], [X_2], \ldots, [X_i])} = K \int_{t_1}^{t_2} [C]_0 \, dt. \quad (19)$$

If $G([P]) = \int \{ d[P] / F([X_i]) \}$, Eq. (19) can be written as

$$G([P]_2) - G([P]_1) = K[C]_0 (t_2 - t_1). \quad (20)$$

Later sections show how Eq. (20) can be used in the determination of catalyst concentrations using the fixed-time and the variable-time methods.

The treatment given above is intended as an introduction to the physical basis of kinetic methods and is not inclusive. The reader is referred to more extensive treatments [10] for additional cases.

C. Single-Component Techniques

A general block diagram of an instrumental system for single-component reaction rate measurements is shown in Fig. 1. The reaction is carried out in a suitable thermostated reaction cell. Reagents and samples can be introduced automatically or by means of syringes of pipets. The reaction is followed by a reaction monitoring system which produces an electrical signal related to the concentration of some species. Spectrophotometric, potentiometric, fluorometric, amperometric, and many other monitoring systems can be employed. The signal modifier is used to provide input signals of the proper levels and impedances for the rate computation system. The signal modifier can also be used to linearize a nonlinear response from the reaction monitor. Signal modifiers can be current-to-voltage converters, amplifiers, voltage dividers, logarithmic amplifiers, domain converters, and so on. The rate computation system contains the circuitry which converts the signal modifier output to a signal proportional

3. Kinetic Methods of Analysis

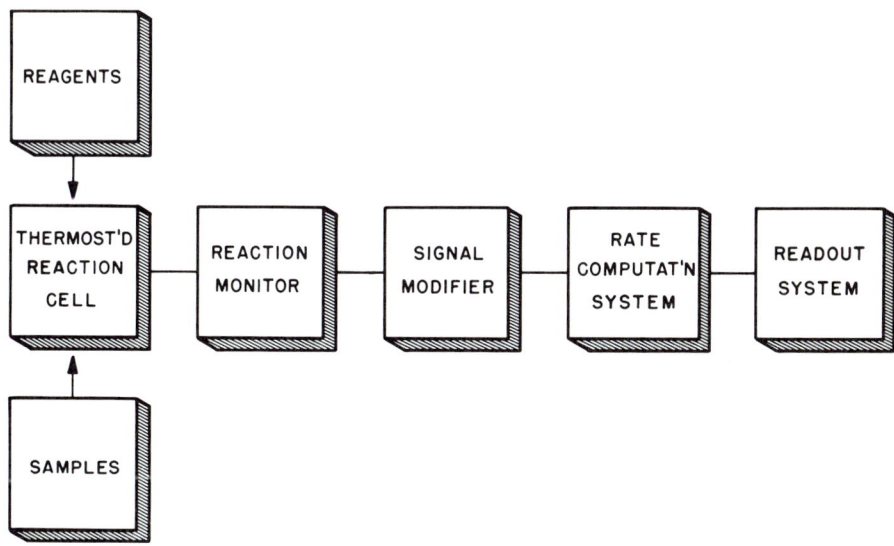

Fig. 1. Block diagram of single-component rate measuring system.

to the rate of the reaction and thus to the concentration of the sought-for constituent. It may also contain the necessary circuitry to allow calibration of the system so that the final readout is in concentration units. For systems where the measured quantity varies linearly with concentration, and the value of the measured quantity is zero when none of the sought-for species is present, direct concentration readout can be accomplished by a simple voltage divider network. Other systems require more complex circuitry.

Rate measurement techniques have been classified into several different approaches by various authors [7, 10, 11]. The most recent classification by Pardue [10] is followed here. The most straightforward approach to rate measurements using instrumental techniques to follow reactions is to obtain electronically the derivative of the electrical signal ($\mathscr{S}$) from the signal modifier. This approach is called the "derivative" or slope method. In the second approach, called the "fixed-time" or constant-time method, the reaction is allowed to proceed for a fixed-time interval, and the change in the signal modifier output ($\Delta\mathscr{S}$) is measured over the time interval. In the third general approach, called the "variable-time" method, the time required (Δt) for the signal modifier output to vary between two fixed limits is measured. Pardue [10] has also considered an additional measurement approach, the "signal-stat" method, which is not considered here since it necessitates using one of the above techniques to obtain the final readout.

1. The Derivative Technique

   a. First-Order or Pseudo-First-Order Reactions. For a linear reaction monitoring system, Eq. (8) is employed directly as the basis of the derivative technique. Solving Eq. (8) for $[A]_0$ gives the desired relationship between the rate of change of the signal modifier output, $\mathscr{S}$, and the initial concentration of A

$$[A]_0 = \left\{\frac{\exp(k_A t)}{\nu x k_A}\right\}\left(\frac{d\mathscr{S}_t}{dt}\right) . \tag{21}$$

There are two ways in which Eq. (21) can be applied. If the derivative measurement is always made at a specific time after the reaction is initiated, the term in braces is a constant and $[A]_0$ may be obtained directly from the measured derivative. A more satisfactory approach, which is also valid for more complex reactions, is to make the derivative measurement early in the reaction before the exponential term in Eq. (21) differs significantly from unity. In this case, the initial rate $(d\mathscr{S}/dt_0)$ is obtained, and Eq. (21) simplifies to

$$[A]_0 = \left\{\frac{1}{\nu x k_A}\right\}\left(\frac{d\mathscr{S}}{dt_0}\right) . \tag{22}$$

The instantaneous slope of the signal vs. time curve remains within 1% of the true initial rate for only 1% of the total reaction. Hence the derivative measurement should be made before the relative concentration of A changes by 1% for Eq. (22) to be valid. Usually, however, the instantaneous derivative is not measured because of instrument response limitations or noise averaging in the readout system. In this case the instantaneous slope can be averaged for 2% of the total reaction within an accuracy of 1%. During the initial portion of the reaction, where Eq. (22) applies, the slope is approximately constant within 1%, and pseudo-zero-order kinetics apply.

   In all the systems that follow, the computation step is greatly simplified if initial rates are measured. Also the measurement of initial rates avoids problems which might be caused by side reactions occurring late in the reaction.

   If the braced quantities in Eqs. (21) and (22) are evaluated experimentally, the rate computation system can be adjusted to read out directly in concentration units.

3. Kinetic Methods of Analysis

If the reaction monitoring system is nonlinear, Eq. (13) is the basis of the measurement. Rearrangement of (13) gives the relationship between the rate of change of the signal and $[A]_0$

$$[A]_0 = \left\{ \frac{\exp(k_A t)}{xk_A (\partial F([P])/\partial [P])_c} \right\} \left(\frac{d\mathscr{S}_t}{dt} \right). \qquad (23)$$

To obtain a direct proportionality between the derivative of the signal and the initial concentration, two conditions are necessary: First, measurements must be made during the initial stages of the reaction before A has undergone significant reaction (less than 1% for 1% accuracy). In this case Eq. (23) reduces to

$$[A]_0 = \left\{ \frac{1}{xk_A (\partial F([P])/\partial [P])_c} \right\} \left(\frac{d\mathscr{S}_t}{dt} \right)_0. \qquad (24)$$

The second condition is that the derivative must be measured at the same absolute value of product concentration (the same signal level). If this second condition is satisfied, the term in braces will be constant from run to run and $[A]_0$ is directly proportional to the measured derivative.

b. Enzyme-Catalyzed Reactions. Enzyme activities can be determined from Eq. (15) if the signal is a linear function of the product concentration. Solving Eq. (15) for $[E]_0$ yields

$$[E]_0 = \left\{ \frac{K_m + [S]_t}{\nu k_2 [S]_t} \right\} \left(\frac{d\mathscr{S}_t}{dt} \right). \qquad (25)$$

At any specific substrate concentration $[S]_t$, the rate is proportional to the enzyme activity. The proportionality constant in Eq. (25) is considerably simplified if conditions are arranged such that $[S]_t \gg K_m$. In this case, the initial reaction rate is obtained and

$$[E]_0 = \left\{ \frac{1}{\nu k_2} \right\} \left(\frac{d\mathscr{S}}{dt} \right)_0. \qquad (26)$$

If k_2 and ν are known, the rate can be used to calculate $[E]_0$ directly without employing standards. Also, Eqs. (25) and (26) predict that the response curve ($\mathscr{S}$ vs. t) will remain linear as long as $[S]_t \gg K_m$. Thus determinations of enzyme activities can often be made very precisely since conditions can usually be arranged such that the slope of the response curve remains constant for a relatively long period of time. During the initial stages of the reaction, while Eq. (26) is valid, the time at which the slope measurement is made is uncritical.

With a nonlinear reaction monitor, Eq. (16) applies. Again a simplification results if conditions are arranged such that $[S]_t \gg K_m$ when the derivative is measured. Under these conditions, the enzyme activity is given by

$$[E]_0 = \left\{ \frac{1}{k_2 (\delta F[P]/\delta [P]_t)_c} \right\} \left(\frac{d\mathscr{S}}{dt} \right)_0 . \qquad (27)$$

If measurements are made at a constant absolute value of signal (constant product concentration), the term in braces is constant, and $[E]_0$ is directly proportional to the measured slope.

The determination of substrate concentrations by the derivative technique also follows from Eqs. (15) and (16). For a linear reaction monitor, Eq. (15) is usually applied under conditions where $[S]_t \ll K_m$. In addition, great simplification occurs if measurements are made before significant amounts of substrate are consumed ($[S]_t \geq 0.99 [S]_0$). If both of the above conditions are satisfied, Eq. (15) reduces within 1% to Eq. (28):

$$[S]_0 = \left\{ \frac{K_m}{\nu k_2 [E]_0} \right\} \left(\frac{d\mathscr{S}}{dt} \right)_0 . \qquad (28)$$

For the determination of substrate concentrations with a nonlinear reaction monitor, three conditions must be satisfied: First, the substrate concentration must be smaller than K_m to simplify Eq. (16). Secondly, initial rates should be measured. If these conditions are satisfied, Eq. (16) becomes

$$[S]_0 = \left\{ \frac{K_m}{k_2 [E]_0 (\delta F([P])/\delta [P]_t)_c} \right\} \left(\frac{d\mathscr{S}}{dt} \right)_0 . \qquad (29)$$

The third condition is that measurements must be made at a constant product concentration for linearity to be observed between the measured slope and the initial substrate concentration.

c. Other Catalyzed Reactions. For simplicity, initial rate measurements are usually made for complex catalytic reactions. Equation (18) shows that if all reactant concentrations are equal to initial concentrations when the derivative is measured, a direct proportionality exists between

3. Kinetic Methods of Analysis

the catalyst concentration and the measured rate. Again, if the reaction monitor is nonlinear, the measurement must also be made at a specific value of product concentration.

d. *Advantages of the Derivative Technique.* The derivative method has several advantages when compared with the fixed-time and variable-time approaches to rate measurement. In principle it is the most straightforward instrumental approach. The result of the measurement is an electrical signal proportional to the reaction rate. Second, the derivative technique is the most rapid means of obtaining reaction rate data. Measurements can be made in a very short time after the reaction is initiated. A third advantage, when the derivative technique is used with linear reaction monitors, is the fact that the user can easily ascertain if the reaction is truly in its initial stages and the conditions of Eq. (22) are being satisfied. In practice, the slope of the rate curve is constant during the initial stages of the reaction, but begins to decrease as significant concentration changes occur. With the derivative technique, these deviations from initial conditions are readily apparent, whereas with the fixed-time and variable-time methods, such deviations are not readily seen. With nonlinear reaction monitors, the slope of the rate curve continuously changes and this latter advantage is lost.

e. *Limitations and Instrumental Problems.* Despite the preceding list of advantages, there are several problems with the derivative technique which have limited its utility in practical kinetic analyses. The most significant of these limitations is the extreme susceptibility of derivative circuits to noise. Because of the high sensitivity required of the reaction monitor in order to follow small concentration changes, noise can be quite severe and is often enhanced in derivative circuits. Spectrophotometric detection in particular can give noisy response curves, which can lead to meaningless derivative signals. Some of the principal noise sources in spectrophotometric detection systems are light-source fluctuations and drifts, shot noise from the photomultiplier, solution noise due to dust particles and incomplete mixing, 60-Hz noise, and amplifier noise in the electronic circuits. Potentiometric detection is normally much quieter than spectrophotometric detection, although not nearly as widely applicable. Many of the techniques developed to circumvent noise problems are discussed in later sections.

Another limitation of the derivative technique is the necessity of taking measurements at a constant signal value when nonlinear reaction monitors are used. It is shown later that the variable-time approach can be used to advantage with such nonlinear transducers.

f. Factors Affecting Precision and Accuracy. There are many factors which influence the precision and accuracy of results obtained using the derivative technique. The present discussion follows the block diagram of Fig. 1 and focuses on the factors affecting each general block of the system. It will be assumed that the system is calibrated by running a standard or standards which are treated identically to the samples.

The reproducibility of the sample and reagent introduction system as well as sample preparation errors will influence both the precision and accuracy of the final readout. Since samples and standards are treated identically, the absolute accuracy of the sampling system is unimportant. However, the reproducibility has a direct effect on the readout. Manual introduction of samples and reagents by pipets and syringes can normally be made within a few tenths percent reproducibility. Complete automation of the sample introduction system can increase the precision.

Any factors which influence the reaction in the cell will influence the final readout. Since all of the equations developed contain rate constants, all of the factors which affect the rate constant must be carefully controlled. Temperature variations in the cell are probably most critical, affecting both precision and accuracy. Other factors such as pH, ionic strength, solvent effects, and so on, are usually quite well regulated by reproducible handling and introduction of samples and reagents. Chemical species present in the sample which influence the reaction rate and are not duplicated in the standards will directly affect the accuracy of the readout.

The absolute accuracy of the reaction monitor system is generally unimportant since the system is calibrated with standards and the rate of change of the reaction monitor signal is ultimately measured. Thus many factors which affect the reaction monitor, but do not influence the rate, will be unimportant. For example, in spectrophotometric detection, the presence of other absorbing species, dirty cells, etc., will be unimportant as long as these factors remain constant during the reaction and do not affect the rate. The reproducibility of the reaction monitor, however, is of direct influence on the final result. Drifts in light-source intensity and shifts in electrode potentials are two examples of factors which can lead to inaccuracy and imprecision in the readout.

The reproducibility of the signal modifier has a direct influence on the final readout. If the signal modifier is a current-to-voltage converter, for example, the absolute accuracy of the conversion is relatively unimportant. However, the conversion factor must remain constant from run to run for accurate and precise results to be obtained. Such factors as changing component values and operational amplifier drifts are likely to be encountered at this stage.

3. Kinetic Methods of Analysis 123

The rate or concentration computation system has a direct influence on the readout. The accuracy and reproducibility can generally be evaluated independently of the rest of the system by applying synthetic signals which simulate response curves from chemical reactions. In the derivative technique, the reproducibility of the slope-measuring circuit will influence both the accuracy and precision of the final readout. If initial reaction rates are not measured, Eqs. (21), (23), and (25) indicate that the measurement time must be reproduced from run to run. Imprecision in timing can lead to errors and imprecision in the readout. If initial rates are measured, the time of the measurement, within the initial pseudo-zero-order portion of the reaction, is not critical. In most cases the measurement must be delayed relative to the initiation of the reaction due to incomplete mixing, induction periods, and so on. With nonlinear reaction monitors, measurements must be made at a particular absolute signal value. Thus any variation in the value of $\mathscr{I}$ at which measurements are made will influence the precision and accuracy. As mentioned previously, any noise in the input of the derivative circuitry is generally enhanced in the output. Thus care should be taken to reduce the amount of noise.

The accuracy and precision of the readout device itself will, of course, directly influence the result. Digital readout devices, such as digital voltmeters, have become quite popular because of the high resolution possible. With digital voltmeters conversion errors and nonlinearities are usually much smaller than any of the preceding sources of inaccuracy and imprecision.

2. Fixed-Time Method

The fixed-time approach, unlike the derivative technique, is an integral technique; although, as mentioned previously, conditions are usually arranged so that measurement times and concentration changes are small enough that a good approximation to the instantaneous reaction rate results.

a. <u>First-Order or Pseudo-First-Order Reactions.</u> If the product is monitored by a linear reaction monitor, solving Eq. (7) for $[A]_0$ gives the desired result

$$[A]_0 = \left\{ \frac{\exp(k_A t_1)}{\nu x [1 - \exp(-k_A \Delta t)]} \right\} \Delta \mathscr{I}. \tag{30}$$

Equation (30) expresses the initial concentration of A in terms of the experimentally adjustable and measurable parameters of the fixed-time measurement. There are two ways in which the fixed-time method can be implemented using Eq. (30) as the physical basis. Since Δt is maintained constant, if all measurements are begun at the same time after the reaction is started, t_1, the quantity in braces will remain constant from run to run, and a linear relationship will be observed between the measured quantity ($\Delta \mathscr{I}$) and the initial concentration ($[A]_0$). This is true for any fixed-time interval during the reaction. Thus fixed-time measurements need not be obtained during the initial stages of the reaction for first- or pseudo-first-order reactions. However, considerable simplification of Eq. (30) can be made if the measurement is made before 2% of A has reacted. In this case, which is equivalent to making the measurement before $\mathscr{I}$ changes by $0.02 \nu x [A]_0$, Eq. (30) reduces within 1% to

$$[A]_0 = \left\{ \frac{1}{\nu x k_A \Delta t} \right\} \Delta \mathscr{I}. \tag{31}$$

If the reaction monitor signal is nonlinearly related to the product concentration, the fixed-time method is very cumbersome and leads to nonlinear relationships between the measured quantity ($\Delta \mathscr{I}$) and the unknown concentration, as can be seen from Eq. (13). If it is assumed that measurement times are small enough that $(\Delta \mathscr{I})_t / \Delta t$ represents the instantaneous reaction rate, Eq. (13) predicts a linear relationship between $\Delta \mathscr{I}$ and $[A]_0$, for a constant Δt, only if all measurements are made at the same value of product concentration (same value of $\mathscr{I}$). However, because Δt is maintained constant and $\Delta \mathscr{I}$ is measured, measurements must necessarily be made at different signal levels, and nonlinearities will usually result. It will be seen later that the variable-time procedure can be used to advantage in situations involving nonlinear reaction monitors.

b. <u>Enzyme-Catalyzed Reactions.</u> For enzyme-catalyzed reactions, Eq. (17) must be used with the fixed-time method. Solving Eq. (17) for the enzyme activity ($[E]_0$) yields

$$[E]_0 = \left[-K_m \ln \left(\frac{[S]_2}{[S]_1} \right) - \Delta[S] \right] / k_2 \Delta t. \tag{32}$$

where $[S]_2$ and $[S]_1$ are the substrate concentrations at the end (t_2) and beginning (t_1) of the measurement time ($\Delta t = t_2 - t_1$). If the relationship $[S]_2 = \Delta[S] + [S]_1$ is substituted into (32) and the logarithmic term expanded in a Maclaurin series, the result is

$$[E]_0 = \left\{ -K_m \left[\frac{\Delta[S]}{[S]_1} - \left(\frac{\Delta[S]}{[S]_1} \right)^2 / 2 + \ldots \right] - \Delta[S] \right\} / k_2 \Delta t. \tag{33}$$

Several approximations in Eq. (33) are necessary in order to use the fixed-time method for the determination of enzyme activities. First, if

3. Kinetic Methods of Analysis

the relative change in substrate concentration ($\Delta[S]/[S]_1$) is very small during the measurement time, the higher-order series terms can be ignored. If the relative concentration change is 2% or less, Eq. (34) holds to better than 1% [13]:

$$[E]_0 = \frac{-1}{k_2 \, \Delta t} \left(\frac{K_m}{[S]_1} + 1 \right) \Delta[S]. \tag{34}$$

Equation (34) can be further simplified if the substrate concentration at the beginning of the measurement ($[S]_1$) is made much larger than K_m. Under these conditions the enzyme activity is given by

$$[E]_0 = \left\{ \frac{-1}{k_2 \, \Delta t} \right\} \Delta[S]. \tag{35}$$

The net effect of the above assumptions is that the fixed-time method increases in dynamic range as $[S]_1$ becomes greater than K_m.

For a linear reaction monitor which follows the product concentration, substitution of $\Delta[P] = -\Delta[S]$ and $\Delta \mathscr{S} = \nu \Delta[P]$ in Eq. (35) gives the relationship between the measured signal ($\Delta \mathscr{S}$) and the enzyme activity

$$[E]_0 = \left\{ \frac{1}{\nu k_2 \, \Delta t} \right\} \Delta \mathscr{S}. \tag{36}$$

Because of the complexities of the above relationships and the necessity that measurements be made at a constant product concentration for linearity, a nonlinear reaction monitor would undoubtedly be avoided with the fixed-time method. The variable-time method can be used for enzyme activities with nonlinear reaction monitors, as is pointed out in the next section.

The determination of substrate concentrations follows from the same mathematical development. Rearrangement of Eq. (33) yields

$$-K_m \left[\frac{\Delta[S]}{[S]_1} + \left(\frac{\Delta[S]}{[S]_1} \right)^2 /2 + \cdots \right] - \Delta[S] = k_2 [E]_0 \, \Delta t. \tag{37}$$

If conditions are arranged such that $K_m > 100[S]_1$, then $\Delta[S]$ is negligible and Eq. (37) reduces within 1% to

$$-K_m \ln \frac{[S]_2}{[S]_1} = k_2[E]_0 \, \Delta t. \tag{38}$$

Equation (38) has the form of a pseudo-first-order reaction, and is ideally suited for the fixed-time method [13]. If Eq. (38) is expressed in terms of the initial concentration of substrate, $[S]_0$, and the change in substrate concentration over the measurement interval, $\Delta[S]$, Eq. (39) results,

$$\Delta[S] = -[S]_0 \exp(-K't_1)(1 - \exp(-K'\Delta t)) \tag{39}$$

where $K' = k_2[E]_0/K_m$.

For a linear reaction monitor which follows the product concentration, $\Delta \mathscr{A} = \nu \Delta[P] = -\nu \Delta[S]$. Substitution for $\Delta[S]$ and rearrangement yields

$$[S]_0 = \left\{ \frac{\exp(K't_1)}{\nu[1-\exp(-K'\Delta t)]} \right\} \Delta \mathscr{A}. \tag{40}$$

As in the case of pseudo-first-order reactions, Eq. (40) reveals that the fixed-time method can be used to determine substrate concentrations without ensuring that measurements are made during the initial stages of the reaction as long as the assumptions used to simplify Eq. (39) are valid. If all measurements are begun at t_1 and ended after a constant time interval (Δt), the quantity in braces will be constant from run to run. Again, however, considerable simplification results if all measurements are made near zero reaction time.

c. Other Catalyzed Reactions. Catalyst concentrations can be determined using the fixed-time method, but strict adherence to pseudo-zero-order kinetics (initial reaction rates) is usually necessary to obtain linear calibration curves. If the fixed-time measurement is confined to the pseudo-zero-order portion of the reaction, the denominator of Eq. (19) is a constant. After evaluating the integral, Eq. (19) becomes

$$\frac{\Delta[P]}{F([X_1]_0, [X_2]_0, \ldots, [X_i]_0)} = K[C]_0 \, \Delta t \tag{41}$$

where $F([X_i]_0)$ is a function of all the initial reactant concentrations. If a linear reaction monitor is employed, the catalyst concentration is

3. Kinetic Methods of Analysis

given by

$$[C]_0 = \left\{ \frac{1}{\nu K F([X_1]_0, [X_2]_0, \ldots, [X_i]_0) \, \Delta t} \right\} \Delta \mathscr{I}. \tag{42}$$

Nonlinear reaction monitors would probably be avoided unless the combined response characteristics were linear during the initial stages of the reaction [10].

d. *Advantages of the Fixed-Time Method.* There are several reasons why the fixed-time approach is advantageous in many situations. First, in contrast to the variable-time method, the measured quantity in the fixed-time method is directly proportional to the concentration of the rate-limiting species under appropriate conditions. Thus no additional computational steps are necessary to obtain direct linear readout. Also, measurement times remain constant from sample to sample in a given procedure. Thus the fixed-time method is easily incorporated into a completely automated sample handling and measurement system. Another advantage of the fixed-time method arises when first- or pseudo-first-order reactions are used. For these reactions, and for the determination of substrates with enzymatic reactions, the measurement time need not correspond to the initial pseudo-zero-order part of the reaction (less than 2% reaction). Thus the fixed-time method is capable of handling a wide dynamic range of concentrations for first- or pseudo-first-order reactions without introducing nonlinearities in calibration curves. Measurement times for these reactions can be made fairly long to average noise fluctuations and instrumental drifts. Another advantage, which also leads to increased noise averaging, arises because integration techniques can be utilized with the fixed-time method. Because the reaction monitor signal is the measured quantity, integration of the signal can lead to increased signal-to-noise ratios, in contrast to the variable-time method where the signal is not the measured quantity.

Specific computing circuits which take advantage of these capabilities are discussed in a later section.

e. *Limitations and Instrumental Problems.* There are also several limitations associated with the fixed-time method. First, the method is not well suited to the determination of enzyme activities or catalyst concentrations. Nonlinearities can occur if measurements are made outside the pseudo zero order initial portion of the reaction. This requirement leads to a limited dynamic range for the fixed-time method with these chemical systems [13]. Secondly, the method is not suited for nonlinear response curves. Detection systems such as potentiometric and conductometric systems often lead to nonlinear response curves which cannot be easily adapted to fixed-time procedures. Another limitation is that the monitored change in signal ($\Delta \mathscr{S}$) varies from run to run. Thus the signal-to-noise ratio of the rate measurement is variable depending on the reaction rate.

f. Factors Affecting Accuracy and Precision. All of the factors mentioned in discussing the derivative technique, with the exception of those influencing the rate or concentration computation system, also influence the fixed-time method and thus are not repeated here. In addition, the reproducibility of the fixed-time interval and the accuracy and reproducibility of measuring $\Delta \mathscr{S}$ will influence the overall precision and accuracy of the measurement. The timing error can usually be made negligibly small in comparison to other factors by using a highly stable crystal oscillator time base in conjunction with scaling circuits as described in later sections. The computation of concentration can usually be made directly from $\Delta \mathscr{S}$ by voltage divider circuits or amplifier circuits. The reproducibility and linearity of these circuits will directly affect the accuracy and precision.

Inaccuracies can also result in the determination of catalysts if the measurements are made outside the initial portion of the reaction.

As with the derivative technique, the precision and accuracy of the rate computation and readout system may be evaluated with signal sources which simulate response curves.

3. The Variable-Time Method

The variable-time method, like the fixed-time method, is an integral method which, for short measurement times and small concentration changes, gives a result approaching the instantaneous reaction rate. Like the fixed-time method, certain reactions and reaction monitors do not require that initial rates be measured for linear calibration curves to be obtained. Other reactions, however, will lead to nonlinearities if measurements are not made during the initial stages of the reaction.

a. First-Order or Pseudo-First-Order Reactions. In the variable-time procedure, the integral form of the first-order rate expression must be used. Solving Eq. (5) for $[A]_0$ results in the following expression:

$$[A]_0 = \left\{ \frac{\Delta[P] \exp(k_A t_1)}{x[1 - \exp(-k_A \Delta t)]} \right\} . \tag{43}$$

3. Kinetic Methods of Analysis

In the variable-time procedure $\Delta[P]$ is held constant and Δt is measured and related to the initial concentration, $[A]_0$. Since both Δt and t_1 vary depending on $[A]_0$, Eq. (43) predicts a nonlinear relationship between the reciprocal of Δt and $[A]_0$. If measurements are restricted to the first 2% of the reaction, and the measurements are begun very near zero reaction time such that $\exp(k_A t_1) \approx 1$, Eq. (43) reduces to

$$[A]_0 = \left\{ \frac{\Delta[P]}{xk_A} \right\} \frac{1}{\Delta t} \,. \tag{44}$$

Thus, in contrast to the fixed-time procedure, strict adherence to pseudo-zero-order kinetics is necessary for linear calibration curves with first- or pseudo-first-order reactions. If these approximations hold, Eq. (44) predicts that the quantity in brackets will remain constant from run to run, and $1/\Delta t$ will be directly proportional to $[A]_0$.

For a linear reaction monitor, $\Delta[P]$ can be replaced by $\Delta \mathscr{S}/\nu$ in Eq. (11), resulting in Eq. (45)

$$[A]_0 = \left\{ \frac{\Delta \mathscr{S}}{\nu x k_A} \right\} \frac{1}{\Delta t} \,. \tag{45}$$

Again, subject to the same restrictions as Eq. (44), a linear relationship between $[A]_0$ and reciprocal time is predicted.

For a nonlinear reaction monitoring system, Eq. (12) is applicable. Solving Eq. (12) for the initial concentration yields

$$[A]_0 = \left\{ \frac{f(\mathscr{S}_2) - f(\mathscr{S}_1)}{x \exp(-k_A t_1)[1 - \exp(-k_A \Delta t)]} \right\}. \tag{46}$$

If the measurements are begun near zero reaction time and restricted to the first 2% of the reaction, Eq. (46) reduces to

$$[A]_0 = \left\{ \frac{[f(\mathscr{S}_2) - f(\mathscr{S}_1)]}{xk_A} \right\} \frac{1}{\Delta t} \,. \tag{47}$$

Equation (47) reveals a distinct advantage of the variable-time procedure. Since the measurement is made between two fixed reaction monitor levels, $f(\mathscr{S}_2)$ and $f(\mathscr{S}_1)$, nonlinearities in the relationship between $\mathscr{S}$ and $[P]$ do not affect the accuracy, and a linear relation will exist between reciprocal time and $[A]_o$ as long as the approximations used to obtain Eq. (47) are valid.

b. *Enzyme-Catalyzed Reactions.* The variable-time method is ideally suited for the determination of enzyme activities. Rearrangement of the integral form of the rate law [Eq. (17)] yields an expression for $[E]_o$

$$[E]_o = \left[-K_m \ln\left(\frac{[S]_2}{[S]_1}\right) - \Delta[S] \right] \left(\frac{1}{k_2 \Delta t} \right) . \qquad (48)$$

Equation (48) is an exact expression if the simple Michaelis-Menten mechanism is valid. Since $[S]_2$, $[S]_1$, and thus $\Delta[S]$, are held constant in the variable-time method, a linear relationship always exists between the enzyme activity and the reciprocal of the time interval. Again, however, considerable simplification can be made if the relative change in substrate concentration, $\Delta[S]/[S]_1$, is very small during the measurement and $[S]_1 >> K_m$. Under these conditions Eq. (48) becomes

$$[E]_o = \left\{ \frac{\Delta[S]}{k_2} \right\} \frac{1}{\Delta t} . \qquad (49)$$

It should be noted that these restrictions are not necessary for linear calibration curves to be obtained. Thus the variable-time method is capable of measuring a wide dynamic range of enzyme activities with linear calibration curves [13].

If a linear reaction monitor is used to monitor the product concentration, Eq. (50) results

$$[E]_o = \left\{ \frac{\Delta \mathscr{A}}{\nu k_2} \right\} \frac{1}{\Delta t} . \qquad (50)$$

If the reaction monitor output is a nonlinear function of product concentration, Eq. (51) is applicable

$$[E]_o = \left\{ \frac{[f(\mathscr{A}_2) - f(\mathscr{A}_1)]}{k_2} \right\} \frac{1}{\Delta t} . \qquad (51)$$

Equations (50) and (51) predict linear relationships between the enzyme activity and $1/\Delta t$ for nonlinear reaction monitors, as well as for linear transducers. Again, measurements need not be made during the initial stages of the reaction for linear relationships to hold.

3. Kinetic Methods of Analysis

The determination of substrate concentrations is not as well suited for the variable-time procedure as for the fixed-time method. Equation (40) predicts a nonlinear relationship between reciprocal time and $[S]_o$ unless pseudo-zero-order kinetics prevail during the measurement interval and measurements are begun very near zero reaction time. Nonlinear reaction monitors can likewise be handled subject to the same restrictions.

c. Other Catalyzed Reactions.
The determination of catalyst concentrations is well suited for the variable-time method. Equation (20) can be rearranged to give

$$[C]_o = \{G([P]_2) - G([P]_1)\} \frac{1}{\Delta t\, K} . \tag{52}$$

Since variable-time measurements are always made between two fixed product concentrations, $1/\Delta t$ will be proportional to $[C]_o$, no matter how complex the actual rate expression. Again measurements need not be made during the initial stages of the reaction, although simplified expressions result. Linear and nonlinear reaction monitors can be handled equally well.

d. Advantages of the Variable-Time Method.
The variable-time method was shown in the preceding section to be advantageous in many situations. For the determination of enzyme activities and catalyst concentrations, a direct linear relationship between the reciprocal of the measurement time and concentration is obtained. For these determinations, the variable-time method is capable of handling a wide range of catalyst concentrations without inaccuracies. Another major advantage is that there is no necessity for the instrumental monitoring technique to be linear in concentration. Thus nonlinear transducers can be utilized directly without additional linearization circuitry. Another advantage over the derivative method, which results for catalytic reactions, is that relatively large concentration changes can be set to decrease the influence of noise without introducing nonlinearities. Since the variable-time method always involves a constant change in signal, $\Delta \mathscr{S}$, the signal-to-noise ratio of the reaction monitoring system is constant from run to run and independent of the reaction rate. Finally, it should be noted that, using precise and stable timing circuits, Δt can be measured with very high resolution and stability [14].

e. Limitations and Instrumental Problems.
The major limitation of the variable-time method is the reciprocal relationship between the measured quantity (Δt) and the concentration of the sought-for constituent. This limitation necessitates an additional computation step (reciprocal time) in automated variable-time systems if direct concentration or rate readout is desired. The variable-time method is also not as well suited for first-

or pseudo-first-order reactions or for the determination of substrate concentrations as is the fixed-time method. For these reactions, measurements must be completed near zero reaction time to maintain linear calibration curves. Finally, it should be noted that measurement times can vary significantly if a wide range of rates is being measured by the variable-time method. Thus the variable-time procedure is not as easily incorporated into completely automated systems as is the fixed-time method.

Several instrumental systems which deal with the above problems are discussed in later sections.

   f. Factors Affecting Precision and Accuracy. In addition to those factors mentioned previously, the variable-time method is influenced by the stability of the fixed signal interval, $\Delta \mathscr{I}$, the precision of the timing device used to measure Δt, and the precision of the computation system used to calculate the reciprocal of the measurement time ($1/\Delta t$). These factors and the accuracy and precision of the readout may be evaluated with synthetic rate curves.

Inaccuracies in the variable-time method can also occur for first-order reactions and substrate determinations if measurements do not correspond to the first 2% of the reaction.

4. Summary and Comparison of Single-Component Techniques

In choosing between the three single-component approaches, the type of reaction, the sought-for species, and the characteristics of the reaction monitor are the most important factors. The derivative technique is the most straightforward approach and gives the desired readout of instantaneous reaction rate. For noisy signals, however, the two integral approaches are more reliable. If the reaction monitor is linearly related to concentration, the fixed-time approach is superior for pseudo-first-order reactions and the determination of substrate concentrations using enzyme reactions. Reaction rate analyses of enzyme activities and the determination of other catalysts are best suited for the variable-time method. If the reaction monitor is nonlinearly related to concentration, the variable-time approach is of great advantage. The two integral approaches are seen to be complementary and should both be available if various types of reactions and instrumental techniques are to be utilized for reaction rate methods.

D. Techniques for Simultaneous Determination of Mixtures

Differential rate methods for the simultaneous determination of closely related mixtures have become extremely useful in analytical chemistry. While a great deal of effort has been devoted to new techniques and chemical

3. Kinetic Methods of Analysis   133

systems, little attention has been focused on complete automation of the procedures. Some of the most common differential rate techniques, such as those involving graphical extrapolation [7,9], are difficult to automate without the aid of a small digital computer. Much future research will probably be devoted to this task. The method of proportional equations [7,9], however, is well suited for automation with normal circuitry. Thus this discussion will be limited to the method of proportional equations as applied to first- or pseudo-zero-order reactions. The reader is referred to more extensive treatments for discussions of other differential rate techniques [7,9]. Section IV discusses some of the applications of the small digital computer in kinetic analyses, including differential rate methods.

Figure 2 illustrates in block diagram form the apparatus needed for automated differential kinetic analyses using the method of proportional equations. As in single-component techniques, the reaction is carried out in a thermostated cell and monitored by a suitable instrumental method. The reaction monitor output is treated as desired by a signal modifier before the measurement circuitry. As discussed later, the measurement circuitry can sample the signal modifier output at specific times, or measure the rate of the reaction by any of the single-component techniques discussed previously. The output of the measurement circuit is the input to the computer circuit, which solves the simultaneous equations that are needed to obtain initial concentrations.

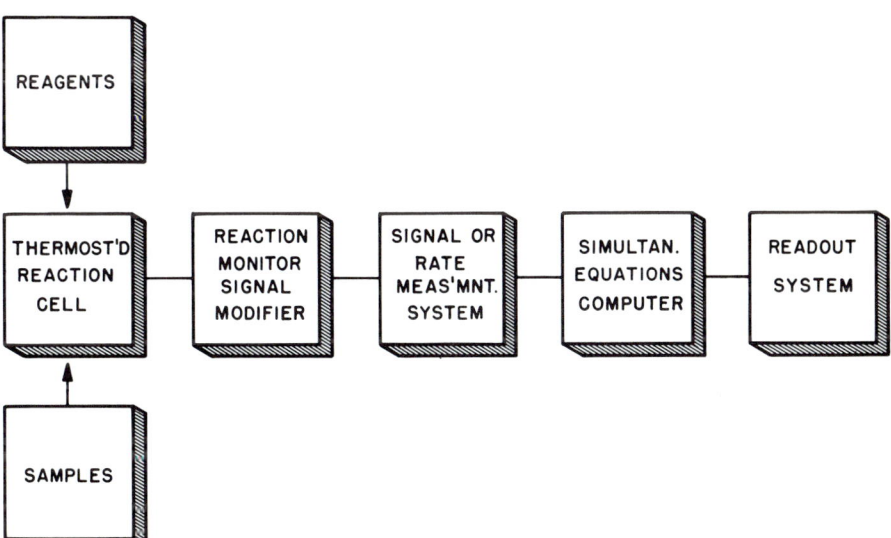

Fig. 2. Block diagram of automated differential rate measuring system.

1. First-Order or Pseudo-First Order-Reactions

The method of proportional equations can be implemented in several ways. If two reactants A and B react with a common reagent R under pseudo-first-order conditions to form a common product P, the reactions may be written

$$A + R \xrightarrow{k_A} xP$$

$$B + R \xrightarrow{k_B} yP$$

where k_A and k_B are pseudo-first-order rate constants. The rate of formation of the product is given by

$$\frac{d[P]}{dt} = xk_A[A] + yk_B[B]. \tag{53}$$

Equation (53) may be written in terms of the initial concentrations of A and B by making use of relationships similar to Eq. (6) for each species:

$$\frac{d[P]_t}{dt} = xk_A[A]_0 \exp(-k_A t) + yk_B[B]_0 \exp(-k_B t). \tag{54}$$

The concentration of product at any time may be obtained by integration of Eq. (54)

$$[P]_t = x[A]_0 [(1 - \exp(-k_A t)] + y[B]_0 [1 - \exp(-k_B t)]. \tag{55}$$

Equations (54) and (55) can be utilized as the basis for differential rate methods. Equation (55) is most commonly applied. Since the exponential terms are constant at any specific time, t_1, the amount of product formed at t_1 can be written as

$$[P]_{t_1} = K_{A1}[A]_0 + K_{B1}[B]_0 \tag{56}$$

where $K_{A1} = x[1 - \exp(-k_A t_1)]$ and $K_{B1} = y[1 - \exp(-k_B t_1)]$. At some later time, t_2, the amount of product formed is given by

3. Kinetic Methods of Analysis 135

$$[P]_{t_2} = K_{A2}[A]_0 + K_{B2}[B]_0. \tag{57}$$

The four constants are usually determined experimentally by measuring the amount of product formed at times t_1 and t_2 upon reaction of known amounts of pure A and B. The mixture of A and B is then reacted with R, and $[P]_{t_1}$ and $[P]_{t_2}$ are measured. These results are then substituted in Eqs. (56) and (57), and the equations are solved simultaneously to give the initial concentrations, $[A]_0$ and $[B]_0$.

For automation purposes, instrumental techniques are employed to obtain an electrical signal $\mathscr{S}$ proportional to the amount of product formed. In this case, the reactants can form different products with R as long as the electrical signal obtained is related to both products. For the case in which two different products are formed and the instrumental technique produces a signal linearly related to the concentration of each product, the relationship between $\mathscr{S}$ and concentration may be written as

$$\mathscr{S} = \nu_A[P_A] + \nu_B[P_B] + \mathscr{S}_0 \tag{58}$$

where $\mathscr{S}_0$ takes into account any instrumental offsets. If each product is formed by an independent first- or pseudo-first-order reaction, the value of $\mathscr{S}$ at times t_1 and t_2 may be expressed as

$$\mathscr{S}_{t_1} = K'_{A1}[A]_0 + K'_{B1}[B]_0 + \mathscr{S}_0 \tag{59}$$

$$\mathscr{S}_{t_2} = K'_{A2}[A]_0 + K'_{B2}[B]_0 + \mathscr{S}_0 \tag{60}$$

where $K'_{A1} = \nu_A x[1 - \exp(-k_A t_1)]$, $K'_{B1} = \nu_B y[1 - \exp(-k_B t_1)]$, etc. Again, if the four constants have been previously evaluated, $[A]_0$ and $[B]_0$ may be determined by measurement of $\mathscr{S}_{t_1}$ and $\mathscr{S}_{t_2}$ and simultaneous solution of Eqs. (59) and (60). This technique is quite similar to the fixed-time method, except that two measurements must be made for each reaction, and an additional computation step is necessary to solve Eqs. (59) and (60) simultaneously. Several authors have discussed the choice of the optimum analysis times, t_1 and t_2, to minimize errors in simultaneous determinations [7, 9, 15].

Measurement of the reaction rate at two times during the reaction of A and B with R may also be used for simultaneous analysis of mixtures. At time t_1 Eq. (54) may be written

$$\frac{d[P]_{t_1}}{dt} = G_{A1}[A]_0 + G_{B1}[B]_0 \tag{61}$$

where $G_{A1} = xk_A[\exp(-k_A t_1)]$, etc.

Likewise at t_2

$$\frac{d[P]_{t_2}}{dt} = G_{A2}[A]_0 + G_{B2}[B]_0. \tag{62}$$

The constants are evaluated by measuring the rate of reaction of known amounts of pure A and B at the two times. Analysis of the mixture is accomplished by measuring the rate of reaction of the mixture at t_1 and t_2. These results are then used in Eqs. (61) and (62), which are simultaneously solved to give the initial concentrations.

If a linear instrumental technique is employed, the rate of change of the resulting electrical signal is measured, in which case Eqs. (63) and (64) result:

$$\frac{d\mathscr{A}_{t_1}}{dt} = G'_{A1}[A]_0 + G'_{B1}[B]_0 \tag{63}$$

$$\frac{d\mathscr{A}_{t_2}}{dt} = G'_{A2}[A]_0 + G'_{B2}[B]_0 \tag{64}$$

where $G'_{A1} = \nu_A G_{A1}$, etc. The derivative technique, described earlier, could be used to measure the slope of the response curve at the two times.

Nonlinear reaction monitors would probably be avoided with these techniques which require measurements at specific times, unless linearization circuits were used prior to the measurement, or a small digital computer were available to solve the resulting equations.

2. Pseudo-Zero-Order Reactions

Initial reaction rates may also be utilized with the method of proportional equations for simultaneous analysis of mixtures [7, 12]. The mixture can be reacted under two different reaction conditions, by changing

3. Kinetic Methods of Analysis 137

some outside parameter on which the rate depends (pH, temperature, solvent, etc.) or by using two different reactions. The initial reaction rates may be expressed

$$\text{initial rate}_{(1)} = Q_{A1}[A]_o + Q_{B1}[B]_o \tag{65}$$

$$\text{initial rate}_{(2)} = Q_{A2}[A]_o + Q_{B2}[B]_o \tag{66}$$

where the subscripts 1 and 2 refer to the two different conditions or reactions. As before, the four constants can be evaluated experimentally by reacting known amounts of pure A and B and measuring the initial rates.

Initial reaction rate measurements can be made by any of the three techniques previously described (derivative, fixed time, or variable time). If the variable-time method is utilized, nonlinear signal sources may be employed with no loss of accuracy or increase in the complexity of the circuits.

3. Comparison of First-Order and Pseudo-Zero-Order Techniques

The techniques described here are but a few of the many methods which have been used in the analysis of mixtures by differential rate techniques. However, the method of proportional equations is the only technique to date which has been completely automated [16].

The initial rate method has the advantage that data can be obtained very rapidly in comparison to the time for complete reaction. However, with the initial rate technique, two reactions must be run under different conditions in order to obtain the necessary data. With the pseudo-first-order methods, the necessary data are obtained with only one reaction. The initial rate method has the additional advantage that side reactions or nonidealities which may occur later in the reaction will often not influence the initial stages of the reaction. Also, any ratio of reactant concentrations may be used when initial rates are measured [7], whereas the pseudo-first-order techniques require a large excess of reagent R in comparison to the reactants A and B.

The first-order technique has the advantage that instrumentation for measuring the electrical signal $\mathscr{S}$ at times t_1 and t_2 is less complex than instrumentation for measuring and computing initial rates.

4. Factors Affecting Accuracy and Precision

Many of the factors which affect the accuracy and precision of differential kinetic analyses are similar to those which influence single-component techniques. However, most of these same factors are more critical in differential techniques. For example, because two reactants are involved, temperature variations will influence both rate constants and generally lead to greater inaccuracies in differential rate methods than in single-component methods. Also, since simultaneous equations must be solved to obtain concentrations, any inaccuracy in determining one of the components will affect the analysis of the other. In addition, the first-order techniques require a stable time base in order to carry out all measurements at times t_1 and t_2. An extensive discussion of the effect of errors in various parameters on the concentration error in differential rate methods has been presented [7]. The computer for solving the simultaneous equations can also cause errors, particularly if analog computation is used.

III. INSTRUMENTATION FOR KINETIC METHODS

Some of the recent developments in automated instrumentation systems for kinetic methods are presented in this section. It will be useful first to consider general ideas about data treatment, the domains of data, and the processing of data to obtain the desired final result; in this case a number proportional to or equal to the concentration of the rate-limiting species. These concepts will prove useful here in discussing the movement of data in and out of the analog and digital domains and in discussing specific analog and digital circuitry for kinetic methods. Reaction rate methods often place severe strain on the sensitivity and stability of reaction monitoring systems. Likewise, reactants must be introduced with high precision since rates are highly concentration dependent. Thus part of this discussion focuses attention on recent efforts to develop instrumental techniques and sampling systems uniquely suited for the requirements of kinetic methods. Automated systems for the processing of information from the reaction monitor to obtain the reaction rate or concentration of a single species in solution are then discussed in terms of the analog and digital devices which make up these computation systems. Circuitry for implementing the derivative, the fixed-time, and the variable-time methods is considered. In the final section, some of the techniques which have been developed for automating differential rate methods for the simultaneous analyses of two or more species are discussed.

A. General Considerations

Physical phenomena which a scientist studies are normally observed by quite indirect means. Most often input transducers are used to convert some physical or chemical characteristic of the phenomenon of interest into an electrical signal which can be processed and treated by electronic

3. Kinetic Methods of Analysis

means to produce a number related to the original phenomenon. Although natural phenomena occur as discrete, individual actions, most of these actions cannot be individually resolved with present state of the art techniques. Thus the results of many individual events are normally averaged and eventually processed to produce a number related to the average of the phenomena over the observation time.

1. Data Domains

In analyzing and describing modern instrumentation systems for scientific measurements, the concepts of data domain analysis [17, 18] have proven extremely useful. Data domain analysis provides a unified method to describe the way in which encoded information proceeds through a complex instrument. At any instant in time the desired physical or chemical information can be represented by some property of an electrical signal. Similar electrical signals are grouped together in "data domains." Three distinct data domains exist for electrical signals [17, 18]. In the "analog" (A) domain, the magnitude of the electrical signal (current or voltage) carries the encoded information. Most of the input transducers presently used in chemical analysis convert physical or chemical information to analog signals because of the averaging of discrete events mentioned earlier. Analog transducers and circuits respond in a continuous manner to input information. A familiar example of a transducer which, as normally used, produces an analog signal is the photomultiplier tube, which converts radiant power (the number of photons per second) to a proportional current.

In the "time-interval" (Δt) domain, the information is encoded in the time relationship between pulses, whose amplitude is of little importance. In this domain, information may be encoded as the frequency of pulses, their width, or the phase angle between two signals. An example of data encoded in the time-interval domain is the output of a Geiger tube used in radioactivity measurements. The electrical signal produced is a series of pulses whose amplitude is unimportant, but whose frequency is related to the number of disintegrations of radioactive material per unit of time. Signals in the Δt domain, because they are less amplitude dependent than analog signals, are less susceptible to noise interferences.

In the third domain of electrical signals, information is encoded as logic levels representing discrete numbers. This domain is known as the digital (D) domain. The digital domain is often confused with the time interval domain and, in fact, these domains are similar in some respects. In both domains the amplitude of the electrical signal is unimportant as long as the two states are distinguishable. The digital domain, however, is distinguished by dealing with a specific, discrete number of pulses or arrangement of logic level signals. A signal can often be converted from the Δt domain to the D domain merely by opening a gate for a specific length of time. In this case a discrete number of pulses will have occurred during the gating time. These pulses can then be processed as a serial digital number, converted to a parallel number, coded in binary coded decimal form, and so on. Thus the Geiger counter output can be converted to a digital signal if a gate is opened for a precise time interval, i.e., 1 sec. There are other characteristics which distinguish the Δt and D domains. In the Δt domain, the frequency of pulses can vary continuously with variations in input information. Once information has been converted to digital form, however, the numbers obtained are discontinuously related to the phenomenon.

The circuits used for processing data in the digital domain are quite similar to those used in the time-interval domain and here these circuits are considered together as digital circuits.

2. Domain Conversions

In any electronic measurement at least two, and often many more, domain conversions are necessary [17] — one to transfer the data into one of the electrical domains, and one to transfer the data out of the electrical domain for readout purposes. Interdomain converters have the basic characteristics of measurement devices in that they transform one characteristic of the data into another. Each interdomain conversion in an instrument involves a difference detector and a reference standard, which is the property or the characteristic being transformed [17]. Each conversion is also subject to errors which depend on the quality of the reference standard and the difference detector. Hence it is often desirable to reduce the number of domain conversions in an instrument by designing systems in which the most direct route between the input domain and the desired output domain is taken.

Typical interdomain converters include voltage-to-frequency (V/F) converters for analog-to-time interval conversions, dual slope, ramp and successive approximation analog-to-digital converters, and count rate meters for time interval-to-analog conversions.

Conversions within the same domain can often be made by very simple circuits. For example, current-to-voltage and voltage-to-current conversions are often made with only a resistor. Operational amplifier circuits are also used for simple conversions within the analog domain with improved signal-to-noise ratio and impedance matching characteristics over simple resistive devices.

3. Kinetic Methods of Analysis 141

Likewise, time-interval signals can be simply converted from one subdomain to another. For example, a simple comparator or Schmidt trigger circuit can accomplish a frequency-to-pulse duration conversion.

Conversions within the digital domain are usually accomplished with counters or shift registers. A counter will allow count serial data to be transformed into parallel data or vice versa. Binary coded serial data can be converted to parallel digital form and vice versa with a simple shift register.

Many of the conversion techniques mentioned here are illustrated in later sections when instruments for kinetic analyses are described in detail.

B. Reaction Monitor Systems

The reaction monitor in a kinetic measurement system encodes information related to the concentration of a reactant or product as an appropriate electrical signal in one of the three domains mentioned previously. The reaction monitor may be a one-stage transducer, such as an electrode sensitive to a species in solution, or a complex instrument such as a spectrophotometer. Although many different reaction monitoring techniques have been used for specific kinetic studies, four are most commonly employed in kinetic methods of analysis: spectrophotometric, fluorometric, potentiometric, and amperometric techniques.

To follow the initial stages of chemical reactions and to obtain large enough signals for accurate and precise measurement, a detection system with extremely high sensitivity is required. As is seen in a later section, rate computation systems have been developed which are highly accurate and precise, and capable of detecting small signal changes. Hence the reaction monitoring system must also be capable of high stability since any drifts in the detection system which occur during the measurement will add or subtract from the actual reaction rate. As a general rule of thumb, the system drift should be at least 100 times lower than the lowest rate to be measured for 1% accuracy.

Unfortunately, the most widely used and versatile reaction monitoring technique, spectrophotometry, is also the most complex and subject to the most problems with noise and instrumental drifts. Thus, much of the effort in developing more sensitive and stable reaction monitoring systems has gone into the spectrophotometric technique.

1. Spectrophotometric Systems

When single-beam spectrophotometric monitoring is used, a physical parameter, light absorption at a specific band of wavelengths, is used as a concentration measure. Thus chemical concentration is encoded as a decrease in radiant power, and the first domain conversion is made. A monochromator or filter selects the wavelength band to be measured, and thus does not perform a domain conversion. The transducer, usually a photomultiplier tube, converts the resultant radiant power to a proportional current or, in the case of photon counting, to a train of current pulses. The photomultiplier can act, therefore, as either a physical-to-analog converter or a physical-to-time interval converter. Vacuum phototubes can also be used with some advantages if high light levels are available. With either of these approaches, the output of the transducer is in one of the electrical domains and ready for processing by electronic circuitry. Electronic signal modifier and rate computation systems are discussed in later sections.

Another important factor in spectrophotometric reaction rate methods is that concentration changes, and thus absorbance changes, are usually very small. Under these conditions, deviations from Beer's law are not often significant. Hence relatively large spectral bandwidths can be employed with little loss in accuracy and much improvement in signal-to-noise ratio due to the increased light level. Filter photometers can also be used for the same reasons.

Although many commercial spectrophotometers are available, only a few possess the required sensitivity and stability for routine use in kinetic methods. For this reason, several researchers have developed new systems with characteristics particularly suited for kinetic analyses. With spectrophotometric detection, light-source fluctuations, detector noise and drifts, and temperature variations in the reaction cell are among the most significant problems.

a. Light-Source Stabilization. The effect of light-source variations can be minimized by several methods. Double-beam systems can partially cancel fluctuations by comparison of reference and sample beam intensities. However, since most instruments employed for kinetic methods are not often called upon for spectral scanning, most users have discarded the double-beam approach and have developed less expensive stabilization techniques for light-source regulation.

The simplest technique for stabilizing the light source against short-term fluctuations is to regulate the line voltage since line-voltage fluctuations can cause changes in light intensity comparable to the measured

3. Kinetic Methods of Analysis

reaction rate. Weichselbaum and coworkers [19] have described an electronic regulator circuit for such stabilization. The circuit compensates for changes in light bulb resistance and is thus superior to circuits which regulate only the applied voltage or the applied current. A reference voltage is applied to the noninverting input of an amplifier and is compared with the voltage at the inverting input. The voltage at the inverting input is the sum of the voltage drops across two resistors. The voltage across one resistor is proportional to the current through the bulb, while that across the other is a fraction of the voltage across the bulb. By changing resistance ratios, the system can be varied from current feedback to voltage feedback and thus can be optimized for changes in light-source resistance. These same authors have also reported that additional light intensity variations can result from heat flow around the light bulb itself. The use of baffling to eliminate convection currents across the optical path was reported to give the most satisfactory results. The stabilized light-source spectrophotometer was reported to have drift stability of better than 0.003 absorbance units (A.U.) per hour, a noise level which produced a rate error of less than 0.001 A.U. per minute and a photometric accuracy of 0.01 A.U. at 1.0 and 0.001 A.U. near zero absorbance.

One disadvantage of the above control system is that it controls the electrical input to the lamp rather than the actual lamp intensity. Several systems have been described using optical feedback techniques to control the lamp intensity. Usually in these techniques the lamp intensity is monitored with a second photodetector, and a signal from this detector is fed back to the lamp power supply. An optical feedback system for this purpose, originally described by Loach and Loyd [20], is diagrammed in Fig. 3. The light source requires 28 V at 5.5 A and is driven from a programmable supply. An additional 28-V supply is used to set the operating point. The control amplifier, P2A, is a summing amplifier, which sums the voltage from a dc balance potentiometer and that derived from the reference photomultiplier. The dc balance potentiometer is adjusted so that the output of P2A is 0 V at the desired lamp intensity. Any fluctuations in lamp intensity will cause the output of P2A to deviate from 0 V and drive the programmable lamp supply to eliminate the change. Amplifier P55 is used to drive a meter so that the error can be observed. Although data on the long-term stability of this system are not given, short-term drifts of less than 2×10^{-5} A.U. per second were reported. Utilizing signal-averaging techniques, absorbance changes of about 5×10^{-5} A.U. were isolated with a sensitivity of at least 1×10^{-5} A.U.

Because long-term stability is often an important factor in kinetic methods, Pardue and coworkers have described spectrophotometer systems using optical feedback techniques for long-term regulation [21, 22]. The first system described was a filter photometer [21], which is diagrammed

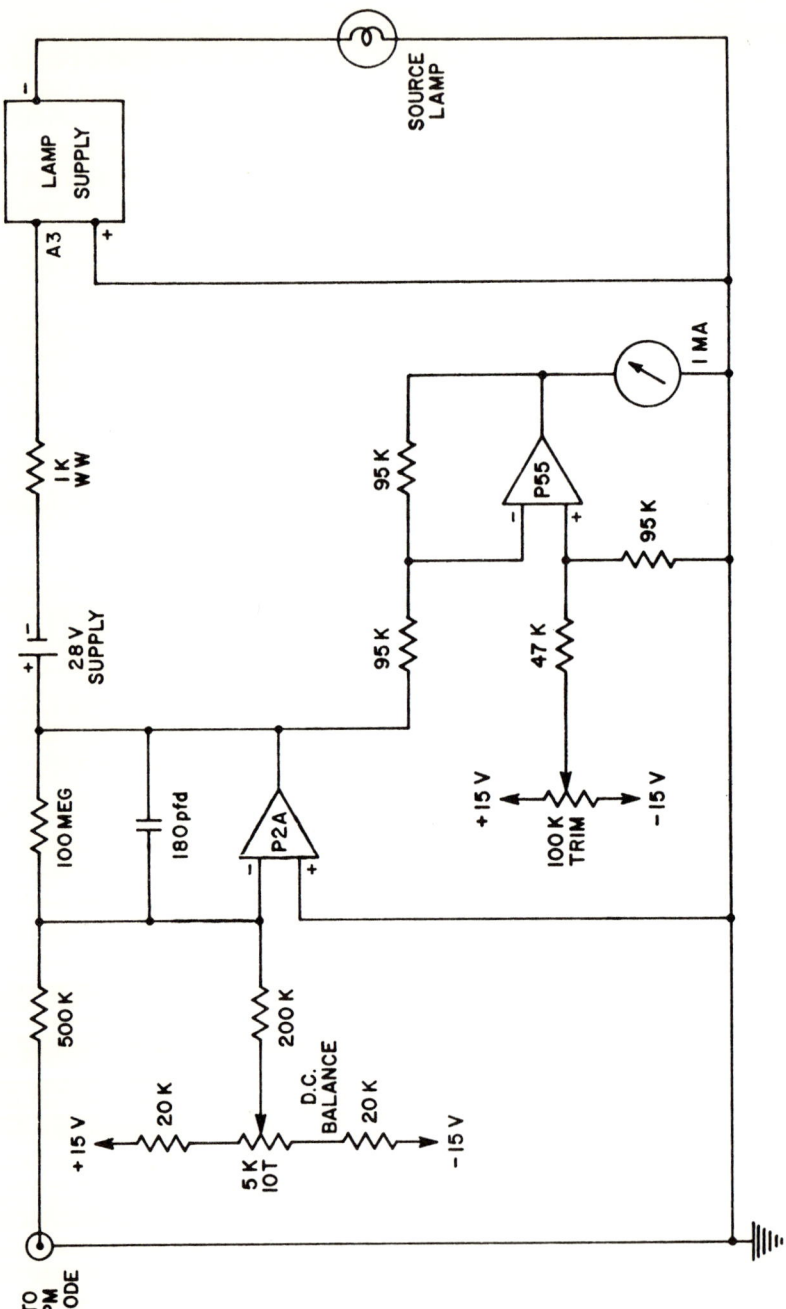

Fig. 3. Circuitry for optical feedback light-source stabilization. Reprinted from Ref. [20], p. 1710, by courtesy of the American Chemical Society.

3. Kinetic Methods of Analysis

in Fig. 4. In this system, two phototubes are used to provide both the readout signal and the signal for feedback stabilization. The control amplifier, OA1, can operate as a current-to-voltage converter or as a current integrator. The current from the phototube is summed at the input to OA1 with the current from a reference cell, which is arranged to be of opposite polarity from the phototube current. Any difference in the reference and phototube currents is either integrated or converted to a voltage and drives a programmable power supply to change the lamp output. Only when the intensity of the bulb produces a photocurrent equal to the reference current is the output of OA1 constant and the system stable. With this system, short-term stabilities of about 0.01% T/sec were reported. This corresponds to a stability of 4×10^{-5} A.U./sec near zero absorbance and about 4×10^{-4} A.U./sec near 1 A.U. The worst case of long-term fluctuation was reported to be about 0.03% T/hr. A warmup time of about 30 min was observed for stable output. These same authors also compared constant voltage operation of the light source to the optical feedback approach. The optical feedback system was found to enhance the stability by about a factor of 10.

Pardue and Deming [22] have described a second system using a mofified commercial grating spectrometer for higher resolution and a similar optical feedback circuit for stabilization. The long-term stability was observed as a function of wavelength, and average drifts of about 0.013% T/hr were reported at 425 and 522 nm. The long-term drift at 600 nm was observed to be somewhat higher, averaging about 0.036% T/hr. Short-term stability was also reported and over a few minutes was well within 0.01% T.

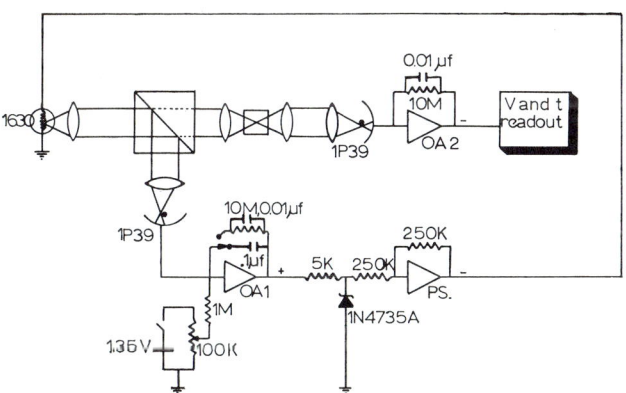

Fig. 4. Diagram of stabilized photometer. Reprinted from Ref. [21], p. 902, by courtesy of the American Chemical Society.

b. Detector Noise and Drift. Transducers for modern spectrophotometric systems are usually vacuum phototubes or photomultiplier tubes. Noise and drift from the transducer can be quite troublesome in kinetic methods, and special care is normally taken to ensure low-noise operation. There is some controversy [19, 23, 24] about which detector provides the highest signal-to-noise ratio, the phototube or the photomultiplier. If comparisons are made at the same light level, the photomultiplier is capable of higher signal-to-noise ratio, because its internal amplification is so high, often about 10^6, that the Johnson noise of the load resistor is insignificant [24]. When a phototube is used at low light levels, Johnson noise becomes the limiting factor and external amplification does not help. If comparisons are made at the same anode current, however, the phototube has the higher signal-to-noise ratio since the light level is much higher to obtain equivalent output current with the phototube. Thus in systems where the operator can control the light level, such as absorption spectrophotometers, it is recommended that intense light sources be used with a phototube detector. As is true in any measurement, if the input signal can be made high enough so that little or no amplification is needed, signal-to-noise ratios can be obtained which are higher than those obtained with amplification. For systems where the light level cannot be controlled, such as an emission spectroscopy, or the light level is very low, the photomultiplier will give superior results.

Another important consideration in detectors is power supply stability. For phototubes, regulation is uncritical because the current-voltage characteristic is essentially flat in the usual operating region. For photomultipliers, however, the gain is highly dependent on the power supply voltage. As a rule of thumb, the power supply stability should be at least an order of magnitude better than the desired gain stability [25]. For 0.1% gain stability, the power supply should be regulated to 0.01%.

c. Temperature Control. Because of the critical dependence of most reaction rates on temperature, the spectrophotometer cell must be thermostated. It is also highly desirable to include stirring in the reaction cell for efficient mixing. Usually, thermostated water is circulated around the reaction cell to provide relatively constant temperature. The most important consideration in temperature control is the rapid achievement of equilibrium temperatures. Thus several cell designs have been proposed for efficient heat transfer.

Pardue and Rodriguez [21] have described one design in which the glass spectrophotometer cell is sealed into a Lucite jacket. Water from a thermostated bath is in direct contact with the walls of the cell. The authors report that solutions which are 0.6°C different from the equilibrium temperature reach within 0.1°C of the equilibrium temperature in less

3. Kinetic Methods of Analysis

than 1 min. A stirring rod is also provided in the cell and is rotated at 1200 rpm. Complete mixing of solutions requires only 15-20 sec with this unit.

Weichselbaum and coworkers [19] have reported on a different type of cell design, which is illustrated in Fig. 5. Here the reaction chamber is separated from the spectrophotometer cell. Two reaction chambers are

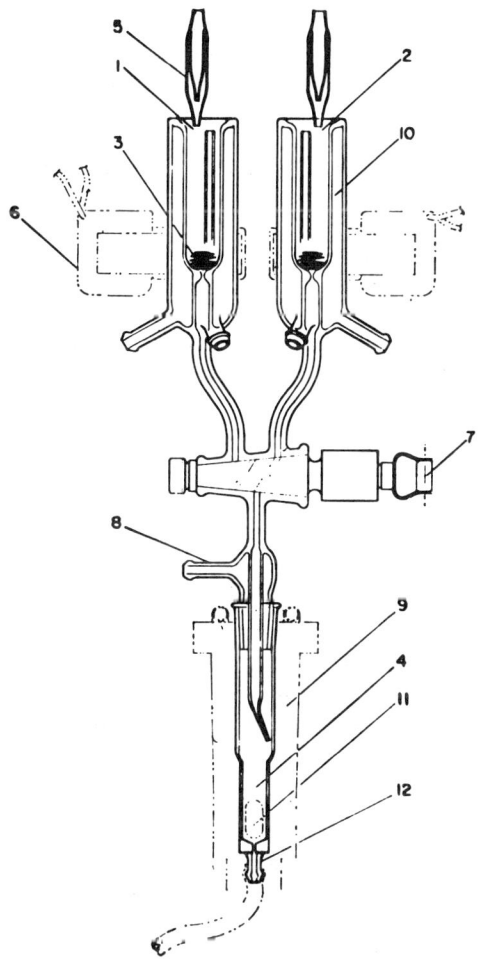

Fig. 5. Diagram of reaction chamber-cuvette system. Reprinted from Ref. [19], p. 735, by courtesy of the American Chemical Society.

provided so that one reaction mixture can be prepared while another is in the cuvette. Rapid mixing is carried out in the reaction chamber by a magnetic stirrer. Temperature control is provided in both the cuvette and the two reaction chambers. A later report [26] described analog-digital control circuitry using a proportional thermistor bridge temperature control system instead of the usual temperature-controlled switching circuits. The sample cell and water jacket are described in more detail. Using such a temperature control system, the authors obtained the data in Table 1 during lactic dehydrogenase (LDH) determinations. Equilibrium temperature control was found to be better than $\pm 0.01°C$ at all times. The data demonstrate a rapid approach to the equilibrium temperature, especially when reagents are stored in a bath at the same temperature.

Feil and coworkers [27] have described a temperature-controlled cell specifically adaptable to the Beckman DU spectrophotometer, which provides rapid response and excellent regulation. Temperature control was reported to be better than $\pm 0.03°C$ at 15 and $35°C$ and $\pm 0.005°C$ at $25°C$. The time required for a solution to come within $0.01°C$ of the equilibrium temperature was reported to be less than 1.5 min.

Finally, the most recent two-part series on temperature control in the Journal of Chemical Education [28] should be noted for its excellent discussion of the design considerations and construction details of temperature controllers.

2. Fluorescence Methods

Recently, fluorometric methods have become quite popular for monitoring chemical species in kinetic methods of analysis [28a]. In the fluorescence technique, the emission of radiation, which follows radiation-induced excitation, is used to follow the concentration of a reactant or product in solution. Thus chemical concentration is converted to radiant power. As in the case of absorption spectrophotometry, the photomultiplier transducer can convert the resultant radiant power to a proportional current or to a train of current pulses for photon counting. For dilute solutions, fluorescence intensity is directly proportional to chemical concentration.

Although fluorometric methods are less generally applicable than absorption techniques, there are several reasons why fluorometric techniques are advantageous in certain cases. First, for those species which do show appreciable fluorescence, an increase in sensitivity of at least two orders of magnitude over absorption methods is often observed. For kinetic methods, this increased sensitivity allows the detection of much smaller concentration changes, which in turn allows the initial reaction rate to be measured with high precision and accuracy. A second

TABLE 1

Temperature Measurements during LDH Determinations[a]

Time from complete addition of reagents (sec)	Monitored temperature readings in cuvette	
	Reading (reagents at $30 \pm 0.5°C$)	Reading (reagents at $26.5°C$, stirred in reaction chamber for 30 sec)
34	29.84	28.8
36	29.92	29.32
38	29.98	29.51
40	29.98	29.83
42	30.00	29.94
44	30.01	29.98
46	30.00	29.99
48	30.01	30.00
60	30.00	30.00
120	30.01	30.01

[a] Reprinted from Ref. [26], p. 1918, by courtesy of the American Chemical Society.

advantage is that the fluorescence signal is directly proportional to concentration for dilute solutions, so that electronic linearization techniques are unnecessary. For many reactions in which reactants and products do not show appreciable fluorescence, it is possible to introduce secondary reactions which take place much more rapidly than the reaction of interest to produce a fluorescent product.

Many of the considerations discussed in the previous section concerning temperature control, detector noise, and light-source stability apply directly to fluorometers, although more intense light sources are normally used in fluorescence methods for high sensitivity. Since absolute fluorescence intensities are not normally required in kinetic methods, fairly simple fluorometers, including filter instruments, can be used.

3. Potentiometric, Amperometric, and Other Methods

Instrumentation for potentiometric and amperometric monitoring of reactions is, of course, much simpler than instrumentation for spectrophotometry. In potentiometric methods, chemical concentration is converted directly into the analog electrical domain by the electrode system. Thus the transducer in a potentiometric measurement is a direct chemical-to-analog converter. Concentration cell techniques are normally used in kinetic methods, and a typical cell [29] is shown in Fig. 6. The beaker contains a reference solution, while the reaction is carried out in a small tube. Similar electrodes are placed in each container. The analog output voltage of the cell is related to the concentration difference of the electrode-sensitive species in the two containers. Such concentration cell techniques have provided potentials reliable to ± 0.02 mV, which is well within the range of reliability needed for kinetic measurements. Temperature control can be provided by circulating thermostated water around the beaker.

With amperometric detection, the rotating platinum electrode has been used for kinetic methods [30]. An external source of polarizing voltage is needed, and the electrode system performs a concentration-to-analog current conversion.

Conductometric techniques have been utilized for several fundamental kinetic studies, but have not been generally useful in kinetic methods of analysis. However, the bipolar pulse technique, recently described by

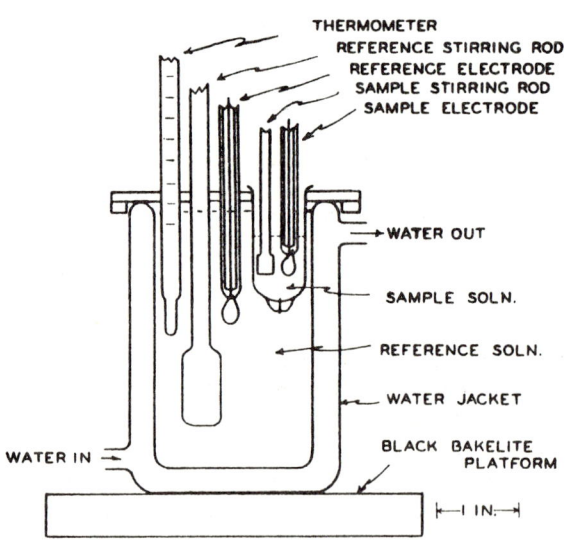

Fig. 6. Concentration cell for potentiometric rate measurements. Reprinted from Ref. [29], p. 1041, by courtesy of the American Chemical Society.

3. Kinetic Methods of Analysis

Johnson and Enke [31], appears highly applicable for continuous monitoring at the sensitivities required in kinetic methods.

C. Signal Modifier Systems

With any of the detection systems discussed, outputs from the transducers must in general be converted and shaped into signals of the proper levels, impedances, and data domains for the rate computation system which follows. If the rate computation system is analog, the signal conditioning is done in the analog domain, whereas hybrid and digital computation systems may require input signals in the time-interval or digital domain.

1. Analog Outputs

With spectrophotometric detection several different signal modifier circuits have been used. In some cases the transmittance signal can be used directly. If measurements are made before the relative transmittance changes by 2%, the error in using the transmittance signal instead of converting to a signal proportional to absorbance is no more than 1% [13]. Likewise, if the variable-time method is used for the rate computation, the transmittance signal can be used directly without introducing error. In these cases a simple operational amplifier current-to-voltage converter is used to convert the photocurrent to an analog voltage prior to the rate computation step. If a signal proportional to absorbance is necessary, a logarithmic amplifier, employing a transistor or diode feedback element, is often used to provide a voltage directly proportional to concentration. With fluorescence monitoring, the current output of the photomultiplier transducer is usually converted to a voltage by a standard operational amplifier circuit.

Potentiometric detection necessarily produces a nonlinear signal versus concentration relationship, although in some catalytic reactions the transducer response and the kinetic characteristics of the reaction combine to give a signal which varies linearly with time [10]. In most cases linearization of the potentiometric response is quite complicated, and nonlinear response curves must be utilized. In these situations, the variable-time method has distinct advantages for the computation of rates or concentrations. Signal-conditioning circuits are often quite simple with potentiometric reaction monitors and may use only a voltage follower to isolate the cell.

With amperometric detection, the current output of the electrochemical cell is inherently linear with concentration. Normally a simple operational amplifier current-to-voltage converter is used to obtain an analog voltage proportional to the concentration of the species being followed.

2. Time-Interval and Digital Outputs

With spectrophotometric monitoring several different possibilities exist. If the computer is to operate directly on the transmittance signal, the photocurrent can be converted to voltage by an operational amplifier and the resulting voltage converted to frequency by a voltage-to-frequency (V/F) converter. Alternately, the analog voltage can be converted directly to the digital domain by an A/D converter. Spectrophotometric detection also offers the possibility of utilizing the photomultiplier as a direct $P/\Delta t$ converter, if photon-counting circuits are employed. This latter route eliminates the V/F or A/D converter and thus eliminates any errors which might result from the conversion step. However, a digital logarithmic converter would be required for direct absorbance readings. If signal conditioning is done in the analog domain, the logarithm of the photocurrent can be obtained prior to V/F or A/D conversion.

With fluorescence monitoring, the photocurrent can be converted to the Δt domain by the current-voltage-frequency route, although the use of photon counting to provide a direct $P/\Delta t$ conversion would seem preferable since the physical parameter (light intensity) is directly proportional to concentration.

With potentiometric and amperometric detectors, the signals are analog and must therefore be converted to Δt or D domain if digital rate computation is employed.

D. Automated Sampling Systems

Because reaction rates are highly dependent on the concentrations of reactants, reproducible sample and reagent introduction is a necessity. For routine applications of kinetic methods, delivery and mixing of samples and reagents should be accomplished automatically. While many reports have appeared concerning instrumentation for the measurement of reaction rates and the calculation of the results, reports dealing with automated sampling systems have been much less frequent. There are several different approaches which have been employed for delivery and mixing of samples and reagents. Continuous flow systems have become quite popular since their commercial introduction. Since these have been discussed by several reviewers [10, 11], this section will focus on sampling systems utilizing discrete discontinuous sampling.

3. Kinetic Methods of Analysis 153

Malmstadt and Hicks [32] have described an automatic injection and refill pipet which operates by remote switching to deliver aliquots of reagents. A motor-driven cam pushing a syringe plunger is utilized in conjunction with delivery and refill valves for delivery of precise aliquots and refilling of the pipet. A 1-ml aliquot could be delivered with a standard deviation of about 0.009 ml. The pipet was used to inject solutions for determinations of glucose by the glucose oxidase procedure [33].

An improved automatic pipet has been described by Malmstadt and Pardue [34]. Although this pipet operates by the same general principle, the reagent and delivery valves are combined in a three-way motor-driven stopcock. Also the pipet can be continuously recycled or made to deliver a preset number of aliquots. A 1-ml aliquot could be delivered with a precision of 0.002 ml in 0.4 sec. Two of these pipets were used to add reagent and sample for glucose determinations using the glucose oxidase rate procedure [35].

An alternative approach which is attractive for many reactions is the stopped-flow technique. Although primarily used for studying the kinetics of rapid reactions, the features of automatic delivery of very small amounts of reactants and rapid mixing make the stopped-flow technique attractive for automated kinetic analysis using slow reactions as well as rapid reactions. Commercial instrumentation has recently become available for stopped-flow spectrophotometry and should continue to improve. One disadvantage of present commercially available stopped-flow systems is the manual manipulations which are required to open and close the valves, which are used for filling the syringes, delivering the reagents, and eliminating the waste solution after the reaction is completed. A completely automated stopped-flow system has been described [36] which requires no manual manipulations. The flow system is motor driven, and the addition of a sample turntable and pickup unit makes possible complete automation. With a moderately rapid reaction, the total time required for the analysis of a single sample was about 0.7 sec. For increased accuracy and precision, the system was programmed to run ten reactions using the same sample. An average of ten results could be obtained in about 10 sec, including the washing of the cell between samples. A system is currently under development in the author's laboratory which utilizes pneumatically operated valves and a pneumatic syringe drive. The stopped-flow unit is completely automatic and can be easily programmed by a digital logic sequencer.

Deming and Pardue [37] have recently described a computer-controlled electromechanical system for automatic reagent addition. The complete computerized instrumentation system is discussed in Section IV. The reagent addition system can be used to deliver up to four different solutions to a spectrophotometer cell. Filling the delivery system from reservoirs, delivery of reagent to the cell, removal of spent solution from the cell, and rinsing of the cell between runs are all accomplished automatically. The delivery system consists of four micrometer syringes, driven by bidirectional stepping motors which are activated by pulses originating from the computer. Relay-actuated solenoid valves, which are also triggered by the computer, are used to operate three-way valves for switching the solution flow. Thus stock solutions can be connected to the syringes for filling purposes, and later the syringes can be connected to the spectrophotometer cell for initiation of reactions. An automatic rinse operation for cleaning the cell and an automatic application of vacuum to remove old solution from the cell are included in the cycle. In addition, a stirring motor is actuated at the time of reagent delivery to provide good mixing of the reagents within the spectrophotometer cell. The authors report that the automated sample handling system is capable of highly accurate and repeatable initiation of reactions. In most cases kinetic measurements using the automated initiation system were reproducible to better than 1%.

Only a few of the developments in automatic sample handling have been mentioned here. Much remains to be done in this important area. With the introduction of the small laboratory computer, complete automation of all steps in the analysis procedure under computer control will soon be within the realm of many laboratories.

E. Computation Systems for Single-Component Rate Methods

1. The Derivative Technique

Since the derivative technique is based on the differential form of the rate law, it is implemented using analog circuitry. Several different computation systems have been developed to extract data related to the slopes of response curves.

Perhaps the simplest system is that developed by Weichselbaum and coworkers [19], which is now commercially available in the Digecon system (Sherwood Medical Industries, Inc., St. Louis, Missouri). This instrument utilizes direct differentiation of the output of a stable, low-noise spectrophotometer by a standard, frequency-degraded operational amplifier differentiator. To obtain the rate of change of absorbance with time, the transmittance signal T is differentiated and referenced to the instantaneous value of transmittance. The resulting signal is related by a constant ($-\log_{10} e$) to the derivative of absorbance with respect to time as shown in Eq. (67):

3. Kinetic Methods of Analysis

$$\frac{d(\log_{10} T)}{dt} = (-\log_{10} e) \left(\frac{1}{T}\frac{dT}{dt} \right). \tag{67}$$

Figure 7 shows the analog circuit which is used to provide the rate readout. The phototube current is first converted to a voltage by a standard operational amplifier signal modifier circuit. This voltage output, proportional to transmittance, is the input to the operational amplifier differentiator of Fig. 7. For simplicity, the capacitor in the feedback loop and the input resistor, which are added to restrict the bandpass of the differentiator, are not shown in Fig. 7. The reader is referred to Fig. 3 of the original article [19] for actual component values. Provision is made for a wide range of filtering capacitors to be introduced before the derivative is taken, which allows filtering to be optimized for the particular reaction being observed.

Weichselbaum and coworkers [19] report that the linearity of the rate measuring system is better than 1% over a range of 1.0 A.U. and that usable readings can be taken to 2.0 A.U.

Four different kinetic methods were investigated by the authors [19] to test the rate measuring instrument. Typical data for the assay of alkaline phosphatase in serum are shown in Table 2. The initial rate data

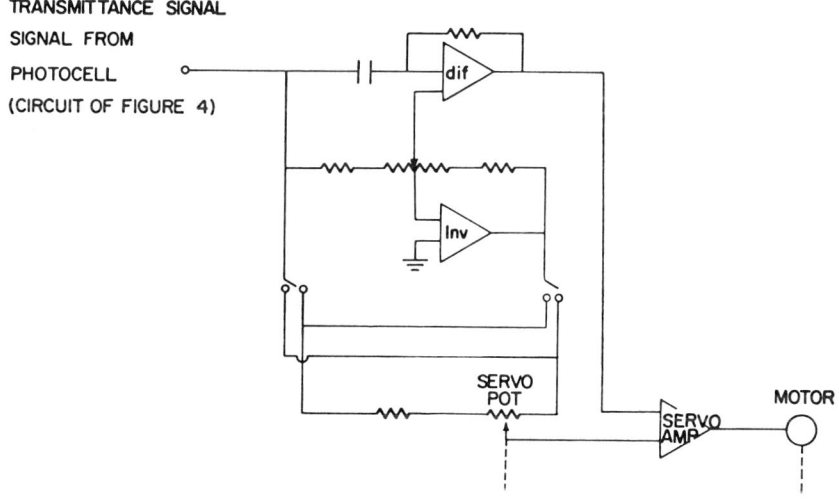

Fig. 7. Schematic diagram of rate of change of absorbance measuring circuit. Reprinted from Ref. [19], p. 730, by courtesy of the American Chemical Society.

TABLE 2

Determination of Alkaline Phosphatase in Human Serum by
by Derivative Method [a]

Serum	Values expressed in Bessey-Lowry-Brock units [b]	
	Initial rate method	Values reported by manufacturer
1	2.7	3.0
	2.7	
	2.6	
	2.5	
	2.5	
2	4.9	5.0 [c]
	5.2	
	4.9	
	5.2	
	5.2	
3	8.2	7.8
	8.4	
	8.3	
	8.3	
	8.3	

[a] Reprinted from Ref. [19], p. 736, by courtesy of the American Chemical Society.

[b] K. Linhardt and C. Walter, Methods of Enzymatic Analysis (H. U. Bergmeyer, ed.), 2nd ed., Academic Press, New York, 1965, p. 799.

[c] Based on calibration with the 5.0-unit standard serum.

are compared with values reported by a standard fixed-time assay for alkaline phosphate. As can be seen excellent reproducibilities were obtained.

Pardue and Dahl [38-40] have developed several systems for implementing the derivative technique based on a different principle. Instead of

3. Kinetic Methods of Analysis 157

obtaining the slope of the response curve by direct differentiation, an indirect procedure is employed. In these instruments, a servo system [38, 39] or an operational amplifier [40] is used to compare the slope of the response curve with a ramp generated by an operational amplifier integrator. The input voltage necessary to cause the ramp to match the slope of the response curve is proportional to the slope as shown in Eq. [68],

$$\frac{de_o}{dt} = -\left(\frac{1}{RC_f}\right) e_i \tag{68}$$

where e_i is the input voltage to the integrator, RC_f is the time constant of the integrator, and de_o/dt is the slope of the response curve. The relationships between the rates of change of reaction monitor signals and the concentrations of rate-limiting species were presented in Section II. A voltage follower is used at the output to isolate the readout (digital voltmeter) from the slope-measuring circuitry. A potentiometer on the follower output allows calibration of the output in concentration units if desired.

The automatic slope-measuring circuitry was applied to the determination of cystine by its catalytic effect on the iodine-azide reaction [40]. In this case a potentiometric technique was used to monitor the iodine concentration. The concentration cell response characteristics and the kinetic characteristics of the reaction were shown to give a cell output signal which varied linearly with time. The slope of the response curve is given by Eq. (69)

$$\frac{d\mathscr{J}_t}{dt} = kk'[\text{cystine}] \tag{69}$$

where k and k' represent proportionality constants. By equating $d\mathscr{J}_t/dt$ in Eq. (69) to de_o/dt in Eq. (68), the cystine concentration is related to the integrator input, e_i, by

$$[\text{cystine}] = \left(\frac{1}{RC_f}\right)\left(\frac{1}{kk'}\right) e_i. \tag{70}$$

Table 3 shows results obtained for cystine concentrations over a tenfold range with the readout calibrated directly in concentration units. Results were obtained about 20 sec after initiation of the reaction and generally showed relative errors and relative standard deviations of less than 1%.

Three systems based on similar principles have been developed and evaluated by Malmstadt and Crouch [41]. Slope-measuring circuits using

TABLE 3

Analysis of Aqueous Cystine Solutions by Derivative Method [a]

Cystine concentration (ppm)	Meter reading R			Relative error (%)	Relative standard deviation (%)
	R_1	R_2	R_3		
1	0.100	0.102	0.102	+1.3	0.94
2	0.199	0.200	0.201	0.0	0.41
4	0.403	0.404	0.401	+0.67	0.31
6	0.606	0.600	0.602	+0.45	0.42
8	0.804	0.802	0.796	+0.08	0.40
10	Calibration (1.000)	1.000	1.000	—	—

[a] Reprinted from Ref. [40], p. 275, by courtesy of Elsevier Publishing Company.

manual, mechanical servo, and operational amplifier slope-matching techniques are described. With the all-electronic system, the output of the reaction monitor-signal modifier circuit is the input, e_u, to the slope-measuring instrument. Two operational amplifiers are used to generate a standard slope, which is controlled by a third amplifier to equal the unknown slope. This circuit operates almost identically to that described by Pardue and Dahl [40] except that the control amplifier has a resistive feedback element instead of capacitive feedback. The resistive feedback causes the output voltage to approach its final value exponentially and, when the exponential term becomes insignificant, the readout voltage, e_r, is given by

$$e_r = -R_s C_s \left(\frac{de_u}{dt} \right) = -R_s C_s \left(\frac{d\mathscr{l}_t}{dt} \right) \tag{71}$$

where $R_s C_s$ is the integrator time constant. The feedback resistor of the control amplifier serves to control the response time of the system and to reduce noise in the output.

3. Kinetic Methods of Analysis 159

Data for the spectrophotometric determination of glucose utilizing the all-electronic slope-measuring system [41] are shown in Table 4. Direct concentration readout was obtained by taking the fraction of e_r necessary for a recorder to read 50 chart divisions for a glucose concentration of 50 ppm. Results were precise and accurate to about 2%. Because the recorder used for the final readout could limit the accuracy of the overall system, an integrating digital voltmeter was later used as the readout device with this slope-measuring system [42]. Improved results were obtained, and relative errors and standard deviations were reduced to below 1% for the same glucose procedure.

This same slope-measuring system was later applied [43] in the determination of phosphate by measuring the rate of formation of the heteropoly blue formed from phosphate, molybdate, and ascorbic acid. Results for the analysis of phosphate in blood serum are shown in Table 5. As can be seen from the data, standard deviations were again within 2%.

As previously mentioned, the major problem with the derivative technique is the enhanced sensitivity to noisy input signals. Figure 8 illustrates the type of noise encountered in the spectrophotometric reaction rate analysis of glucose [41]. As can be seen, small fluctuations on the response curve can cause large fluctuations in the measured slope, even for systems with relatively long time constants. The signals in Fig. 8 can be rather easily eye averaged or averaged electronically with an integrating readout system. For most determinations, however, care must be exercised with averaging techniques to ensure that measurement times correspond to the initial stages of the reaction. Therefore some compromise between response speed and noise must be made. When input signals are relatively noise free, the derivative technique is probably the fastest automatic method for extracting rate information.

2. Fixed-Time Method

Most of the early developments in automated instrumentation for the fixed-time approach were made in conjunction with continuous analysis systems which utilize flowing streams. As many of these systems have been previously discussed [10, 11], only the more recent advances are discussed here. The most recent fixed-time systems have used discrete sampling techniques and have employed either analog or digital integration techniques for high noise immunity. These systems employ two integration cycles with subsequent subtraction of the two integrals. Under the appropriate conditions, excellent noise immunity and linearity between readout and concentration may be obtained.

TABLE 4

Analysis of Aqueous Glucose Solutions by Derivative Method [a]

Glucose (ppm)	Recorded slope, chart divisions	Relative standard deviation (%)
80	78.0, 79.0	0.9
60	60.0, 57.5, 59.0	2.1
50	49.5, 49.0, 50.0	1.0
20	19.5, 20.0	1.8

[a] Reprinted from Ref. [41], p. 351, by courtesy of the Division of Chemical Education, American Chemical Society.

TABLE 5

Determination of Phosphate in Blood Serum by Derivative Method [a]

Phosphorus (mg per 100 ml serum)		Relative standard deviation (%)
Reported	Found	
3.0	2.97, 2.95, 3.02, 2.99	1.2
Unassayed	3.45, 3.52, 3.37	2.1
4.6	4.57, 4.65, 4.55	1.2

[a] Reprinted from Ref. [43], p. 1092, by courtesy of the American Chemical Society.

a. Analog Systems. Cordos and coworkers [44] have described the first of these fixed-time systems, based on the use of operational amplifiers to provide the integration and subtraction. Digital logic circuitry was used to control the timing and sequencing of the analog computation system. Figure 9 illustrates the basic integration and subtraction circuitry

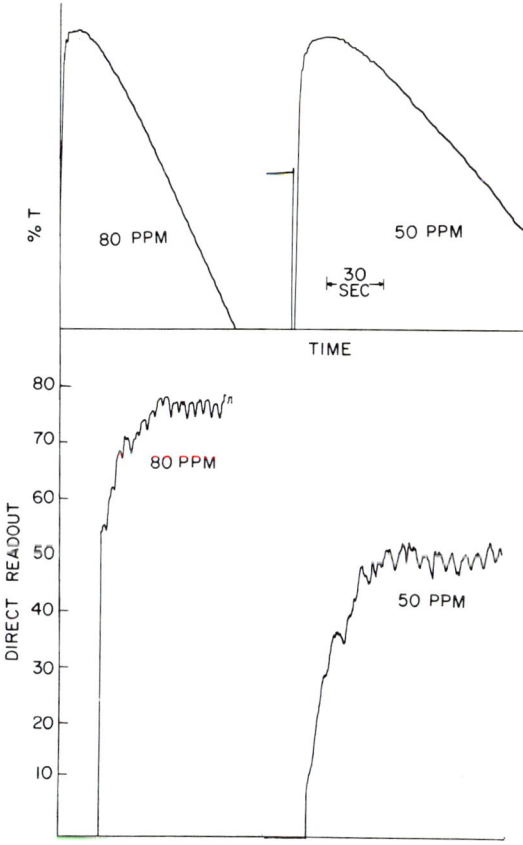

Fig. 8. Recorded rate curves and direct readout of slopes for glucose determinations. Reprinted from Ref. [41], p. 351, by courtesy of the Division of Chemical Education, American Chemical Society.

employed. The signal from the reaction monitor-signal modifier system is directed by relays 1 and 3 through OA1 and OA2 for a fixed-time interval (Δt_1). Thus the signal modifier output is inverted and integrated during the first time interval. During a second identical time interval, relay 3 is opened, and relays 1 and 2 direct the signal to OA2 without inversion. The net effect of the computation is that the output of OA2, after the two-cycle integration, is proportional to the difference in the average values of the signal during each interval as shown in Eq. (72),

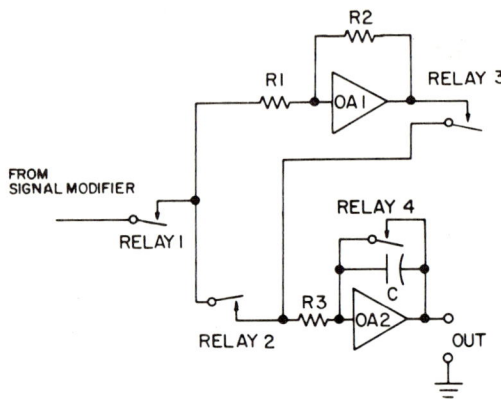

Fig. 9. Schematic diagram of analog fixed-time rate measuring circuit. Reprinted from Ref. [44], p. 1815, by courtesy of the American Chemical Society.

$$\text{output} = \bar{\mathscr{I}}_2 - \bar{\mathscr{I}}_1 = \Delta\bar{\mathscr{I}} \tag{72}$$

where $\bar{\mathscr{I}}_2$ is the average value of $\mathscr{I}_t$ during Δt_2 and $\bar{\mathscr{I}}_1$ is the average of $\mathscr{I}_t$ during Δt_1.

For response curves which are linear during the total measurement interval, ($2\Delta t = \Delta t_1 + \Delta t_2$), $\Delta\bar{\mathscr{I}}$ is related to the slope of the response curve by (44)

$$\Delta\bar{\mathscr{I}} = \frac{(\text{slope})(\Delta t)^2}{R_3 C}. \tag{73}$$

Thus, if measurements are made during the initial stages of the reaction, $\Delta\bar{\mathscr{I}}$ is directly proportional to the concentration of rate-limiting species. For situations where measurements cannot be restricted to the linear region of the response curve, this technique is subject to the limitations previously described in Section II for all fixed-time methods.

To satisfy the reader that the two-cycle integration technique does indeed give identical mathematical relationships to those given in Section II for two-point measurements, the application of the integration technique to a first- or pseudo-first-order reaction can be illustrated easily. If the product of the reaction is monitored by a linear reaction monitor, Eq. (74)

3. Kinetic Methods of Analysis 163

expresses the signal modifier output as a function of time

$$\mathscr{S}_t = \nu x [A]_0 [1 - \exp(-k_A t)] + C \tag{74}$$

where C is a constant offset signal to account for cases in which $\mathscr{S}_t$ is not zero for zero concentration, and all other terms are defined as in Section II. Integration of the inverted signal modifier output during time interval Δt_1 gives the following:

$$\bar{\mathscr{S}}_1 = -\left(\frac{1}{RC}\right) \int_{t_1}^{t_2} (-\mathscr{S}_t) \, dt$$

$$= \frac{\nu x [A]_0}{RC} \left\{ \Delta t_1 - \frac{\exp(k_A t_1)}{k_A} [1 - \exp(-k_A \Delta t_1)] \right\} \tag{75}$$

where RC is the integrator time constant. Direct integration of the signal modifier output during Δt_2 yields

$$\bar{\mathscr{S}}_2 = -\left(\frac{1}{RC}\right) \int_{t_2}^{t_3} (\mathscr{S}_t) \, dt$$

$$= \frac{-\nu x [A]_0}{RC} \left\{ \Delta t_2 - \frac{\exp(-k_A t_2)}{k_A} [1 - \exp(-k_A \Delta t_2)] \right\}. \tag{76}$$

The final readout after the second interval is given by

$$\Delta \bar{\mathscr{S}} = \bar{\mathscr{S}}_2 - \bar{\mathscr{S}}_1 = \frac{\nu x [A]_0}{RC} \left\{ (\Delta t_1 - \Delta t_2) \right.$$

$$\left. + \frac{\exp(-k_A t_2)}{k_A} [1 - \exp(-k_A \Delta t_2)] - \frac{\exp(-k_A t_1)}{k_A} [1 - \exp(-k_A \Delta t_1)] \right\}.$$

$$\tag{77}$$

Since the two integration intervals are chosen to be equal, Eq. (77) reduces to

$$\Delta \bar{\mathscr{S}} = \frac{\nu x [A]_0}{RC k_A} [1 - \exp(-k_A \Delta t)][\exp(-k_A t_2) - \exp(-k_A t_1)]. \tag{78}$$

Equation (78) is directly comparable to Eq. (7) in Section II. During the initial stages of a reaction, the exponential terms in Eq. (28) may be approximated by $e^{-x} = 1 - x$, and Eq. (78) reduces to

$$\Delta \bar{\mathscr{J}} = \frac{\nu x k_A [A]_o}{RC} (\Delta t)^2 . \tag{79}$$

Since $\nu x k_A[A]_o$ is the initial slope of the response curve, Eq. (79) reduces to the simplified form, Eq. (73), when the initial rate approximation is valid. Again, Eq. (78) reveals that linearity between $\Delta \bar{\mathscr{J}}$ and $[A]_o$ will hold for any fixed-time interval during the reaction since Δt, t_1, and t_2 are all held constant from run to run.

The major advantage of the two-cycle integration technique is the high noise immunity which results from averaging the signal during the two cycles. Exact cancellation of noise occurs when the period of the noise is a multiple of the integration time. Noise of other frequencies will not cancel exactly, but in the worst case the integration method gives results with signal-to-noise ratios equal to a two-point fixed-time measurement.

A complete schematic of the fixed-time system [44] including the sequencing circuitry is shown in Fig. 10, where the top part of the diagram shows the analog computation system of Fig. 9. The sequencing system consists of a synchronous four-stage binary counter, an input NAND gate (G), a monostable multivibrator, and a standard clock. The sequencer is begun by activating switch S, which triggers the monostable multivibrator M. The monostable is set to delay the measurement until mixing is

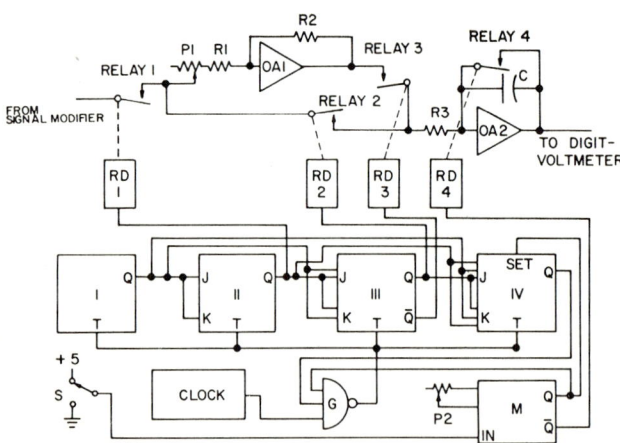

Fig. 10. Diagram of analog fixed-time rate measuring circuit with digital logic system. Reprinted from Ref. [44], p. 1815, by courtesy of the American Chemical Society.

3. Kinetic Methods of Analysis

complete in the reaction cell and any induction period in the reaction has passed. On the initial trigger the Q output of the monostable undergoes a $\underline{1} - \underline{0}$ transition, which closes gate G, preventing any clock pulses from being counted. This same transition "sets" flip-flop IV so that Q_{IV} becomes $\underline{1}$. After the delay time, the monostable returns to the $\underline{1}$ state, which opens the gate to allow clock pulses to reach the counter. Eight pulses from the clock are counted, after which Q_{IV} undergoes a $\underline{1} - \underline{0}$ transition, which again closes the gate. During the eight cycles of the clock, flip-flop II undergoes a $\underline{0} - \underline{1}$ transition twice and is utilized to control relay 1. This directs the signal into the computer for the two equal time intervals. Flip-flop output Q_{III} controls relay 2, while the complementary output $\overline{Q}_{III}$ controls relay 3. During the first measurement interval Q_{III} is $\underline{0}$, while $\overline{Q}_{III}$ is $\underline{1}$. Thus the signal is directed through the inverter before being integrated by OA2. During the second measurement interval, Q_{III} is $\underline{1}$, which closes relay 2. The inverter is thus bypassed during the second interval, and the signal is directly integrated.

The computation system itself was evaluated with a synthetic ramp generated by an analog sweep generator [44]. Data on these synthetic slopes are shown in Table 6. These data show the high precision and accuracy possible with the fixed-time integration system. Results were also obtained on synthetic signals which simulated response curves from fast reactions. Integration times of 50 msec were employed for these rapidly changing signals. The data again indicated relative errors and standard deviations of about 0.2% on these linear sweeps.

Two chemical systems were chosen to illustrate the application of the fixed-time system for kinetic analyses [44]. The determination of glucose by the glucose oxidase procedure gave the results shown in Table 7. For this procedure, direct concentration readout was obtained by holding the difference voltage output of OA2 and adjusting a potentiometer on the output until the voltage corresponded directly to the concentration of a standard glucose solution. Relative errors and standard deviations of about 1% were obtained routinely for glucose determinations in water. Results were also obtained for the determination of phosphate by the heteropoly blue procedure with slightly higher relative errors and standard deviations.

This same fixed-time readout system was later applied to the determination of phosphate utilizing the rapid reaction of phosphate and Mo(VI) to form 12-molybdophosphoric acid [36]. A stopped-flow spectrophotometer was used for rapid mixing and automated sampling. For this analysis, a total measurement time ($2 \Delta t$) of 100 msec was employed, and direct concentration readout was used. Because of the rapid analysis time and high degree of automation, ten results could be obtained in less than 10 sec, allowing increased averaging and thus high precision.

TABLE 6

Fixed-Time Rate Measurements on Synthetic Signals
Simulating Slow Reactions [a]

Input rate (mV/sec)	Digital readout [b] (mV)	Proportionality constant	Relative standard deviation (%)
2.50	111.8	0.0223	0.24
3.60	163.4	0.0220	0.24
5.30	239.7	0.0221	0.33
7.90	357.3	0.0221	0.15
10.80	485.6	0.0222	0.09

[a] Reprinted from Ref. [44], p. 1817, by courtesy of the American Chemical Society.

[b] Averages of ten results; integration time of 10 sec; premeasurement time of 30 sec.

TABLE 7

Fixed-Time Determinations of Glucose [a]

Direct concentration readout [b]	Glucose concentration (ppm)		
	Taken	Relative error (%)	Relative standard deviation (%)
5.0	5.0	0.0	1.6
10.0	10.0	—	1.0
15.1	15.0	+0.7	0.7
20.1	20.0	+0.5	0.8

[a] Reprinted from Ref. [44], p. 1817, by courtesy of the American Chemical Society.

[b] Averages of five results; 10.0-ppm standard used to set readout; integration time of 10 sec; premeasurement time of 30 sec.

Toren, Hicks, and coworkers [45, 46] have described a computer interface for reaction rate measurements which utilizes a similar analog integration technique. The authors chose to develop a hardware interface for rate measurements rather than utilize software techniques because the latter approach would have required the sampling of too many data points to be practical for their time-shared computer system. This computer-based system is discussed in more detail in a later section.

Although these analog computation systems are capable of providing high-quality data, the operational amplifiers utilized in the integration and subtraction circuits are subject to current and voltage offsets and drifts. In particular the integrator must be of rather high quality for accurate results to be obtained. There are also limitations to the input slopes which can be measured with these systems. Parameters must be adjusted to ensure that the integrator does not reach its voltage limit at any time during the measurement and to ensure that a sufficient difference voltage exists for accurate readout. With these instrumental factors in mind, a theoretical range of slopes of about 10^6 can be measured with the circuit of Fig. 10 using only two integration times, 10 sec and 100 msec. Chemical factors, as discussed in Section II, place more stringent limits on this range, particularly for reactions which are not first order.

b. Digital System. More recently a completely digital computation system was described [47] to implement the fixed-time approach. This system is similar in principle to the analog systems just described, except that digital integration techniques are used. In this system the analog reaction monitor output is converted immediately to the time-interval domain by a voltage-to-frequency (V/F) converter. The pulses from the V/F converter are gated into an up-down counter for two identical fixed-time intervals. During one interval the counter counts up, while during the second interval the counter counts down. The net result of the counting process is integration of the signal and conversion to a digital number proportional to the slope of the input signal. The readout of the counter, R, was shown to be related to the input slope, for response curves which are linear during the total measurement interval (2 Δt), by

$$R = (\text{slope}) k (\Delta t)^2 \tag{80}$$

where k is the conversion rate of the V/F converter in hertz per volt. For nonlinear response curves, this system has the same limitations as the analog systems just described.

A complete circuit diagram of the instrument is shown in Fig. 11. Switch S_2 allows the instrument to take measurements continuously in the "continuous measurement" position, or allows only one measurement in

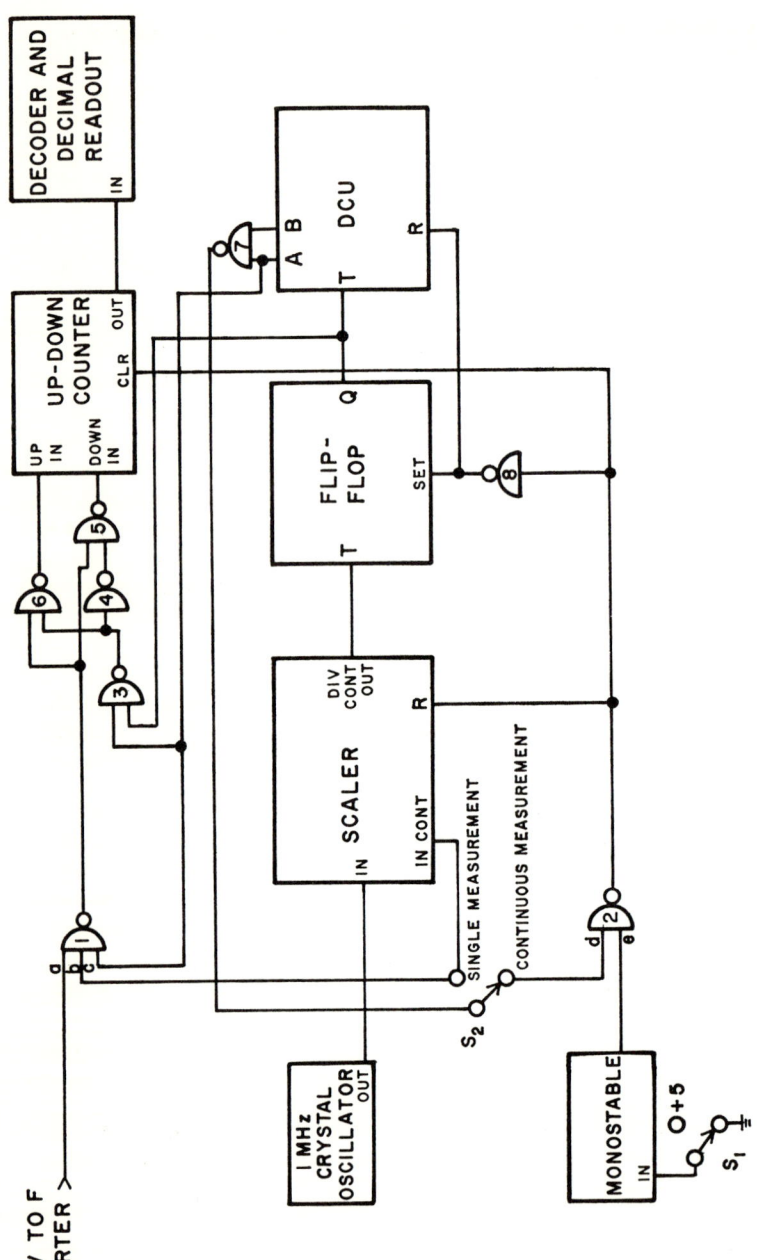

Fig. 11. Diagram of digital fixed-time rate computer. Reprinted from Ref. [47], p. 1057, by courtesy of the American Chemical Society.

3. Kinetic Methods of Analysis

the "single measurement" position. The monostable multivibrator is triggered by S_1 and immediately undergoes a $\underline{1} - \underline{0}$ transition, which resets all circuits to their initial states. When the monostable returns to the $\underline{1}$ state, the 1-MHz crystal oscillator signal is divided by the scaler to obtain clock periods from 10^{-6} to 10 sec. The desired scaler output is the input to a JK flip-flop, which further divides the output by 2 to obtain gating signals of the appropriate time intervals. The first $\underline{1} - \underline{0}$ transition of the scaler occurs after one period of the selected time base, and causes the flip-flop to undergo a $\underline{1} - \underline{0}$ transition at the Q output, which causes the A output of the decade counting unit (DCU) to change from a logical $\underline{0}$ to a logical $\underline{1}$.

Gate 1 controls the transmission of the V/F converter pulses to the computer. When S_2 is in the continuous mode, input b to gate 1 is open and has no effect. Input c to gate 1 is controlled by the A output of the DCU, which also controls gate 3. From the first to second $\underline{1} - \underline{0}$ transition of the scaler, Q is $\underline{0}$ and A is $\underline{1}$. During this period, gate 3 has a $\underline{1}$ output, which opens gate 6 to the pulses from the V/F converter. The output of gate 3 is inverted by gate 4, which closes gate 5 and causes the down counter to be a logical $\underline{1}$. When counting up, the down counter input must be in the $\underline{1}$ state, and the opposite must apply when counting down. Thus from clock pulses 1 to 2, the up-down counter counts up.

From the second to the third $\underline{1}$ to $\underline{0}$ transition of the scaler, both Q and A are $\underline{1}$. During this interval the output of gate 3 is in the $\underline{0}$ state, which closes gate 6 and opens gate 5 to the V/F converter pulses. Thus from clock pulses 2 to 3, the up-down counter counts down.

From the third to fifth $\underline{1} - \underline{0}$ transition of the scaler, A is $\underline{0}$ and B is $\underline{1}$. Thus gates 1 and 3 are closed, isolating the up-down counter and holding the accumulated counts for the readout. In the continuous mode, the fifth $\underline{1} - \underline{0}$ clock pulse resets all circuits, and the cycle repeats itself. In the single measurement mode, however, the output of gate 7 is diverted from the reset circuitry to the scaler input control, which stops the cycle.

This all-digital computation system was evaluated by using synthetic signals to simulate response curves [47]. Data for these signals are shown in Table 8. Except for the extremely slow or fast rates, the data obtained are precise and accurate to a few tenths percent. An improvement over purely analog processing can be seen if the data in the middle of the range tested are compared with those in Table 6. At the extremes of the slopes tested, factors other than the computation system limit the accuracy of the results. For low rates, the analog integrator used to generate the synthetic slopes becomes inaccurate due to voltage and current offsets and drifts. For high rates, the V/F converter is unable to

TABLE 8

Fixed-Time Rate Measurements of Synthetic Slopes[a]

Input slope (mV/sec)	Digital readout[b] (counts)	Relative standard deviation (%)	Relative error (%)
0.1	1,008[c]	4.00	+0.80
0.5	5,024[c]	0.94	+0.48
1	9,974[c]	0.16	−0.26
2	19,982[c]	0.19	−0.09
5	50,050[c]	0.16	+0.16
7	702[d]	0.28	+0.29
10	1,000[d]	0.13	0.00
20	2,000[d]	0.09	0.00
50	Calibration	—	—
70	6,999[d]	0.02	−0.01
100	10,001[d]	0.02	+0.01
200	20,003[d]	0.01	+0.02
500	501[e]	0.27	+0.20
700	701[e]	0.26	+0.14
1000	1,003[e]	0.26	+0.30
2000	2,015[e]	0.21	+0.75

[a] Reprinted from Ref. [47], p. 1059, by courtesy of the American Chemical Society.

[b] Averages of ten results.

[c] Integration time of 20 sec.

[d] Integration time of 2 sec.

[e] Integration time of 0.2 sec.

3. Kinetic Methods of Analysis 171

follow the rapidly changing input signal and thus limits the accuracy. The noise immunity of the system was also evaluated by superimposing sinusoidal noise on the synthetic ramp. Although the precision and accuracy were decreased somewhat by the presence of noise, relative errors and standard deviations were still within 0.2%.

The complete fixed-time rate measuring system was applied to the determination of phosphate using spectrophotometric monitoring of the heteropoly blue reaction [47]. The data obtained are shown in Table 9, and illustrate accuracies and precisions in the 1 to 2% range. Because this computation system is all digital, it is not as convenient to adjust the output for direct concentration readout as with analog systems. To do this, the input to the V/F converter is the most logical place where a continuous adjustment can be made.

The all-digital computation system [47] has several advantages over the analog circuits discussed in the previous section: First, any drifts and nonlinearities in the computation system itself are eliminated. Second, there are no mechanical or analog switches involved in the actual computation circuit. Thus much shorter integration times can be used and faster rates measured without loss of accuracy. There are also several limitations to the system: First, there are still a few analog circuits employed in the reaction monitor-signal modifier system and in the V/F converter. Thus drifts and nonlinearities still occur, although the influence of these is much smaller than with the analog integration system. The digital system also has purely instrumental limitations on the range of rates which can be processed. The readout should be high enough to accumulate 1000 counts for at least 0.1% counting accuracy. With a seven-decade up-down counter the readout cannot be greater than 9,999,999. Thus for any given time base, a dynamic range of about 10^4 is possible. Other factors which limit the dynamic range are the input voltage of the V/F converter and the speed with which the converter can follow input voltage changes. With three different time bases, a dynamic range of 2×10^4 was covered, as shown in Table 8, with high accuracy and precision. As mentioned previously, chemical factors usually limit the dynamic range to less than this in practical analyses.

The all-digital fixed-time computation system is capable of such high accuracy and precision that other factors must be sought to explain why data on actual chemical systems are at least an order of magnitude less accurate and precise than data for simulated rates [47]. With spectrophotometric detection, very small drifts in light-source radiance, shot noise, or changes in photomultiplier supply voltage are enough to cause significant standard deviations in the overall results. That these factors are more important than deviations due to sampling was made

TABLE 9

Fixed-Time Determinations of Phosphate [a]

Digital readout [b]	Phosphorus concentration (ppm)		Relative error (%)	Relative standard deviation (%)
	Taken	Found [c]		
6,730	3	3.01	+0.33	1.48
11,167	5	—	—	2.05
17,800	8	7.97	−0.38	0.66
22,249	10	10.01	+0.10	1.05

[a] Reprinted from Ref. [47], p. 1060, by courtesy of the American Chemical Society.

[b] Averages of five results.

[c] Based on 5-ppm standard; integration time of 40 sec; premeasurement time of 40 sec.

evident by many blank measurements using the fixed-time computer. Such blanks showed an average readout of very nearly zero counts, but displayed standard deviations comparable to those obtained during actual kinetic runs. Highly stabilized photometers, such as those previously described, should help improve the quality of data on real chemical systems.

3. Variable-Time Method

With the variable-time method, the computation system itself must be somewhat more complex than fixed-time computers if complete automation is desired. The instrument must not only detect the preset signal interval ($\Delta \mathcal{I}$) and measure the time required for the signal modifier output to change by this amount but must also compute the reciprocal of this time interval if direct rate or concentration readout is desired. The early developments in automating the variable-time approach employed automatic timing equipment [29, 33, 35], but required a manual conversion

3. Kinetic Methods of Analysis   173

to reciprocal time. More recently, however, completely automatic variable-time systems have appeared. These systems are classified in this section as analog, hybrid, and digital systems.

a. Analog Systems. The first successful completely automatic variable-time measurement system was described by Pardue and coworkers [48, 49]. In the original system, spectrophotometric or amperometric monitoring was employed, but the system is applicable to virtually any detector system. Figure 12 illustrates the detection and computation circuitry employed with amperometric reaction monitoring [49]. The current output of the reaction cell is converted to a voltage by OA1. The output of OA1 is then converted back to a current by the 18K input resistor to the meter relay. The current-voltage-current conversion prevents loading of the cell by the meter relay and provides current gain. The input current to the meter relay is the signal which is used to control the automatic measurement circuits. The meter relay serves as the control device, which detects the preselected signal interval ($\Delta \mathscr{I} = \mathscr{I}_2 - \mathscr{I}_1$) and activates either a normal laboratory timer or a circuit to compute reciprocal time. The reciprocal time computer is shown at the bottom of Fig. 12. When the first set point ($\mathscr{I}_1$) is reached, contacts 5 and 6 are made, and motor M1 is activated. The motor turns the shaft of potentiometer P3 at a constant rate, which increases the input resistance, R_{in}, to OA2 at a steady rate. When the current reaches the second set point ($\mathscr{I}_2$), contacts 12 and 13 are broken, and the motor stops. Since the output voltage, e_o, of OA2 is proportional to the ratio of the feedback to input resistances, e_o is given by

$$e_o = -\left(\frac{R_f e_{in}}{k''}\right) \frac{1}{\Delta t} \tag{81}$$

where e_{in} is the input voltage to OA2 and k'' is the rate of resistance change, assuming $R_{in} = 0$ before the signal reaches $\mathscr{I}_1$. For those cases discussed in Section II, in which $1/\Delta t$ is directly related to the concentration of the rate-limiting species, it follows that e_o will be directly proportional to concentration. Provision is also made in the system for electrical compensation for any nonzero intercept in the relationship between e_o and concentration. This is accomplished by introducing a voltage source of the appropriate magnitude at point J3. Potentiometer P4 allows the user to calibrate the unit for direct concentration readout. A digital voltmeter can then be connected at point J2 for readout purposes.

Table 10 demonstrates the high accuracy and precision which were obtained on an earlier version of the instrument shown in Fig. 12 for glucose determinations in the 20-80 ppm range [48]. Results in column I are based on a single standard, 30 ppm, for calibration of the computer. Two

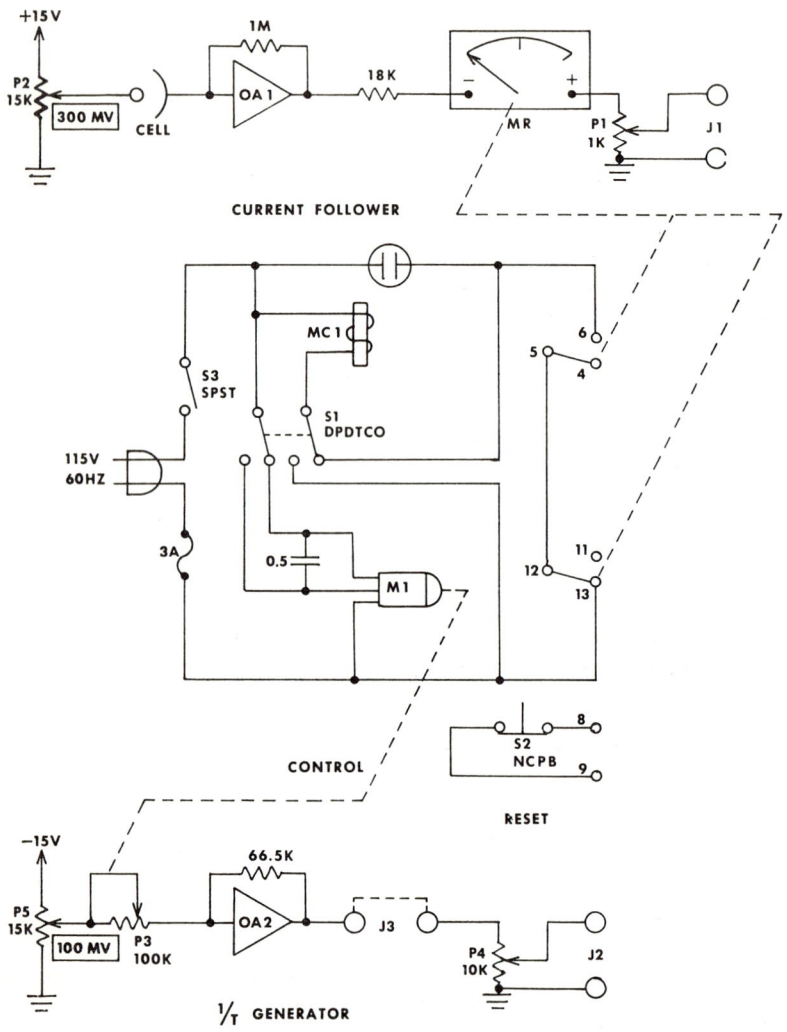

Fig. 12. Diagram of variable-time rate computation circuit. Reprinted from Ref. [49], p. 686, by courtesy of the Division of Chemical Education, American Chemical Society.

3. Kinetic Methods of Analysis

TABLE 10

Determination of Glucose Using Variable-Time Method [a]

Glucose concentration (ppm)			Relative error (%)		Relative standard deviation (%)	
Taken	Found (mV x 10^{-1})					
	I	II	I	II	I	II
20	19.8	20.1	1.0	0.5	0.8	1.6
30	29.4	—	2.0	—	0.3	—
40	39.8	40.6	0.5	1.5	0.6	1.1
50	50.4	50.5	0.8	1.0	0.3	0.8
60	60.0	61.9	0.0	3.1	1.1	0.8
80	81.0	83.0	1.3	3.8	0.8	0.8

[a] Reprinted from Ref. [48], p. 1428, by courtesy of the American Chemical Society.

solutions, 30 and 60 ppm, were used to establish the appropriate nonzero intercept. The results in column II were obtained using 20- and 40-ppm standards for zero compensation and the 30-ppm standard for calibration. The larger relative errors obtained under these latter conditions were attributed to the use of standards near one of the extreme ends of the concentration range, instead of in the middle of the range. Data were also presented for galactose, glucose oxidase, and galactose oxidase determinations [48].

There are several limitations to the circuit shown in Fig. 12. First, both the control system and the reciprocal time computer contain electromechanical components, which limit the response speed and thus the range of rates which can be measured. Also the electromechanical components are subject to higher failure rates than purely electronic devices. A second limitation is the susceptibility of the analog amplifiers to offsets and drifts.

A circuit which partially overcomes the first limitation has been described by Stehl et al. [50] and is illustrated in Fig. 13. In this system, a meter relay is again used to detect the preselected signal interval,

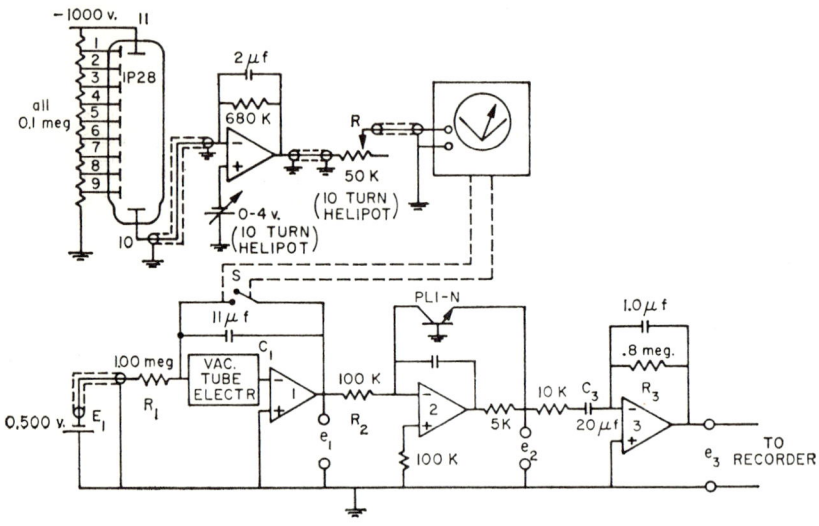

Fig. 13. Diagram of all-electronic variable-time rate computation circuit. Reprinted from Ref. [50], p. 1347, by courtesy of the American Chemical Society.

$\Delta \mathscr{I}$, but the remainder of the circuitry is all electronic. Spectrophotometric detection is illustrated in Fig. 13. The output current of the photomultiplier tube is converted to voltage by the operational amplifier at the top of the figure, and the voltage is converted back to current for driving the meter relay. When the input current to the meter relay reaches the first set point, $\mathscr{I}_1$, switch S is opened. Closure of S occurs when $\mathscr{I}_2$ is reached. Amplifiers 1, 2, and 3 constitute the reciprocal time computer. Amplifier 1 is an integrator which generates a linear ramp when switch S is opened. The output voltage of amplifier 1 reaches a maximum value e_1 just prior to the closure of S. At this point, Eq. (82) gives the relationship between e_1 and the time interval, Δt, that switch S was open

$$e_1 = \left(\frac{-E_1}{R_1 C_1} \right) \Delta t. \tag{82}$$

Amplifier 2 computes the logarithm of e_1, and the output of 2, e_2, is given by

$$e_2 = k \log\left(\frac{e_1}{R_2 I}\right) \tag{83}$$

where k and I are constants at a constant temperature. Amplifier 3 computes the derivative of the output of e_2, according to Eq. (84):

$$e_3 = -R_3 C_3 \left(\frac{de_2}{dt}\right) = \frac{-R_3 C_3 k R_2 I}{e_1} . \tag{84}$$

Substitution of Eq. (82) gives the relationship between the output of amplifier 3 at the point just before closure of switch S and the time interval, Δt

$$e_3 = \left(\frac{R_3 C_3 k R_2 I R_1 C_1}{E_1}\right) \frac{1}{\Delta t} . \tag{85}$$

Since all the factors in parentheses are constant at a given temperature, e_3 is proportional to the reciprocal of the time necessary for the signal to change by $\Delta \mathscr{I}$.

There are again several limitations to the circuit of Fig. 13. The meter relay system used to detect the two set points again limits the response speed so that only relatively slow reactions may be used. Also, the logarithmic amplifier is subject to rather severe temperature effects, unless critical compensation circuits or thermostating of the logarithmic element is employed. The use of a differentiator on the output can cause noise enhancement, although the signals employed in the computer itself should be relatively noise free. Probably the major limitation to this circuit is the transient nature of the output signal. As soon as switch S closes, the output of amplifier 1 returns to zero, and the reciprocal time information is lost. Noise averaging of the final output e_3 cannot be accomplished because of its transient nature.

An alternative analog circuit, which overcomes some of the limitations of the previous systems, has been described by James and Pardue [51]. In this system neither the control system nor the reciprocal time computer contains any mechanical components. The control system utilizes tunnel diodes and silicon-controlled switches to detect the signal interval ($\Delta \mathscr{I} = \mathscr{I}_2 - \mathscr{I}_1$) over which the reciprocal of time is computed. The reader is referred to Fig. 3 of the original report [51] for details of the solid-state switching circuit. The reciprocal time computer itself is illustrated in Fig. 14. As in the previous example, an operational amplifier integrator, OA4, is used to generate a voltage proportional to the

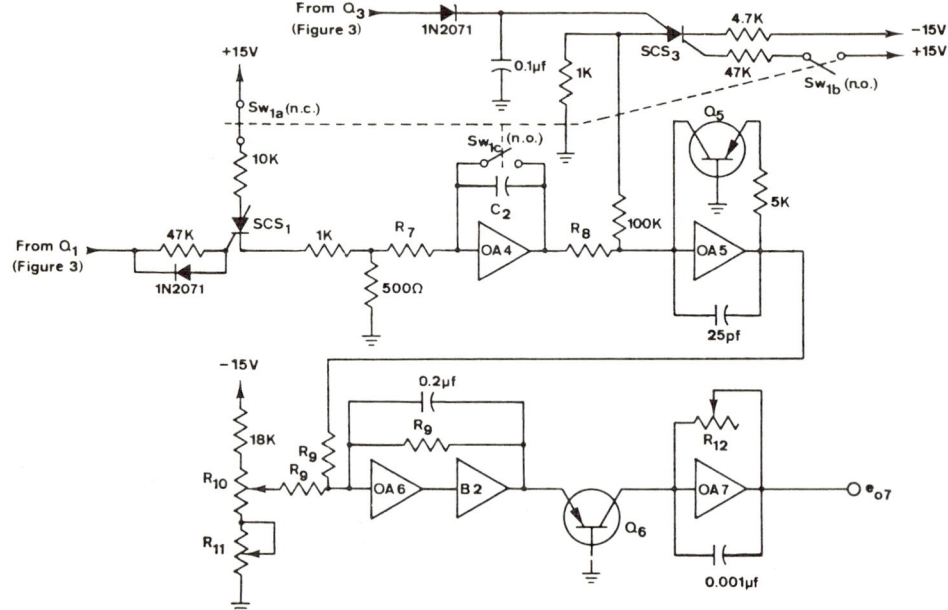

Fig. 14. Circuit of analog reciprocal time generator. Reprinted from Ref. [51], p. 799, by courtesy of the American Chemical Society.

time interval that the signal is between the two set points, $\mathscr{I}_1$ and $\mathscr{I}_2$, by integration of a constant voltage, e_{in}. In this circuit, in contrast to the previous example, the input voltage to the integrator is switched on and off at $\mathscr{I}_1$ and $\mathscr{I}_2$, respectively, which means that the output of OA4 is steady after the signal interval is detected, instead of being immediately shorted. Amplifier OA5 is a logarithmic amplifier which computes the logarithm of the output of OA4. The output of OA5 is summed with a reference voltage e_{ref} at summing amplifier OA6. The final combination of transistor Q_6 and OA7 computes the antilogarithm of the inverted sum, giving a voltage readout proportional to the reciprocal of the time interval as shown in Eq. (86).

$$e_{o7} = \left[R_{12} \left(\frac{-R_8 R_7 C_2 I_1 I_2}{e_{in}} \right) 10^{-e_{ref}/E_o} \right] \frac{1}{\Delta t} \qquad (86)$$

3. Kinetic Methods of Analysis 179

where I_1 and E_0 are characteristic of transistor Q_5, and I_2 is characteristic of Q_6. At a constant temperature, all terms in the brackets of Eq. (86) are constant, and the output of OA7 is directly proportional to reciprocal time. Potentiometer R_{12} is used to calibrate the computer in the desired output units.

The major advantage of the circuit of Fig. 14 over those reciprocal time computers previously discussed is the increased response speed which results from the complete replacement of electromechanical components and switches. Table 11 illustrates results which were taken to demonstrate the short time response capabilities [51]. An external integrator was used to generate a ramp to simulate fast reaction rates. Ramps from 1 to 5 V/sec were measured. The actual time intervals computed ranged from 6 to 18 msec. The data illustrate that the computer is highly reliable even at these short time intervals. Results were also reported for the determination of alkaline phosphatase in blood serum by a spectrophotometric reaction rate procedure. These data indicate that relative errors and standard deviations of 1% or less can be expected with appropriate chemical systems.

The major limitation to the circuit of Fig. 14 is the critical dependence of the logarithmic and antilogarithmic circuits on temperature. Thermostating should be used to reduce temperature effects. The integrating amplifier, OA4, should have low current offset and low voltage drift for accurate conversion of time interval to voltage.

b. **Hybrid Systems.** One system has been described [52] for implementing the variable-time approach which is a hybrid analog-digital system. This system uses all solid-state circuitry for rapid response and eliminates the critical logarithmic elements which limit the analog system of Jamos and Pardue [51]. A block diagram of the hybrid computer is shown in Fig. 15. The control system, the voltage interval sensor, has two set signal levels $\mathscr{S}_1$ and $\mathscr{S}_2$ corresponding to two concentrations of the monitored reactant or product. When the signal from the reaction monitor-signal modifier system reaches $\mathscr{S}_1$, a trigger signal starts a linear sweep generator, which serves as a time interval-to-voltage converter. When the input signal reaches $\mathscr{S}_2$, the linear sweep generator is stopped, and the time interval required for the signal to change by $\Delta\mathscr{S}$ is encoded as a proportional voltage. This voltage is then converted to frequency by a voltage-to-frequency converter. Thus Δt is encoded as the frequency of a pulse train. A digital period meter, triggered when level $\mathscr{S}_2$ is reached, measures the period of the pulse train, which is proportional to the reciprocal of the time interval.

TABLE 11

Variable-Time Measurements on Synthetic Signals
for Short Time Intervals [a]

Integrator input (V)	Computer output (V)	Proportionality constant x 10⁴	Deviation from average (%)	Relative standard deviation (%)
0.0400	0.602	664	+1.4	0.42
0.0501	0.785	638	-2.6	0.39
0.0600	0.925	648	-1.1	0.23
0.0699	1.08	648	-1.1	0.66
0.0900	1.38	652	-0.5	0.51
0.100	1.50	666	+1.7	1.22
0.111	1.66	661	+0.9	1.35
0.120	1.82	660	+0.8	0.39
0.130 [b]	1.97	660	+0.8	0.62
		Average 655		

[a] Reprinted from Ref. [51], p. 801, by courtesy of the American Chemical Society.

[b] Corresponds to time interval of about 6 msec.

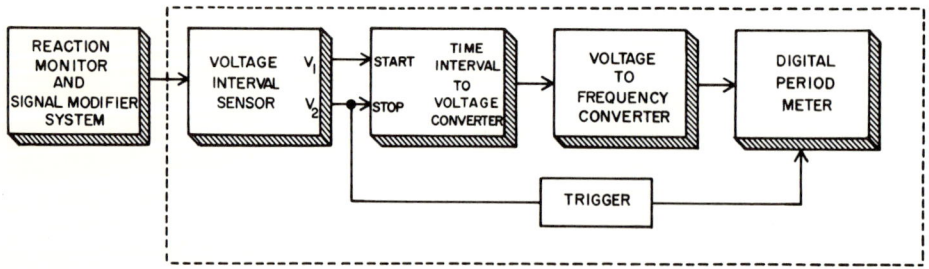

Fig. 15. Block diagram of hybrid variable-time rate measurement system. Reprinted from Ref. [52], p. 880, by courtesy of the American Chemical Society.

3. Kinetic Methods of Analysis

The voltage interval sensor consists of two high-speed voltage comparators and two stable reference voltages, which provide the preselected levels $\mathscr{l}_1$ and $\mathscr{l}_2$. Logic driver outputs from the comparators are used to trigger subsequent circuits, such as the time interval-to-voltage converter illustrated in Fig. 16. When level $\mathscr{l}_1$ is reached, comparator 1 changes from a logical 0 to a logical 1, which opens the analog gate consisting of Q_1 and FET 1 and allows the integrating capacitor, C_f, to begin charging. When level $\mathscr{l}_2$ is reached, the analog gate consisting of Q_2 and FET 2 is opened, removing the integrator input voltage E_{in}. Thus, the output of OA1 increases linearly for a time interval, Δt, when the input signal is between $\mathscr{l}_1$ and $\mathscr{l}_2$ and is held until the next measurement begins. The output of OA1 after level $\mathscr{l}_2$ has been reached is given by Eq. (87)

$$E_o = \frac{-E_{in} \Delta t}{R_{in} C_f} . \tag{87}$$

After level $\mathscr{l}_2$ has been reached, E_o is converted to a proportional frequency by the V/F converter, and the period, T, is displayed on the digital period meter. Equation (88) expresses the relationship between the period and reciprocal time

$$T = \frac{1}{f} = \left(\frac{R_{in} C_f}{k E_{in}} \right) \frac{1}{\Delta t} \tag{88}$$

where k is the conversion rate of the V/F converter in hertz per volt.

For higher resolution, multiple periods can be taken. In this mode, the internal clock of the digital period meter is counted for a scaled number of input periods. If F periods are averaged by selecting a scaling factor of F, the readout becomes

$$\text{readout} = \left(\frac{KFR_{in} C_f}{kE_{in}} \right) \frac{1}{\Delta t} \tag{89}$$

where K is the frequency of the internal oscillator of the counter. The scaling factor, F, and the integrator input voltage, E_{in}, can be adjusted to accommodate a wide variety of input rates with high accuracy. Potentiometer P_2 at the output of OA1 can be used to provide direct concentration or rate readout in the units desired.

Table 12 illustrates typical results obtained using the hybrid computer to measure the rates of synthetic input signals whose slopes varied from

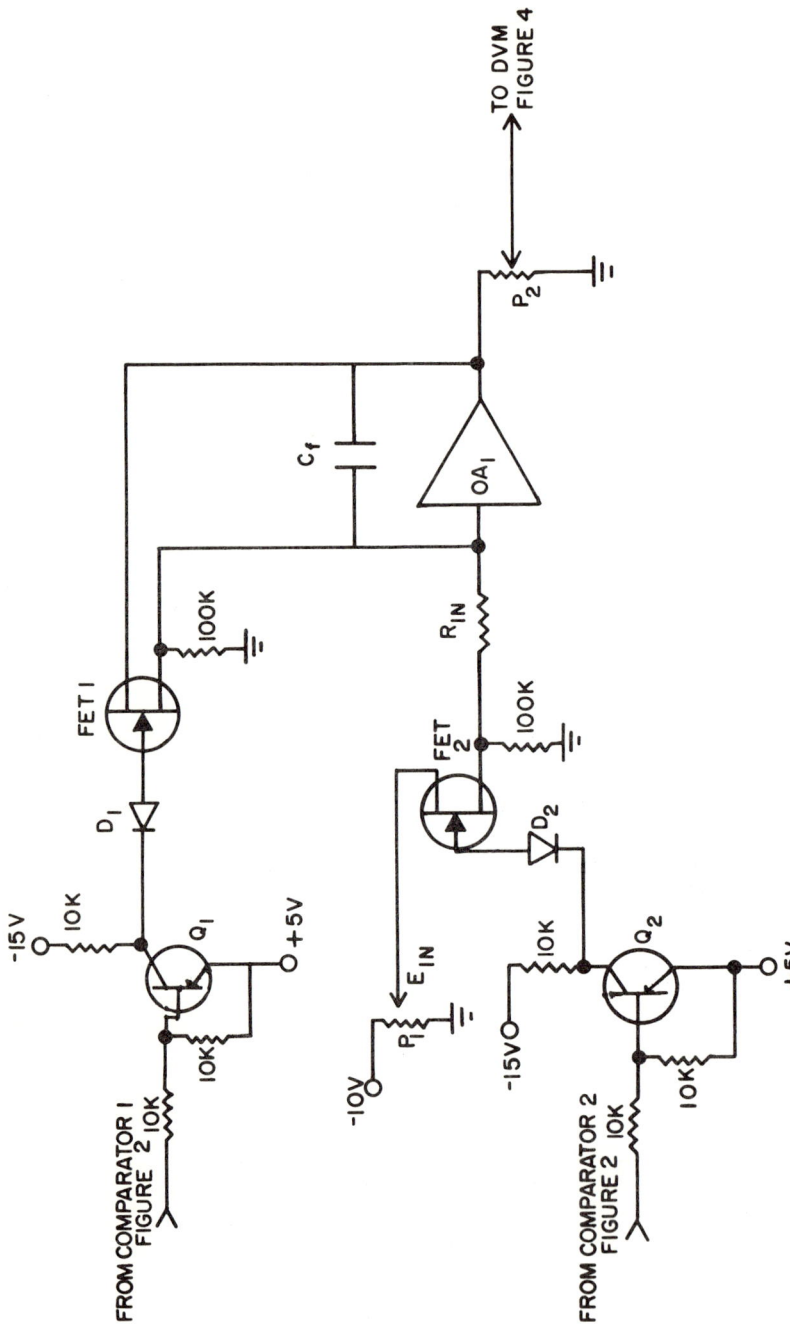

Fig. 16. Time interval-to-voltage converter. Reprinted from Ref. [52], p. 881, by courtesy of the American Chemical Society.

TABLE 12

Variable-Time Rate Measurements with Hybrid Computer[a]

Input rate (mV/sec)	Digital readout[b] (μsec)	Relative standard deviation (%)	Relative error (%)
+5	497	0.15	−0.6
10	1,004	0.25	+0.4
30	2,998	0.22	−0.1
50	5,007	0.21	Calibration
70	7,004	0.13	+0.1
100	9,979	0.06	−0.2
300	29,780	0.34	−0.7
500	49,740	0.27	−0.6
1,000	1,001	0.38	+0.1
5,000	5,020	0.59	+0.4
7,000	7,041	0.87	+0.6
10,000	10,100	1.32	+1.0
20,000	19,840	1.82	−0.8

[a] Reprinted from Ref. [52], p. 882, by courtesy of the American Chemical Society.

[b] Averages of ten results.

5 mV/sec to 20 V/sec [52]. This entire range of input rates was measured with only one change in instrument parameters, and measurement times as low as 5 msec could be handled. Data were also presented to illustrate the application of the hybrid counter to the determination of phosphate based on spectrophotometric monitoring of the heteropoly blue reaction. Relative errors were better than 1%, while relative standard deviations were in the 1-3% range.

Although the hybrid computer eliminates some of the problems associated with the analog systems previously described, it still suffers from several limitations. The operational amplifier integrator used to convert time interval to voltage must be of high quality since current and voltage offsets and drifts will affect the accuracy. Likewise, the analog gates used to short the integrating capacitor and the integrator input voltage must transmit the applied analog signals accurately and reproducibly. Also, the stability of the comparator switching levels will influence the reproducibility which can be obtained. Many of these limitations are eliminated by the all-digital computation system described in the following section.

c. Digital System. Pardue et al. [53] have reported on an all-digital computer for implementing the variable-time approach. In contrast to the hybrid system just described, the digital circuitry does not require an analog time base or a V/F converter. Drift in the computer is limited to the drift of a crystal oscillator, which can be very small. Rapid response speed is maintained, limited by propagation delays in logic gates and the counter, or by the crystal clock frequency, whichever is slower. A significant development is the ability of the system to provide for noise averaging, a capability which has not been previously included in variable-time systems.

Figure 17 shows a block diagram of the system. The voltage interval detector functions as before to detect when the analog signal from the reaction monitor-signal modifier circuit passes two set levels, $\mathscr{A}_1$ and $\mathscr{A}_2$. The reciprocal time computer, which is controlled by the voltage interval detector, consists of the crystal clock, a frequency divider, a 16-bit up counter, a 16-bit down counter and several logic gates. Initially the frequency divider flip-flops are cleared (all set to logical $\underline{0}$), and the two counters are preset (set to logical $\underline{1}$). Prior to the time when level $\mathscr{A}_1$ is reached, the clock is prevented from reaching the divider and the down counter by gate G_2.

Each pulse at the up counter output opens the gated preset input of the up counter and transfers in parallel the information present in the down counter. Since all bits of the down counter are initially preset to the logical $\underline{1}$ state, the up counter is preset by each transfer of information from the down counter. Therefore the output frequency of the up counter, prior to the signal reaching $\mathscr{A}_1$, is merely the clock frequency. When $\mathscr{A}_1$ is exceeded, gate G_2 is enabled, allowing pulses from the clock to reach the divider and the down counter. Each pulse from the divider decrements the contents of the down counter, while each output pulse from the up counter causes transfer of the down counter contents to the up counter. When level $\mathscr{A}_2$ is reached, clock pulses are again prevented from reaching the

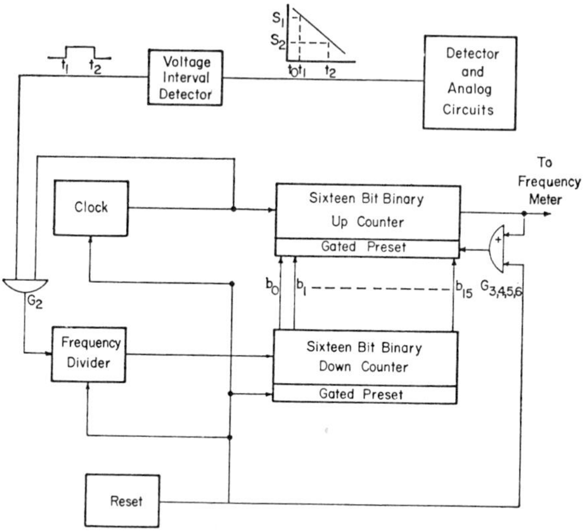

3. Kinetic Methods of Analysis

Fig. 17. Block diagram of digital variable-time rate measuring system. Reprinted from Ref. [53], p. 56, by courtesy of the American Chemical Society.

divider and down counter by gate G_2. The net result is that the down counter contains a count equal to its initial count minus the divided clock frequency times the time interval, Δt, that the signal was between levels $\mathscr{A}_1$ and $\mathscr{A}_2$. The frequency at the output of the up counter, f_u, is related reciprocally to the number of counts in the down counter, and is expressed by

$$f_u = \frac{f_o}{f_D \Delta t + 1} \tag{90}$$

where f_o is the original clock frequency and f_D is the divided clock frequency at the input to the down counter. If the down counter input frequency and Δt are arranged such that $f_D \Delta t \gg 1$, Eq. (91) results

$$f_u \approx \left(\frac{f_o}{f_D}\right) \frac{1}{\Delta t} . \tag{91}$$

Thus the output frequency of the up counter is related directly to reciprocal time and, under the appropriate conditions, to the concentration of the rate-limiting species.

Noise averaging results because of the very low hysteresis of the voltage interval detector [53]. A noise spike which causes the signal to exceed $\mathscr{A}_1$ momentarily will cause the down counter to operate only for the length of time the signal remains above $\mathscr{A}_1$. Similarly, noise pulses which cause the voltage interval detector to disable the down counter will be effective only for the amount of time $\mathscr{A}_2$ is actually exceeded. Random noise pulses would then tend to cancel. Level sensors with hysteresis, such as the comparators used with the hybrid computer described in the previous section, are triggered by the first pulse which exceeds the reference level and remain impervious to further excursions above and below the level.

Experimental results of measurements on simulated reaction rate curves using the all-digital computer are presented in Table 13 [53]. Input slopes ranged from less than 1 mV/sec to almost 100 mV/sec, and the excellent results are apparent. The system was applied to the determination of alkaline phosphatase in blood serum by a spectrophotometric rate procedure. Results are presented for normal and elevated amounts of alkaline phosphatase in serum and showed relative errors and standard deviations of less than 1% in most cases.

The all-digital reciprocal time computer eliminates many of the limitations of previous analog or hybrid systems. Drifts and nonlinearities in the computation system are eliminated. There are no mechanical or analog switches in the computer, which means the response speed can be quite fast so that initial rates of rapid reactions can be measured. The only drifts and nonlinearities which remain are those associated with the reaction monitor, the analog signal-modifier circuits, and the voltage interval sensor.

4. Conclusions

It is apparent from the preceding discussion that automated instrumentation for reaction rate methods has progressed extremely rapidly in the past few years. Completely digital computation systems are now available for both the fixed-time and the variable-time methods. Direct concentration or rate readout can be accommodated with either technique. Noise averaging is possible with either method and has led to improved signal-to-noise ratios. Therefore, it is this author's opinion that the choice between the two integral methods of rate measurement should no longer involve instrumentation considerations, but should be dictated entirely by the characteristics of the reaction, the sought-for species, and the reaction monitor.

TABLE 13

Variable-Time Rate Measurements with Miniature Digital Computer [a]

Input to integrator (mV)	Digital readout [b] (Hz)	Normalized readout [c]	Relative standard deviation (%)	Deviation from mean (%)
2.00	154.1	1541	0.4	−0.06
4.00	307.6	1538	0.2	−0.26
5.00	386.0	1544	0.0	+0.13
10.00	770.5	1541	0.02	−0.06
20.00	154.2	1542	0.0	0.00
40.00	308.1	1541	0.2	−0.06
80.00	618.0	1545	0.0	+0.19
100.00	772.6	1545	0.02	+0.19
200.00	154.2	1542	0.0	0.00
		Average 1542		

[a] Reprinted from Ref. [53], p. 59, by courtesy of the American Chemical Society.

[b] Averages of three determinations.

[c] All values normalized to the value at 20.00 mV.

The derivative method, being a true differential method, must be implemented with analog circuitry, and the limitations associated with analog circuits must be tolerated. Also, the noise enhancement of derivative circuitry leads to further problems. The derivative technique, however, remains the most straightforward approach to rate measurement and is capable of excellent results when signal-to-noise ratios are high.

Future improvements in automated rate measurement systems will probably involve the small digital computer. Several steps in this direction have already been made in several laboratories and are discussed in a subsequent section.

F. Computation Systems for Simultaneous Analysis of Mixtures

Since the computational problems involved in automating differential rate methods are much more complex than those for automating single-component techniques, only a few reports have appeared on computer circuits for solving differential rate equations. This section considers those techniques which can be implemented without the aid of a small laboratory computer. Techniques which involve the small computer are discussed in Section IV.

As previously discussed, many of the techniques which have been devised for in situ simultaneous determination of mixtures require laborious graphical or mathematical data processing in order to solve for the concentrations of the species of interest. Most of these techniques are quite difficult to automate short of using the small computer to solve the equations. The method of proportional equations, however, gives rise to simple linear equations which must be solved simultaneously to arrive at the unknown concentrations. Such equations are readily solved using analog computer techniques.

Pinkel and Mark [16] have presented a simple analog computer circuit for the automatic spectrophotometric analysis of binary mixtures using the method of proportional equations. The computer solves Eqs. (58) and (59) (Section II.D.1) simultaneously. The photocurrent output of the spectrophotometer, proportional to transmittance, is converted by a logarithmic amplifier to a signal proportional to absorbance and thus linearly related to concentration. During the reaction, at two preselected times, t_1 and t_2, the linearized signal $\mathscr{A}_t$ is connected to two sample and hold circuits which act as memory units. The two signals in the memory are proportional to $\mathscr{A}_{t_1}$ and $\mathscr{A}_{t_2}$, respectively. At a certain time after t_2, a programmed timer applies the signal in the memory unit to a standard analog computer circuit for solving simultaneous equations. The memory and computing circuitry are shown in Fig. 18. Operational amplifiers 1 and 2 are integrators, which are used as sample and hold units. A programmed timer sequences all events in the memory and computing circuits. Initially, before the preselected times t_1 and t_2, switches S_2 and S_4 are closed, while all other switches are opened. Thus, the integrating capacitors are initially shorted, and the logarithmic signal from the spectrophotometer is not connected to the memory units. At time t_1, latching relay S_2 opens and remains open. Simultaneously, S_1 closes for a time that is short compared to the overall reaction time. Thus the signal at t_1 is connected to the sample and hold amplifier and held after switch S_1 reopens. Likewise, at t_2, the signal is sampled and held at the output of integrator 2. At some preset time after t_2, relay S_5 is activated and the outputs of both integrators are connected to the

3. Kinetic Methods of Analysis

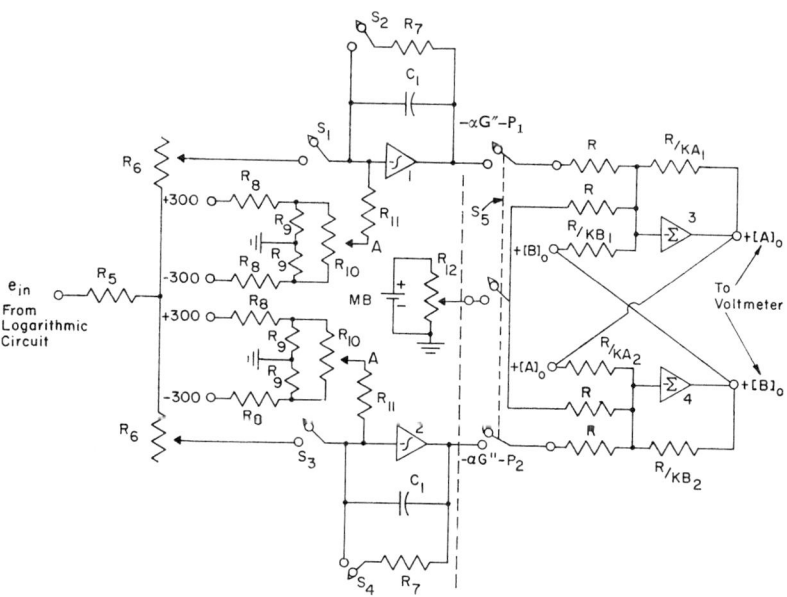

Fig. 18. Memory and computing circuitry for analysis of binary mixtures. Reprinted from Ref. [16], p. 495, by courtesy of Pergamon Press.

computer, which consists of operational amplifiers 3 and 4 and assorted input and feedback impedances. These impedances are selected, presumably by calibration with suitable standard solutions, to make the outputs of amplifiers 3 and 4 read out directly in terms of the unknown concentrations $[A]_o$ and $[B]_o$.

Although the authors do not report the application of the circuitry of Fig. 18 to an actual kinetic analysis, results using simulated rate curves were reported [16]. An electronic function generator was employed to construct rate curves with various values of rate constants and various ratios of initial concentrations. Results obtained on these simulated rate curves were accurate to better than $\pm 2\%$ when compared to the theoretically calculated results. Relative standard deviations of less than $\pm 2\%$ were also reported. It should also be possible to extend this circuit to the analysis of mixtures of more than two components by the addition of more memory amplifiers and computational circuitry.

Toren and Davis [54] have described a similar analog computer circuit for solving the simultaneous equations which result when the method of proportional equations is used for the analysis of mixtures. The computer described by these authors, however, is not meant to operate directly on the data. Rather data representing the values of a parameter or signal at times t_1 and t_2 are set into the computer using potentiometers. Presumably, the addition of memory circuits and sequencing circuitry would allow this computer to operate on-line with no manual transfer of data. Performance data are presented which indicate that the computer itself contributes only about 0.3% error. However, the authors state that the resolution of the data into their components can increase the error many fold. Any errors present in the original data are, of course, magnified when the simultaneous equations are solved. In a later report, Toren and Gnuse [55] applied the computer to the determination of ketone mixtures by a pH-stat method. For mixtures containing approximately 10^{-5} mole of ketone, relative errors and standard deviations of about 1% were obtained for mixtures in which the two rate constants differed by a factor of 5. Relative errors and standard deviations increased to the 3% level if the rate constants differed by only a factor of 3.

IV. SMALL DIGITAL COMPUTERS IN KINETIC ANALYSIS

The use of high-speed digital computers in chemistry has increased tremendously in the past few years. The computer's initial utility to the experimentalist came primarily in the reduction and analysis of data, usually a considerable time after the experiment was completed. The ability to utilize the computer's potential in real time has only recently become feasible with the introduction of the small, relatively low cost, laboratory-oriented computer [45]. In kinetic analyses, on-line computer systems were first applied to assist in data acquisition and high-speed reduction of the data into their final form. While this use of the on-line computer certainly aided the experimenter, the full potential of the computer to interact in the experiment and thus release the experimenter to do nonroutine tasks was not utilized. Only recently have on-line computers begun to take an active part in the experiment itself, performing many routine experimental tasks, processing the data immediately, optimizing experimental parameters, making various decisions, and controlling experimental variables. Applications of the on-line computer to routine kinetic methods and to research in chemical kinetics are now increasing at a rapid rate. Some of the recent developments in these areas are presented in this section.

A. Applications in Single-Component Kinetic Analysis

In single-component reaction rate methods, particularly in the area of enzyme activity measurements, the on-line computer has tremendous potential importance. In hospital laboratories, the increased speed with which data can be acquired and analyzed, and the increased reliability which can result from computer-assisted measurements, should greatly aid in the rapid diagnosis of disease.

James and Pardue [14] have described a system for on-line processing of reaction rate data for analytical purposes. This system utilizes software to provide the calculations necessary for fixed-time or variable-time rate measurements. The program is quite flexible and provides for approximate rate measurements very early in the reaction with subsequent optimization of parameters to yield highly reliable results. The signal-time profile is stored in memory, and the time scale is optimized on the basis of the preliminary rate measurements. Thus a wide dynamic range of rates can be measured with optimum results and no operator decisions. In addition to optimizing the time scale, the program provides for various mathematical manipulations of the raw data prior to the computation step, variable numbers of measurements on each sample, a filtering routine for increasing signal-to-noise ratios, and computation by either the fixed-time or the variable-time approach.

A block diagram of the program described by James and Pardue [14] is shown in Fig. 19. The analog output of a stabilized photometer is converted into the digital domain by a ten-bit A/D converter. Initially, data are acquired at a rate which is rapid compared to the fastest reaction rate expected. An initial approximate rate measurement is made and used to adjust the data acquisition rate for further measurements. This preliminary rate measurement is also used to select the time constant of the digital filter. After the preliminary parameter adjustments, the data can be treated by either the variable-time method or by a pseudo-fixed-time method. With either method, a series of independent measurements is made on each reaction, and the rates or concentrations of the sought-for species are computed. Individual results on each reaction are averaged and printed out on a teletype in the desired format.

The on-line computer system was applied to rate measurements on a synthetic ramp in addition to two analytical determinations [14]. Rate measurements made on an electrically generated ramp with no noise present showed a relative standard deviation of 0.003%. Even when the signal to noise ratio was reduced to 2 by superimposing a 60-Hz sine wave on the synthetic ramp, relative standard deviations of less than 0.5% were obtained, showing that the computer-assisted technique can tolerate high

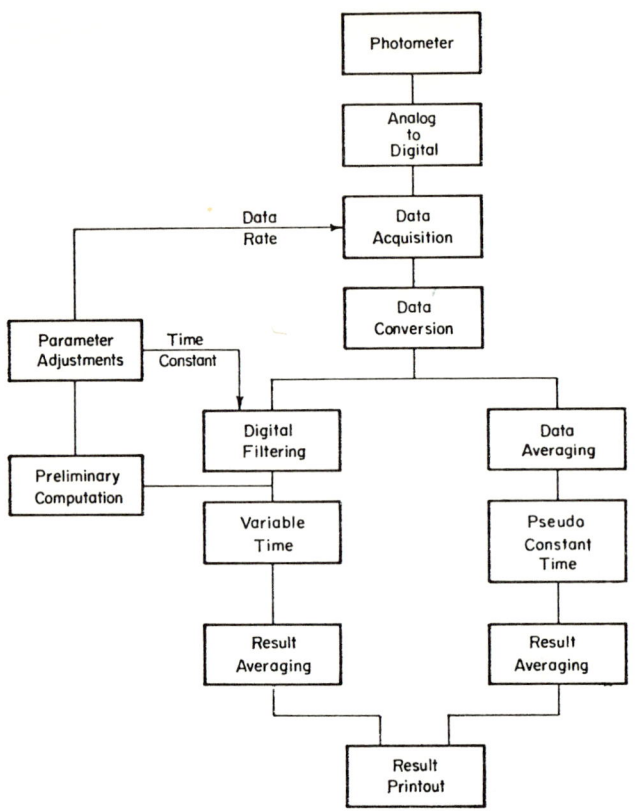

Fig. 19. Diagram of computer program for reaction rate analyses. Reprinted from Ref. [14], p. 1620, by courtesy of the American Chemical Society.

noise levels. Measurements of alkaline phosphatase activity in saline solution and in blood serum were also reported, and the results of the latter study are summarized in Table 14. The maximum variation of an individual normalized measurement from the average was 1.25% with the majority of the results being much better than this. Relative standard deviations of less than 0.6% were obtained, illustrating the extremely high precision which can be obtained in actual enzyme activity measurements. Results were also obtained for the determination of osmium in the 10^{-9} - 10^{-11} M range by catalysis of the Ce(IV)-As(III) reaction. Relative standard deviations were better than 2% for concentrations of 10^{-10} M or higher. Even at the 10^{-11} M level, the precision of repeated

TABLE 14

Computer-Assisted Determinations of Alkaline Phosphatase Activity in Reconstituted Serum by Variable-Time Method[a]

Relative concentration taken	Enzyme activity found[b] (mU/ml)				Average	Normalized activity[c]	Relative standard deviation (%)	Deviated normal values from mean (%)
	I	II	III	IV				
10	255.4	255.8	256.7	—	256.0	256.0	0.26	−0.19
6	150.9	152.3	152.7	—	152.0	253.3	0.62	−1.25
4	102.8	102.5	103.0	103.7	103.0	257.5	0.50	−0.39
2	51.18	51.25	51.22	—	51.22	256.1	0.07	−0.16
1	25.65	25.55	25.75	—	25.65	256.5	0.39	0.00
						Average 256.5		

[a] Reprinted from Ref. [14], p. 1622, by courtesy of the American Chemical Society.

[b] Each result reported is average of seven computations made on each run.

[c] All activities normalized to relative value of 10.

measurements was about 5%. In this latter case, the rate of the osmium-catalyzed reaction was ten times smaller than the uncatalyzed reaction rate, which illustrates that computer-assisted measurements can be made with high reliability under exceedingly unfavorable circumstances.

A different approach to the automation of rate measurements using an on-line computer has been described by Hicks et al. [45]. In this approach, a hardware interface [46] was constructed to compute the reaction rate outside the computer, which frees the computer from much of the data acquisition and allows the computer more time to perform other tasks such as monitoring equipment, analyzing data, and controlling various aspects of the experiment. This approach allows the computer to interact with the experimenter in real time as an integral part of the experiment. A block diagram of the interactive computer system, which is called ELLA (Experimental LINC Laboratory Analytical system), is shown in Fig. 20. The reaction rate interface is a modification of the analog fixed-time computer of Cordos et al. [44], which was discussed in Section III. Solid lines and solid boxes in the diagram show the flow of information and connections between the laboratory equipment and the computer (Laboratory Instrument Computer, LINC). Dotted lines and boxes represent the information flow and actions which involve the operator. The output of the log amplifier, proportional to absorbance, is the input to both the reaction rate interface and a strip chart recorder, which allows the operator to observe the progress of reactions if desired. The computer completely controls the

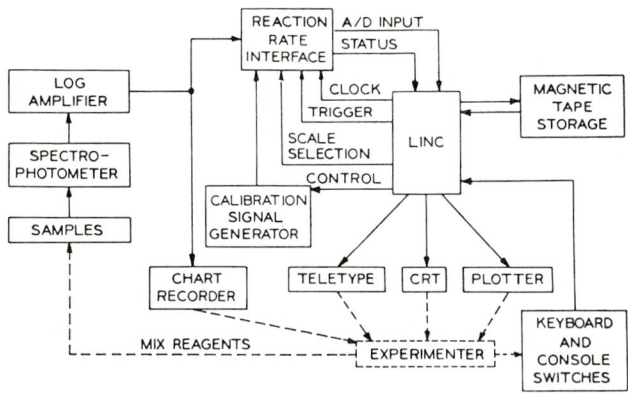

Fig. 20. Block diagram of interactive computer system; arrows indicate flow of information and control. Reprinted from Ref. [45], p. 730, by courtesy of the American Chemical Society.

reaction rate interface and provides automatic scale selection, calibration with a standard ramp, and clock and trigger signals. Experimental results can be obtained from a variety of output devices. In addition to strip chart recordings, the data can be printed on the teletype, displayed in graphical or alphanumeric form on the CRT, or plotted on an incremental plotter. During the experiment, the data can be reduced and analyzed in various forms on command from the experimenter. Magnetic tape storage allows the data to be stored permanently. On command from the operator, results of previous experiments stored on magnetic tape can be retrieved and compared with the results of an experiment in progress. The extreme flexibility of this interactive approach to computer automation is apparent from the above discussion. This system can be applied equally well to nonroutine kinetic studies as to routine analytical determinations. Some of the non-routine applications of the ELLA system and other approaches to computer automation are discussed in a subsequent section.

To demonstrate the applicability of the ELLA system to enzymic determinations, several evaluation studies were performed [46]. Enzyme dilution studies were performed for lactic dehydrogenase and alkaline phosphatase determinations. On-line plots of the output voltage of the reaction rate interface as a function of relative enzyme activity were performed by the ELLA system. Standard errors of estimate were reported to be 2.0% for lactic dehydrogenase and 1.1% for alkaline phosphatase using manual sample handling procedures. Determinations of alkaline phosphatase activity could be made at the rate of 50 to 70 samples per hour. As a further test of the ELLA system, blood serum samples were analyzed for alkaline phosphatase and compared to results obtained using the SMA-12/60 (Technicon Corporation, Tarrytown, N.Y.) system. A relative correlation of 0.3% between the two methods was obtained. A similar comparison for lactic dehydrogenase determinations did not correlate as well, probably because the methods employed for activity measurements with the SMA-12/60 are not rate methods, whereas those employed with the ELLA system measure the actual reaction rate.

Recently, an on-line computer has been utilized to process data from a stopped-flow spectrophotometer for single-component reaction rate analyses [56]. Conventional stopped-flow systems utilize oscilloscopic readout and subsequent data treatment from photographs of the oscilloscope traces. The use of fast reaction rate measurements for quantitative analysis is attractive because of the potential rapid rate at which analytical data can be acquired. Clearly, manual treatment of data from stopped-flow experiments eliminates this potential advantage. In the system described [56], the spectrometer output is interfaced directly to a small laboratory computer. Data can be acquired and processed using real-time averaging and smoothing routines, and a variety of output options is provided by the software. For the analysis of a single component undergoing a first- or pseudo-first-order reaction, a modification of the derivative method was utilized. According to Eq. (21) (Section II.C), the slope of the response curve at any time is directly proportional to the initial concentration of the rate-limiting species. With the on-line computer, the proportionality constant between the slope of the absorbance vs. time curve and the initial concentration can be evaluated at several different times during a single reaction using a standard solution. Unknown concentrations can then be determined from slopes measured at the times for which the proportionality constants are stored in the computer. With such a program, multiple slope measurements can be provided on a single reaction for averaging purposes. In addition, because of the rapid data acquisition and processing, repetitive runs can be made on the same sample with very little additional expenditure of time, providing ensemble averaging for high precision. For nonanalytical uses of stopped-flow measurements, the program provides for the display of a variety of kinetic parameters.

Data are presented for several applications of the computer-interfaced stopped-flow spectrophotometer [56]. The reaction between iron(III) and thiocyanate was used for the kinetic determination of iron. Utilizing a stabilized spectrometer, which is also described, iron determinations were reported with relative standard deviations and relative errors generally below 1%. Strontium was determined by the exchange reaction between the strontium complex of trans-1,2-diaminocyclohexane-N,N,N,N-tetraacetate ($CyDTA^{4-}$) and hydrogen ion, using lead (II) as a scavenger [57, 58]. Results were obtained under conditions in which the average signal change corresponded to about 3% T, and the signal-to-noise ratio was estimated to be 20. A relative standard deviation of 1.47% was obtained, demonstrating that the on-line ensemble averaging and data smoothing routines are highly effective in improving precision.

B. Applications in Simultaneous Kinetic Analysis of Mixtures

One of the most promising areas where the small computer should be able to provide both a substantial saving of data reduction time and an improvement in data quality is in multicomponent kinetic analysis. Routine uses of multicomponent techniques are undoubtedly limited by the large amount of time required for solution of the simultaneous equations or for graphical data treatment. Likewise, multicomponent methods can give rise to rather large errors unless extremely high-quality data are obtained. In simultaneous determinations of mixtures, manual data handling approaches have used a small number of data points or graphical methods for isolating the responses due to each component. With an on-line digital computer, many more data points can be acquired for analysis, and

3. Kinetic Methods of Analysis

separation of the responses due to each component is unnecessary. Thus vastly improved data can be obtained by making use of the entire response-time curve and allowing the computer to calculate the initial concentrations which best fit the experimental curve.

The approach discussed above has been used for the analysis of binary and ternary mixtures by an on-line regression technique [59]. If two components react with a common reagent under pseudo-first-order conditions to form a common product P, Eq. (55) (Section II.D.1) expresses the concentration of product at any time t during the reaction. When an instrumental method is used to follow the amount of product formed during the reaction, Eqs. (59) and (60) apply at particular times t_1 and t_2. Since the instrumental offset $\mathscr{I}_0$ remains constant during the reaction, solution of Eqs. (59) and (60) simultaneously is sufficient to determine $[A]_0$ and $[B]_0$. With manual processing of data, determination of $\mathscr{I}_t$ at two times during the reaction is usually all that is feasible. With the on-line computer, however, it is possible to determine $\mathscr{I}_t$ hundreds of times during a given reaction and to find the values of $[A]_0$ and $[B]_0$ which best fit the entire curve of $\mathscr{I}$ vs. time. In the approach presented, the sum of the squares of the error between the measured parameters at any time and assigned values of $[A]_0$, $[B]_0$, and $\mathscr{I}_0$ are minimized with respect to $[A]_0$, $[B]_0$, and $\mathscr{I}_0$. A weighting factor is used to account for differences in the weight of the measured parameter as the reaction proceeds. Mixtures of more than two components can also be analyzed by the same technique.

To utilize the proportional equations technique, the rate constants for each component must be accurately known or determined prior to the analysis by calibration. To test the effect of errors in rate constants, perfect simulated data were generated for a two-component mixture under a typical set of experimental conditions [59]. Each rate constant was first varied independently and the error in $[A]_0$ and $[B]_0$ determined. Errors in the larger rate constant were shown to be more significant than errors in the rate constant of the slower reacting component. Likewise, both rate constants were varied and a vector surface of the error function was determined. Least-squares error was found to be highest when both rate constants had errors in the same direction. In addition, the concentration error was determined as a function of rate constant ratio. For equal initial concentrations, the method exceeds 1% error only when the rate constant ratio is 1.1:1 or lower. For a 10:1 ratio of k_A/k_B, the concentration error exceeds 1% only when the concentration ratio of $[A]_0/[B]_0$ is lower than 10^{-3}. Such error calculations demonstrate that the least-squares data treatment can yield substantial improvement over the conventional method of proportional equations [7], which utilizes only a few data points.

Data were also obtained for the analysis of binary and ternary mixtures of metals using the acid dissociation reaction of the metal complexes of CyDTA with Pb(II) as a scavenger [59]. Results for binary mixtures of strontium and calcium were presented for concentration ratios of Sr/Ca from 0.113 to 11.3 at absolute concentrations below 10^{-4} M. The largest error was 12% with most results being considerably better than this. Comparison of the on-line analysis results to those previously obtained [58] by manual reading of photographs of oscilloscope traces and subsequent computer processing showed a general improvement in the error by a factor of 2 for strontium and 10 for calcium. Analyses of ternary mixtures of strontium, calcium, and magnesium were also reported, and the results are shown in Table 15. Relative standard deviations are generally below 2%, with 7% being the highest reported for a minor amount of the slowest reacting component, magnesium. Errors for major components average about 4%, but errors for minor components can be fairly large if large amounts of other components are present with similar or higher rate constants. The authors report that the entire procedure, from initiation of the reaction to printout of concentration, requires only about 3 min. The results shown in Table 15 are extremely impressive, being obtained under conditions where it would be very difficult to obtain reliable data using conventional methods. Such use of on-line computer techniques should open up many new applications of multicomponent techniques in practical situations.

C. Applications in Fundamental Studies

The place where computer-aided experimentation promises to bring about the most profound changes is in the research laboratory. The complete automation of routine analytical procedures is relatively straightforward and only a small extrapolation from today's common practice. Much less emphasis has been placed on the automation of research laboratories since nonroutine fundamental studies involve widely different experimental conditions, decision making based on what is known about the system, decision making based on experiments just concluded, and the establishment of the goals of the project, which may change as the investigation proceeds. Within any fundamental investigation, however, many routine experiments are carried out. In principle, part of any investigation can be automated by the straightforward methods used to automate routine procedures. However, it is now feasible with the small computer to provide a much higher degree of automation and interaction. Certain thought and decision-making processes can be automated if experimental goals are well established and sufficient information is made available. In fundamental kinetic studies of chemical reactions, several first steps have been made to introduce the computer as an integral part of the investigation.

TABLE 15

On-Line Kinetic Analysis of Ternary Mixtures [a]

Number of runs	Metal	Standard concentration (M x 10⁵)	Average found concentration (M x 10⁵)	Standard deviation	% Relative standard deviation	% Error
3	Sr	4.98	5.16	0.051	1.0	+3.6
	Ca	4.41	4.27	0.017	0.4	-3.2
	Mg	4.90	4.59	0.162	3.5	-6.3
4	Sr	2.49	2.52	0.154	6.1	+1.2
	Ca	0.882	0.961	0.022	2.3	+9.0
	Mg	2.45	2.46	0.021	0.41	+0.41
4	Sr	2.49	2.69	0.054	2.0	+7.8
	Ca	0.882	0.968	0.014	1.5	+9.8
	Mg	0.980	0.937	0.0067	0.72	-4.4
4	Sr	0.996	1.18	0.029	2.5	+18
	Ca	2.21	2.32	0.034	1.5	-4.8
	Mg	2.45	2.51	0.032	1.3	+2.6
4	Sr	0.996	1.02	0.051	5.0	+2.3
	Ca	2.21	2.36	0.030	1.3	+6.6
	Mg	0.980	0.605	0.043	7.1	-37.7

[a] Reprinted from Ref. [59], p. 1355, by courtesy of the American Chemical Society.

The ELLA system described by Hicks et al. [45] incorporates many of the features which are necessary to automate a fundamental reaction rate study. Several of the more routine features have been previously mentioned. For nonroutine applications, the software has been designed for maximum flexibility and a minimum of control by the experimenter. The flexibility is provided by a wide variety of options which are available to the experimenter, such as output options, data fitting programs, and so on. Minimum experimenter control is provided by having hardware operated under computer control. For example, rates of reaction are measured by a hardware interface operated under computer control. The rate measuring system can be calibrated by a standard sweep generator controlled by the computer. Thus if a calibration is desired, the experimenter pushes a button, and the calibration software causes the computer to control the rate generator and finally display the ideal and calculated rate values. Thus, by building into the system hardware to perform specific functions, the experimenter is required to perform fewer supervisory tasks.

In addition to the calibration procedure, the system has an on-line reading program which begins when information about the sample is entered and a button is pushed telling the system when to read. All the manipulations needed to obtain the initial reaction rate are then performed under computer control. An on-line analysis package allows the experimenter to examine processed data while the experiment is still in progress. Various storage, output, and data-fitting options are also provided. The data fitting option allows data to be fitted to theoretical curves and compared with previous results.

The authors report that several significant results have been obtained from the ELLA system [45]. First, the feasibility of operating complex hardware under computer control was demonstrated. This frees the computer from a large part of the data acquisition and allows more time for analysis and control. On-line fitting of data to theoretical plots is another significant development. The on-line analysis allows the experimenter to see the results as the experiment proceeds so that decisions can be made and carried out immediately. In addition, on-line analysis allowed the authors to note a weakness in the standard method of plotting enzyme kinetic data. Normally such data are plotted by a rearrangement of Eq. (14) (Section II.B.2). If reciprocal rate is plotted against the reciprocal of the substrate concentration, a linear plot should be obtained from which K_m and k_2 can be determined. The simple Michaelis-Menten treatment applies only in the absence of complicating factors such as inhibition by products, substrates, or other species. Deviations from the simplified mechanism are not easily recognized when data are hand plotted and made to fit an assumed form. However, on-line analysis of data obtained from

3. Kinetic Methods of Analysis

a study of the glutamic oxaloacetic transaminase system revealed that the data fit the simplified equation quite poorly. Taking substrate inhibition into account, a new method of plotting was developed which allowed a much better fit of the data.

The ELLA system allows the automation of many steps in a reaction rate experiment. However, in the initial report [45], the steps of solution preparation, initiation of reactions, and decision making on what experiments to perform are still left to the operator. Deming and Pardue [37] have described a highly automated system in which routine procedures as well as the decision-making procedures of experimental design and interpretation of data have been automated. The system described consists of a small computer, which serves as the controller, an analytical instrument, which converts chemical information into the electrical domain, a data acquisition system, and an electromechanical reagent addition system, which makes possible automated initiation and control of experiments.

The instrumental system was evaluated by carrying out a characterization of the hydrolysis of p-nitrophenyl phosphate by the enzyme alkaline phosphate using photometric methods to follow the reaction [37]. The evaluation was divided into three phases with each succeeding phase involving more extensive automation. In the first phase, the system was evaluated while performing routine procedures. The possibilities of experimental design and automated data interpretation were not utilized. In the second more advanced phase, the ability of the computer to design experiments was demonstrated. In the final phase, the system demonstrated the capability of performing data interpretation and experimental design as well as routine procedures. The system was supplied with information already known about the chemical system, the goal of the study, a method for obtaining the goal, and decision-making criteria for determining when sufficient information had been obtained to achieve the goal. Acting entirely by itself, the instrumental system demonstrated the ability to complete the study and report the data to the experimenter.

During the first phase of the study, such characterizations as the linearity of the pseudo-zero-order hydrolysis reaction, the dependence of the reaction rate on buffer concentration, the influence of incubation time on the rate, and dependence of the rate on the amount of enzyme added and the amount of substrate added were carried out by preprogramming various parameters for each experiment. For example, in a study of the rate dependence on incubation period, the experimenter provides information on the desired concentrations of buffer, enzyme, and substrate and the incubation periods to be studied. The appropriate amounts of water, buffer, and enzyme solutions are then added automatically to the spectrometer cell and mixed. At the end of the chosen incubation period, substrate is added by the system and data acquisition begun. Results obtained as reaction rate versus incubation period were found to exhibit a slight dependence of rate on incubation time. Data interpretation was done manually.

In the second phase of evaluation [37], the type of inhibition which the product, inorganic phosphate, exhibits was determined. The controller was asked to demonstrate the ability to design the experiment on a modest scale. The computer was given the concentrations of stock solutions and the range over which the substrate and inhibitor concentrations should be varied. The controller then designed a matrix of 25 experiments, consisting of five concentrations of substrate and five concentrations of inhibitor, and carried out the experiments in a random fashion. Plots of (rate)$^{-1}$ vs. [S]$^{-1}$ for each inhibitor concentration, [I], as well as plots of (rate)$^{-1}$ vs. [I]$^{-1}$ for each [S] were obtained automatically. Competitive inhibition was indicated on the basis of these plots.

In the third phase, the design and interpretive abilities of the system were demonstrated in studies of phosphate inhibition and the pH dependence of the reaction. In these studies, the relationship between the dependent variable (rate) and the independent variable (phosphate concentration or pH) is nonlinear. In the usual experiment, the independent variable is varied in equal increments and the dependent variable obtained at each value of the independent variable. For nonlinear relationships, such methods yield curves that are overdetermined in certain ranges and underdetermined in others. With the computer-controlled instrumental system, a preliminary experiment can be carried out with equal increments of the independent variable. The results of this experiment can then be used to design new experiments that will ensure properly spaced experimental points. For example, in the phosphate inhibition study two criteria are given: The first is the maximum difference in phosphate concentrations for any two points. This criterion thus establishes the conditions for the initial set of experiments, and rates are obtained for equal increments of phosphate concentration. The second criterion is to set a maximum difference in the dependent variable, reaction rate, between any two points. These criteria then serve as the goal of the characterization and as the condition for determining when the goal has been reached. The initial set of data points obtained under the first criterion is scanned by the controller to determine which points do not satisfy the second criterion. If two points exceed the maximum rate difference, a new experiment is carried out automatically in which the phosphate concentration is set midway between the two points. After this new experiment, the expanded group of data points is again scanned to detect pairs of points which still exceed the limit. Additional experiments are then carried out until all points satisfy the two criteria. For the phosphate inhibition study, 15 experiments were

3. Kinetic Methods of Analysis

necessary. A second study was also carried out to determine the pH maximum of the reaction rate. Again an initial experiment was carried out at equal pH increments. The results of the initial experiment were used to design new experiments to determine the maximum rate with higher resolution. Such procedures were then repeated until the preselected resolution was obtained.

Several advantages of such an automated research system can be cited [37]. First, the automated system increases tremendously the efficiency of the experimentalist. Deming and Pardue [37] state that the progress of a project can be measured in hours instead of days. Likewise the setbacks in a project, which presumably would not be eliminated, are also measured in hours because of the rapid availability of the results. Another advantage is the highly accurate and reproducible system for initiation of experiments. Experimental results were found to be reproducible to better than 1% in many cases.

V. FUTURE PERSPECTIVES

From the preceding discussion it is apparent that much progress has been made in automated analyses based on reaction rate measurements. Recent advances in digital instrumentation have improved greatly the reliability of rate computation systems, and increased the speed with which rate computations can be made. However, much work remains to be done in developing complete instrumental systems for rate methods. Improvements in automated sample handling and the development of highly stable reaction monitoring systems are just two of the areas where advances are necessary before the full potential of the reaction rate approach is realized. Also, work with the small digital computer has just begun and many future advances in this important area can be foreseen. The development of specific hard-wired computation systems to be operated under computer control would seem to be a particularly advantageous approach.

Likewise, a profitable area for future work is in the theory of automated kinetic methods. To put rate methods on a sound theoretical basis, it is desirable to know all of the factors influencing signal-to-noise ratios and errors in kinetic procedures. Once these factors are known, complete instrumental systems can be designed to achieve optimum results.

Another area of increasing importance is the complete understanding of the mechanisms of the reactions utilized in rate procedures. Fundamental studies should enable the user to choose reaction conditions so as to minimize interferences or to maximize rate differences in multicomponent analyses.

Finally, commercial interest in reaction rate methods has picked up because of the importance of these techniques in the clinical chemistry area. New instrumentation systems have been introduced by several companies, and many advances in rate measurement technology should be the result.

Development in instrumentation systems, theory, and reaction mechanisms should stimulate many new applications of kinetics.

ACKNOWLEDGMENTS

The author gratefully acknowledges the aid of Dr. Jim Ingle, who read the entire manuscript and contributed many useful ideas, and Dr. Paul Beckwith and Dr. C. G. Enke, who made many helpful comments during the preparation of this manuscript.

REFERENCES

[1]. G. A. Rechnitz, Anal. Chem., 36:453R (1964).

[2]. G. A. Rechnitz, Anal. Chem., 38:513R (1966).

[3]. G. A. Rechnitz, Anal. Chem., 40:455R (1968).

[4]. G. G. Guilbault, Anal. Chem., 42:334R (1970).

[5]. G. G. Guilbault, Anal. Chem., 38:527R (1966).

[6]. G. G. Guilbault, Anal. Chem., 40:459R (1968).

[7]. H. B. Mark, Jr., and G. A. Rechnitz, Kinetics in Analytical Chemistry, Wiley-Interscience, New York, 1968.

[8]. K. B. Yatsimerskii, Kinetic Methods of Analysis, Pergamon, Oxford, 1966.

[9]. H. B. Mark, Jr., L. J. Papa, and C. N. Reilley, in Advances in Analytical Chemistry and Instrumentation (C. N. Reilley, ed.), Vol. 2, Wiley-Interscience, New York, 1963, pp. 255-385.

3. Kinetic Methods of Analysis

[10]. H. L. Pardue, in Advances in Analytical Chemistry and Instrumentation (C. N. Reilley and F. W. McLafferty, eds.), Vol. 7, Wiley-Interscience, New York, 1968, pp. 141-207.

[11]. W. J. Blaedel and G. P. Hicks, in Advances in Analytical Chemistry and Instrumentation (C. N. Reilley, ed.), Vol. 3, Wiley-Interscience, New York, 1964, pp. 126-140.

[12]. H. B. Mark, Jr., Anal. Chem., 36:1668 (1965).

[13]. J. D. Ingle, Jr., and S. R. Crouch, Anal. Chem., 43:697 (1971).

[14]. G. E. James and H. L. Pardue, Anal. Chem., 41:1618 (1969).

[15]. H. A. Laitinen, Chemical Analysis, McGraw-Hill, New York, 1960, pp. 454-460.

[16]. D. Pinkel and H. B. Mark, Jr., Talanta, 12:491 (1965).

[17]. C. G. Enke, Anal. Chem., 43:69A (1971).

[18]. H. V. Malmstadt and C. G. Enke, Digital Electronics for Scientists, Benjamin, New York, 1969, pp. 279-289.

[19]. T. E. Weichselbaum, W. H. Plumpe, Jr., R. E. Adams, J. C. Hagerty, and H. B. Mark, Jr., Anal. Chem., 41:725 (1969).

[20]. P. A. Loach and R. J. Loyd, Anal. Chem., 38:1709 (1966).

[21]. H. L. Pardue and P. A. Rodriguez, Anal. Chem., 39:901 (1967).

[22]. H. L. Pardue and S. N. Deming, Anal. Chem., 41:986 (1969).

[23]. R. H. Müller, Anal. Chem., 41:108A (1969).

[24]. J. D. Ingle, Jr., and S. R. Crouch, Anal. Chem., 43:1331 (1971).

[25]. M. L. Franklin, G. Horlick, and H. V. Malmstadt, Anal. Chem., 41:2 (1969).

[26]. T. E. Weichselbaum, R. E. Adams, and H. B. Mark, Jr., Anal. Chem., 41:1913 (1969).

[27]. P. D. Feil, D. G. Kubler, and D. J. Wells, Jr., Anal. Chem., 41:1908 (1969).

[28]. M. VanSwaay, J. Chem. Educ., 46:A515, A565 (1969).

[28a]. G. G. Guilbault, in Fluorescence, Theory, Instrumentation, Practice, (G. G. Guilbault, ed.), Dekker, New York, 1967, pp. 297-358.

[29]. H. V. Malmstadt and H. L. Pardue, Anal. Chem., 33:1040 (1961).

[30]. H. L. Pardue, Anal. Chem., 35:1240 (1963).

[31]. D. E. Johnson and C. G. Enke, Anal. Chem., 42:329 (1970).

[32]. H. V. Malmstadt and G. P. Hicks, Anal. Chem., 32:445 (1960).

[33]. H. V. Malmstadt and G. P. Hicks, Anal. Chem., 32:394 (1960).

[34]. H. V. Malmstadt and H. L. Pardue, Anal. Chem., 34:299 (1962).

[35]. H. V. Malmstadt and H. L. Pardue, Clin. Chem., 8:606 (1962).

[36]. A. C. Javier, S. R. Crouch, and H. V. Malmstadt, Anal. Chem., 41:239 (1969).

[37]. S. N. Deming and H. L. Pardue, Anal. Chem., 43:192 (1971).

[38]. H. L. Pardue, Anal. Chem., 36:633 (1964).

[39]. H. L. Pardue, Anal. Chem., 36:1110 (1964).

[40]. H. L. Pardue and W. E. Dahl, J. Electroanal. Chem., 8:268 (1964).

[41]. H. V. Malmstadt and S. R. Crouch, J. Chem. Educ., 43:340 (1966).

[42]. S. R. Crouch, Ph.D. Thesis, University of Illinois, Urbana, Illinois, 1967.

[43]. S. R. Crouch and H. V. Malmstadt, Anal. Chem., 39:1090 (1967).

[44]. E. M. Cordos, S. R. Crouch, and H. V. Malmstadt, Anal. Chem., 40:1812 (1968).

[45]. G. P. Hicks, A. A. Eggert, and E. C. Toren, Jr., Anal. Chem., 42:729 (1970).

[46]. E. C. Toren, Jr., A. A. Eggert, A. E. Sherry, and G. P. Hicks, Clin. Chem., 16:215 (1970).

[47]. J. D. Ingle, Jr., and S. R. Crouch, Anal. Chem., 42:1055 (1970).

[48]. H. L. Pardue, C. S. Frings, and C. J. Delaney, Anal. Chem., 37:1426 (1965).

[49]. H. L. Pardue, M. F. Burke, and D. O. Jones, J. Chem. Educ., 44:684 (1967).

[50]. R. H. Stehl, D. W. Margerum, and J. J. Latterell, Anal. Chem., 39:1346 (1967).

[51]. G. E. James and H. L. Pardue, Anal. Chem., 40:796 (1968).

[52]. S. R. Crouch, Anal. Chem., 41:880 (1969).

[53]. H. L. Pardue, R. A. Parker, and B. G. Willis, Anal. Chem., 42:56 (1970).

[54]. E. C. Toren, Jr., and J. E. Davis, Anal. Letters, 1:289 (1968).

[55]. E. C. Toren, Jr., and M. K. Gnuse, Anal. Letters, 1:295 (1968).

[56]. B. G. Willis, J. A. Bittikofer, H. L. Pardue, and D. W. Margerum, Anal. Chem., 42:1340 (1970).

[57]. D. W. Margerum, J. B. Pausch, G. A. Nyssen, and G. F. Smith, Anal. Chem., 41:233 (1969).

[58]. J. B. Pausch and D. W. Margerum, Anal. Chem., 41:226 (1969).

[59]. B. G. Willis, W. H. Woodruff, J. R. Frysinger, D. W. Margerum, and H. L. Pardue, Anal. Chem., 42:1350 (1970).

Chapter 4

ANALOG COMPUTER SIMULATION OF KINETIC MODELS

Jiří Janata

Imperial Chemical Industries Limited
Petrochemical and Polymer Laboratory
The Heath
Runcorn, Cheshire
England

I.	INTRODUCTION	210
II.	CLOSED KINETICS	211
	A. Reactions of First Order	215
	B. Reactions of Second Order	223
III.	OPEN KINETICS	233
IV.	BIOCHEMICAL APPLICATIONS	243
	A. Pharmacokinetics	244
	B. Enzyme Kinetics In Vivo	245
V.	MISCELLANEOUS APPLICATIONS	248
	A. Kinetic Analysis	248
	B. Simulation of Log(K)-pH Profile	250
	C. Kinetics of Chemisorption	251
	D. Nonisothermal Kinetics	254
	REFERENCES	257

Copyright © 1973 by Marcel Dekker, Inc. All Rights Reserved. Neither this work nor any part may be reproduced or transmitted in any form or by any means, electronic or mechanical, including photocopying, microfilming, and recording, or by any information storage and retrieval system, without permission in writing from the publisher.

I. INTRODUCTION

Analog computers are particularly suitable for the solution of ordinary differential equations. It is therefore not surprising that they have found a widespread use in the simulation of kinetic systems in both chemistry and biology. There are several features of analog computers which make their application to reaction kinetics very attractive: First of all they are much cheaper than even a small digital computer and their accuracy is quite adequate for most kinetic purposes. They are easy to operate and the preparation of a program takes only a fraction of the time required for the preparation of an equivalent digital program. Any changes which have to be made in the program can also be done easily by adjusting the values of computing elements representing the parameters of the kinetic equations. This ensures a close interaction between the operator and the problem programmed on the computer. Furthermore, parts of the analog program can be easily identified with parts of the real system which this program represents. Finally, the actual execution of a program takes normally 10-30 sec but in a REP-OP mode (repetitive operation), which is a standard facility of all commercial computers, it can be as short as 10-30 msec. This high speed of execution, which is independent of the size of the program, stems from the fact that all variables are being operated on in parallel as opposed to a digital computer where individual operations are performed in sequence.

The high speed of an analog computer and the storage and control facilities of a digital computer are combined in a <u>hybrid computer</u>. Since an ordinary analog computer is quite adequate for most of the kinetic problems there have been only a few applications of hybrid computers to reaction kinetics. As they are more suitable for solution of partial differential equations than any other type of computer it is likely that they will be used more often for solution of kinetic problems in heterogeneous and nonisothermal kinetics, electrochemical kinetics, and so on. The term "hybrid computer" is sometimes used improperly for an analog computer with <u>parallel logic</u>. This is an extra facility of an analog computer which enables a high speed control (reseting, updating, optimization) and logical decisions to be made. Some of the most common parallel logic elements together with ordinary analog computing symbols are summarized in Tables 1 and 2.

There are many programs available which will allow an ordinary digital computer to be programmed as an analog computer. They have evolved from <u>digital differential analyzers</u> which were designed to behave like analog computers but the mathematical operations were carried out digitally. These programs are block oriented; i.e., each operational element (e.g., multiplier, integrator, etc.) is a digital subroutine

4. Simulation of Kinetic Models 211

represented by the name and the ordinary analog symbol and has standard input(s) and output. The program preparation is done in the same way as on an analog computer; each block is assigned an execution number. All operations are performed in sequence and computed values of each block are stored in a specified location in the computer's memory. After completing the first round, execution returns to the first block and starts again with the time value incremented in a specified way. Results are presented in the form of a printout and/or on an incremental plotter or are eventually converted into an analog signal and displayed in the usual way. Needless to say, the major advantage of an analog computer, i.e., programming speed, is lost through this arrangement. However, in some cases where speed is not critical this can be offset by some additional advantages. First of all any number of more or less exotic computing blocks (e.g., dividers, log-antilog, logical elements, etc.) can be programmed. All blocks can be made noninverting. This and less stringent scaling requirements make the programming easier. The computing accuracy is also increased. It is evident that this approach, which is now common in engineering and process control, will become even more popular, particularly with increasing use of small digital computers. The more recent conversion programs are compatible with common computer languages and are even available as a standard software from the manufacturers. A detailed discussion of this subject is given in Volume 1 of this series and in review articles where some kinetic applications can also be found [1-6].

A detailed discussion of analog computation techniques can be found in textbooks [7-10] and review articles [11, 12]. There are several reviews dealing specifically with applications of analog computers to reaction kinetics [13-15]. Reviews of engineering and process control applications as well as of biological applications are listed in the Appendix.

The aim of this chapter is to show some typical as well as some less common applications of analog computers in the reaction kinetics. Apart from review articles and textbooks, two main sources of references were used, Chemical Abstracts and Computer Abstracts, covering the period 1958-1970.

II. CLOSED KINETICS

There is no analog computer treatment of a general kinetic case although some attempts of the general solution of homogeneous [16] and homogeneous nonisothermal [17] kinetics have been made. It is quite obvious that the advantages of a general treatment would be outweighed by its complexity and particularly by the heavy demand on nonlinear computing

TABLE 1

Summary of Common Analog Computing Elements

Operation	Meaning
Multiplication by constant 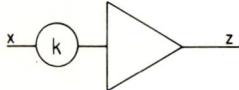	$z = -kx$
Summation 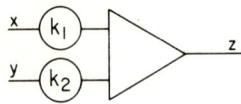	$z = -(k_1 x + k_2 y)$
Integration with initial condition 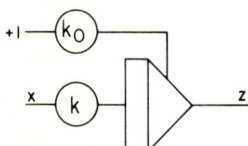	$z = -k \int_0^t x\,dt + x_0$
Integration with summation 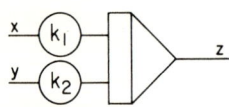	$z = -\int_0^t (k_1 x + k_2 y)\,dt$
Multiplication 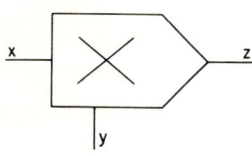	$z = \pm xy$
Function generator 	$z = F(x)$

4. Simulation of Kinetic Models 213

TABLE 2

Summary of Common Parallel Logic Symbols

	Meaning[a]
Comparator	For $v_1 + v_2 > 0$ $A = 1$ $v_1 + v_2 < 0$ $A = 0$
Switch	For $A = 1$ switch conducting $A = 0$ switch nonconducting
Relay	For $A = 1$ $x = v_1$ $A = 0$ $x = v_2$
AND - Gate	For $A = B = C = 1$ $D = 1$ For either of A, B, C = 0 $D = 0$
OR - Gate	For A or B or C = 1 $D = 1$ $A = B = C = 0$ $D = 0$
NOT	For $A = 1$ $B = 0$ $A = 0$ $B = 1$

[a] The digits 1 and 0 imply logic 1 and logic 0, respectively.

elements [17]. Another argument against a generalized treatment is that the analog programming is a relatively simple operation so that a program can always be tailored to suit a particular problem. Furthermore, the problem must always be written with the available computer in mind.

The most common area in chemical kinetics is that of homogeneous isothermal reactions. The rate equations then are usually written as a system of ordinary differential equations and their solution assumes a routine format.* It is customary in analog computations to begin with the separation of the highest derivative on the left-hand side of the differential equation, which is also the usual way of writing the chemical rate equations. The next step is the scaling of variables and of time. The resulting machine equations are then solved by performing the individual mathematical steps, e.g., summing, multiplying, integrating, and so on, using appropriate analog computing blocks arranged in accord with the program. The resulting voltage (e.g., concentration)-time curves are then displayed on an oscilloscope or on an X, Y recorder. There are two common cases to be encountered: First, there are the experimental curves to be compared with a hypothetical model. This is done by a "curve-fitting" procedure, i.e., by matching computed concentration-time curves with the experimental runs. If an overlap is achieved, it can be assumed that the mathematical model corresponds to the chemical reaction. However, some caution must be exercised here for it has been shown, for example, on the isomerization of cyclohexene to butylcyclopentenes [18], that the same set of experimental results can be described by completely different kinetic equations.

The other situation arises when a definitive mathematical model has been conceived and the results are sought to be displayed in graphical form. By adjusting the potentiometers corresponding to rate constants and/or initial concentrations, it is often possible to find optimum conditions for a given reaction. It is obvious that such an exercise is particularly valuable for the planning of experiments.

There is nothing complicated in designing an analog computer program. One is only limited by his own imagination and by the computing capacity of his computer. Numerous hypothetical models have been designed and discussed in many review articles.

---

*As the general analog computing procedures were dealt with in Volume 1 only a brief description relevant to kinetic applications is presented here.

4. Simulation of Kinetic Models 215

A. Reactions of First Order

1. Consecutive Reactions

Let us consider the reaction

$$A \xrightarrow{k_1} B \xrightarrow{k_2} C. \tag{I}$$

This reaction can be described by the following set of differential equations:

$$\frac{dA}{dt} = -k_1 A \tag{1a}$$

$$\frac{dB}{dt} = k_1 A - k_2 B \tag{1b}$$

$$\frac{dC}{dt} = k_2 B \tag{1c}$$

where for $t = 0$, $A = A_0$ and $B = C = 0$.

Since first-order kinetics are independent of the initial concentrations, the proportionality constant between the concentrations and corresponding voltages need not to be known. Therefore, we can write

$$\frac{dx}{dt} = -k_1 x \tag{2a}$$

$$\frac{dy}{dt} = k_1 x - k_2 y \tag{2b}$$

$$\frac{dz}{dt} = k_2 y \tag{2c}$$

where for $t = 0$, $A = x = 1$ machine unit and $y = z = 0$.

The <u>machine unit</u> (M. U.) is the maximum operating voltage of the computer and in most cases is ± 10 V. It is improbable that the time scale of an experiment is identical with the time scale of the computation (usually 10-30 sec). It is therefore necessary to generate the <u>computer time</u>, which is related to the <u>real time</u> by the equation

$$t = \alpha T \tag{3}$$

where α is an arbitrary time constant, t is the real time, and T is the computer time. Equations (2a)-(2c) must be rewritten in this new time scale as follows:

$$\frac{dx}{dT} = -k_1 \alpha x \tag{4a}$$

$$\frac{dy}{dT} = k_1 \alpha x - k_2 \alpha y \tag{4b}$$

$$\frac{dz}{dT} = k_2 \alpha y. \tag{4c}$$

where for T = 0, x = 1 and y = z = 0.

The procedure discussed above is called <u>time scaling</u>. It can readily be seen that the identical set of voltage-time curves (the computer output) will be obtained regardless of the magnitude of k_1 and k_2 (as long as the ratio k_1/k_2 remains constant). The spread of the curves in time is effected by a single parameter, the time constant α, for

$$\text{potentiometer setting} = k_1 \alpha \ (\text{or } k_2 \alpha, \ldots).$$

The X axis (time base) will then be calibrated in milliseconds or hours, etc., depending on α.

To solve Eqs. (4a)-(4c) we have to perform three integrations. The original equations can also be formulated as

$$\frac{dA}{dt} = -k_1 A \tag{5a}$$

$$\frac{dB}{dt} = k_1 A - k_2 B \tag{5b}$$

$$C = S - A - B. \tag{5c}$$

4. Simulation of Kinetic Models

where for $t = 0$, $A = A_0 = S$ and $B = C = 0$, and where S is the total concentration of the reacting species. Only two integrators are needed in this case. Computing diagrams corresponding to Eqs. (4) and (5) are given in parts (a) and (b), respectively, of Fig. 1. The solutions via Eqs. (5a)-(5c) would be more advantageous if there were a shortage of integrators on the computer. Apart from programming flexibility this also illustrates why no two programs are alike, which is true particularly for more complicated reaction schemes.

It may be useful to reiterate some aspects of analog programming on this model using Eqs. (4a)-(4c) and part (a) of Fig. 1. First we assume that there is a voltage $+x$ at the output of integrator 1. Next we multiply it on amplifier 2 with potentiometer setting $k_1\alpha$ to obtain $-k_1\alpha \cdot x$. This quantity, multiplied by -1 (amplifier 3), is equal to $-dx/dT$ which is fed back to the input of integrator 1. Thus, the first loop is completed and constrained to the solution of Eq. (4a). The second loop is built in the same way, assuming $+y$ at the output of integrator 4. Integration of the output voltage of amplifier 5 yields $+z$. The initial condition on integrator 1 is set to -1 M.U. The time base is generated by integrating -1 M.U. and multiplying it by α.

A simple example of this type of reaction is the hydrolysis of succinylcholine chloride (SCC) to succinylmonocholine chloride (SMCC) followed by further hydrolysis to succinic acid [19]. The course of this reaction has been followed spectrophotometrically at the wavelength where both SCC and SMCC absorb. This total absorbance has also been computed by multiplying the voltages corresponding to SCC and SMCC with respective absorptivities and then summing. The rate constants obtained by curve fitting of computed and measured total absorbances (Fig. 2) agreed well with those reported previously from chromatographic determinations.

Bonath et al. [20] found that chlorination of the di-o-tolylcarbonate ester at 100 and 200°C under UV irradiation proceeds stepwise

$$\underset{X}{\text{Ph}}-\text{O-C(=O)-O}-\underset{Y}{\text{Ph}} \xrightarrow[v, Cl_2]{k} \text{products}$$

$(X, Y):$ $(CH_3; CH_3) \xrightarrow{k_1} (CH_3; CH_2Cl) \xrightarrow{k_2} (CH_2Cl; CH_2Cl) \xrightarrow{k_3}$
$(CHCl_2; CH_2Cl) \xrightarrow{k_4} (CHCl_2; CHCl_2) \xrightarrow{k_5} (CHCl_2; CCl_3) \xrightarrow{k_6}$
$(CCl_3; CCl_3).$

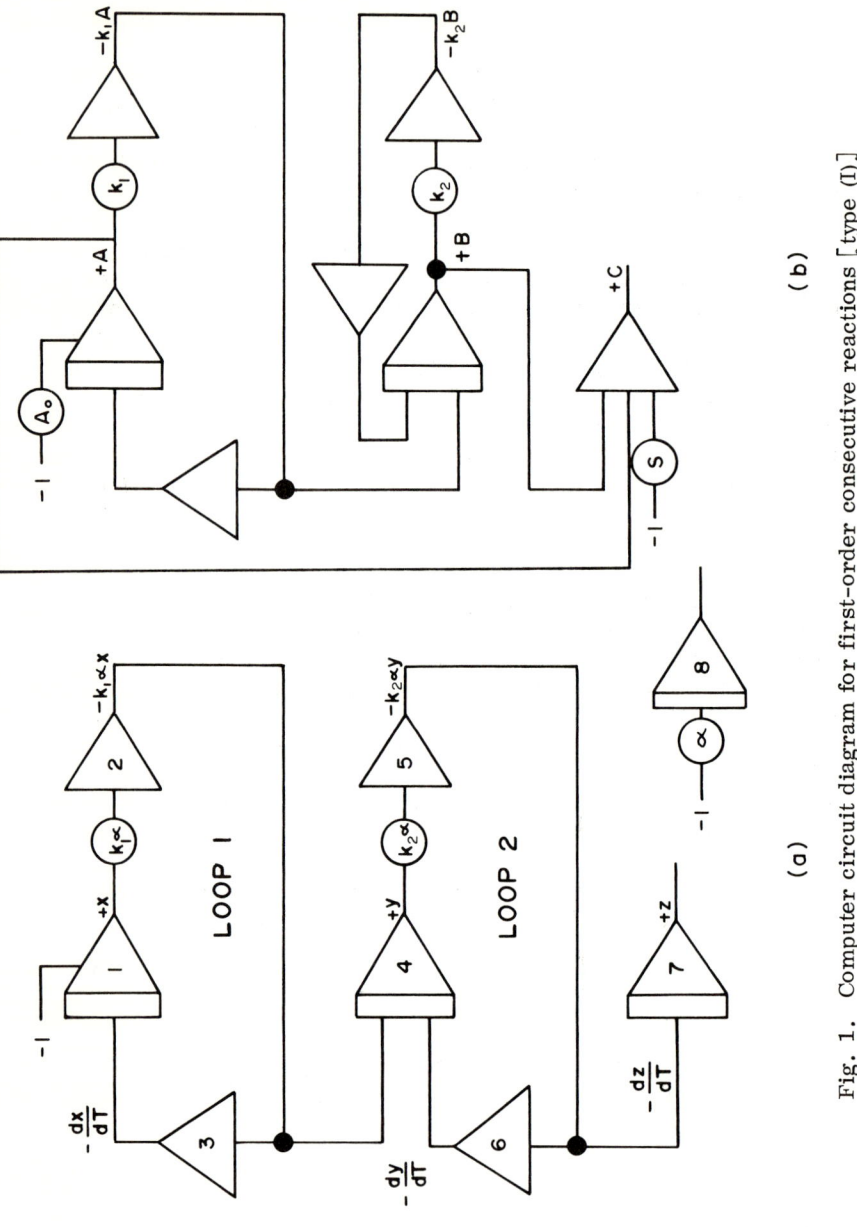

Fig. 1. Computer circuit diagram for first-order consecutive reactions [type (I)] according to: (a) Eqs. (4a)–(4c); (b) Eqs. (5a)–(5c).

4. Simulation of Kinetic Models

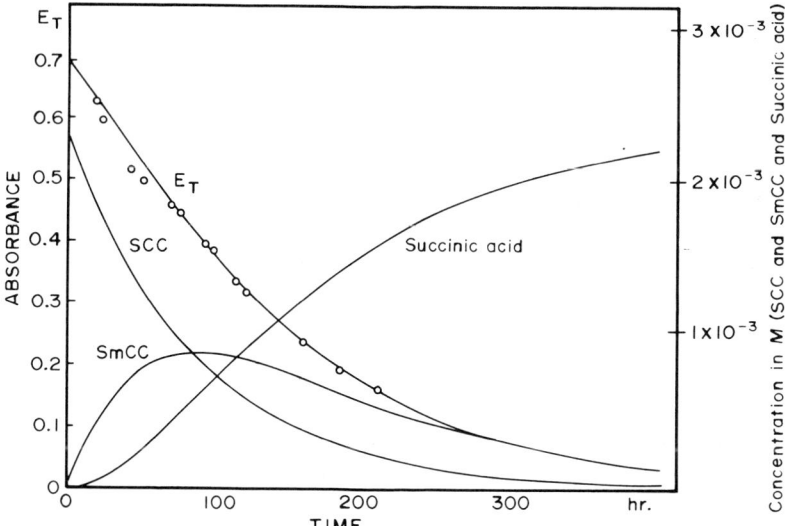

Fig. 2. Typical concentration-time curves for hydrolytic decomposition of succinylcholine chloride in 0.4 M acetate buffer. The initial concentration of SCC was 2.34×10^{-3} M and the rate constants were determined as $k_1 = 3.14 \times 10^{-6}$ sec^{-1} and $k_2 = 2.87 \times 10^{-6}$ sec^{-1}. Reprinted from Ref. [19], p. 1057, by courtesy of the Pharmaceutical Society of Japan.

No (CH_3; CCl_3) or (CH_2Cl; CCl_3) derivatives and only less than 5% of (CH_3; $CHCl_2$) were found on analysis of products. The concentration-time curves were obtained from glc analysis of the reaction mixture. As the amount of chlorine was always kept in excess, the reaction scheme can be described by the sequence of pseudo-first-order reactions:

$$A \xrightarrow{k_1} B \xrightarrow{k_2} C \xrightarrow{k_3} D \xrightarrow{k_4} E \xrightarrow{k_5} F \xrightarrow{k_6} G. \quad\quad (II)$$

The differential equations and corresponding computer program are similar to those discussed above.

We have been considering so far only the most common situations encountered in reaction kinetic studies, i.e., calculations where time is the independent variable. Concentrations can be plotted, however, against variables other than time, as it is shown in the paper by Butterfield et al. [21]. The reactions of concern are the catalytic hydrogenation of methyl linolenate:

$$\text{methyl linolenate} \xrightarrow[H_2]{k_1} \text{methyl linoleate} \xrightarrow[H_2]{k_2}$$

$$\text{methyl oleate,} \qquad \text{(III)}$$

and the reduction of linolenic acid by hydrazine:

$$\text{linolenic acid} \xrightarrow[N_2H_4]{k_1} \text{linoleic acid} \xrightarrow[N_2H_4]{k_2} \text{oleic acid} \xrightarrow[N_2H_4]{k_3}$$

$$\text{stearic acid.} \qquad \text{(IV)}$$

Apart from the usual concentration-time curves it was desirable to obtain plots of the concentration versus the average number of double bonds remaining per molecule. The voltage to be fed into the X axis of an X, Y recorder was derived from the outputs of the integrators for A, B, and C (Fig. 1a) (where A, B, C correspond to the above-mentioned esters), multiplied by the respective number of double bonds (Fig. 3). The dependence of composition on the number of double bonds is shown in Fig. 4. This method can be particularly useful in hydrogenation of oils of variable starting composition. Since the values of k_1, k_2, and k_3 are known, it is only necessary to set the potentiometers corresponding to the initial concentrations of the individual acids to produce a plot for any mixture. This possibility has been demonstrated for the hydrogenation of linolenic acid, linseed oil fatty acids, soybean oil fatty acids, and safflower oil fatty acids [21]. Another problem encountered is a prediction of the time when the reacting mixture reaches a desired composition. This possibility has been verified again experimentally by the same authors for the hydrazine reduction of linolenic acid-oleic acid mixture.

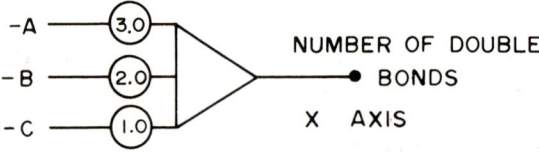

Fig. 3. Computer circuit diagram for generation of double bonds.

4. Simulation of Kinetic Models

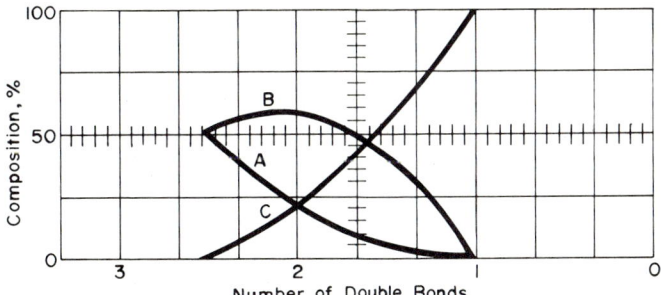

Fig. 4. Composition of the reaction mixtures [reaction (III)] as the function of remaining double bonds; $A = B = 50\%$ and $k_1/k_2 = 2$. Reprinted from Ref. [21], p. 31, by courtesy of the American Oil Chemists' Society.

2. Competitive Reactions

The reaction scheme

$$A \xrightarrow{k_1, k_2, k_3} \{B, C, B_1\} \xrightarrow{k_4, k_5} C \xrightarrow{k_6} D \qquad (V)$$

is a combination of competitive and consecutive reactions. The differential equations describing the concentration-time functions are as follows:

$$\frac{dA}{dt} = -pA \tag{6a}$$

$$\frac{dB}{dt} = k_1 A - k_4 B \tag{6b}$$

$$\frac{dB_1}{dt} = k_3 A - k_5 B_1 \tag{6c}$$

$$\frac{dC}{dt} = k_2 A + k_4 B + k_5 B_1 - k_6 C \tag{6d}$$

$$\frac{dD}{dt} = k_6 C \tag{6e}$$

where $p = k_1 + k_2 + k_3$. The corresponding rate diagram is shown in Fig. 5. It should be noted that the program as it stands requires a dual setting of the rate constants k_2, k_4, k_5, and k_6 and of the sum p. It

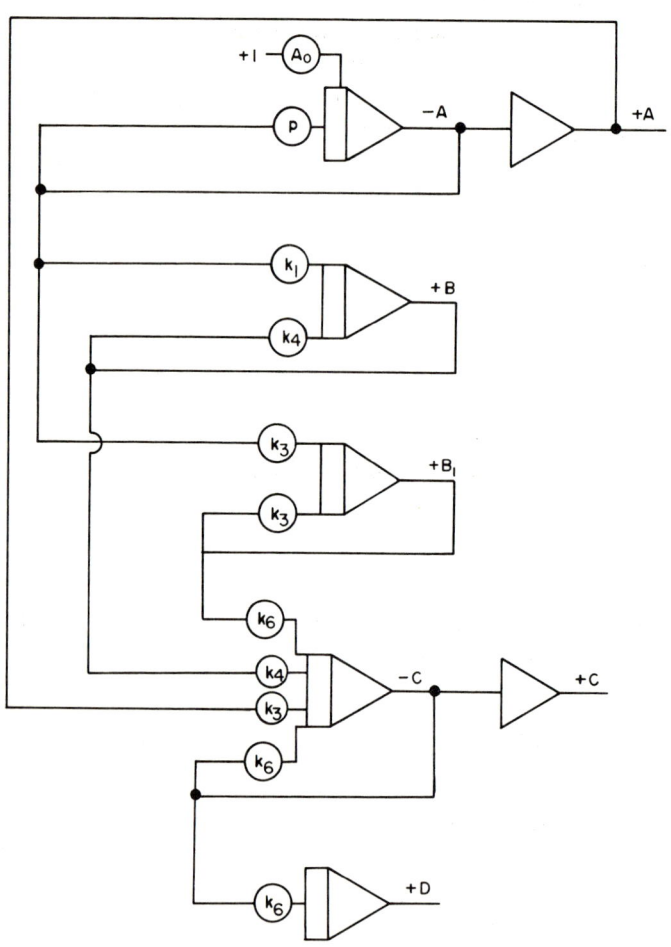

Fig. 5. Computer circuit diagram for first-order competitive consecutive reactions [type (V)].

4. Simulation of Kinetic Models

would be possible to rewrite the program in such a way that only one potentiometer would represent each of these rate constants. This would, of course, require additional inverters to be used. Obviously, one has to compromise between economy and the convenience of the program.

The scheme (V) has been used in study of high-pressure catalytic heterogeneous hydrogenation of fatty acid esters [5, 6]: Linolenate, linoleate, isolinoleate, oleate, and stearate concentrations correspond to A, B, B_1, C, and D, respectively. Many examples of hypothetical first-order reactions can be found in Refs. [7] and [8].

B. Reactions of Second Order

1. Reversible Reactions

Let us consider the following reversible reactions:

$$A + B \underset{\underline{k}_1}{\overset{k_1}{\rightleftharpoons}} C + E$$

$$C + B \underset{\underline{k}_2}{\overset{k_2}{\rightleftharpoons}} D + E. \tag{VI}$$

The differential equations are (E is in a large excess)

$$\frac{dA}{dt} = -k_1 AB + \underline{k}_1 C \tag{7a}$$

$$\frac{dB}{dt} = -k_1 AB - k_2 CB + \underline{k}_2 D + \underline{k}_1 C \tag{7b}$$

$$\frac{dC}{dt} = k_1 AB - \underline{k}_1 C - k_2 CB + \underline{k}_2 D \tag{7c}$$

$$\frac{dD}{dt} = k_2 CB - \underline{k}_2 D \tag{7d}$$

where for $t = 0$, $A = A_0$, $B = B_0$, and $C = D = 0$. To convert the equations above into a form suitable for the analog computer we have to perform a scaling of variables. In other words we have to find the relationship between concentrations and the corresponding computer voltages. This step is superfluous in first-order kinetics where rate constants do not depend on the initial concentrations. There are several ways to approach variable scaling. The most convenient procedure for kinetic purposes can be outlined as follows: First, new variables are introduced

$$A = ax; \quad B = by; \quad C = cz; \quad D = dw \tag{8}$$

where $a = A_{max}$, $b = B_{max}$, $c = C_{max}$, and $d = D_{max}$. As the maximum concentrations of A and B are their respective initial concentrations and as $C_{max} \leq A_{max}$ and $D_{max} \leq A_{max}$, we can write

$$A = ax; \quad B = by; \quad C = az; \quad D = aw \tag{9}$$

where $a = A_{max} = A_o$ and $b = B_{max} = B_o$. This ensures that the machine variables x, y, z, and w will not exceed 1 M.U. Combining Eqs. (7) and (9) then yields

$$\frac{dx}{dt} = -k_1 bxy + \underline{k_1} z \tag{10a}$$

$$\frac{dy}{dt} = -k_1 axy - k_2 ayz + \underline{k_1}\frac{a}{b}z + \underline{k_2}\frac{a}{b}w \tag{10b}$$

$$\frac{dz}{dt} = k_1 bxy - \underline{k_1} z - k_2 byz + \underline{k_2} w \tag{10c}$$

$$\frac{dw}{dt} = k_2 byz - \underline{k_2} w. \tag{10d}$$

Time scaling is now performed in the usual way, yielding the final machine equations:

$$\frac{dx}{dT} = -k_1 \alpha bxy + \underline{k_1}\alpha z \tag{11a}$$

$$\frac{dy}{dT} = -k_1 \alpha axy - k_2 \alpha ayz + \underline{k_1}\alpha\left(\frac{a}{b}\right)z + \underline{k_2}\alpha\left(\frac{a}{b}\right)w \tag{11b}$$

4. Simulation of Kinetic Models

$$\frac{dz}{dT} = k_1 \alpha bxy - \underline{k}_1 \alpha z - k_2 \alpha byz + \underline{k}_2 \alpha w \tag{11c}$$

$$\frac{dw}{dT} = k_2 \alpha byz - \underline{k}_2 \alpha w \tag{11d}$$

where for $t = 0$, $x = y = 1$ and $z = w = 0$.

The computing diagram corresponding to Eqs. (11a)-(11d) is given in Fig. 6. A new computing element comes into play whenever second- (or higher) order kinetics is encountered, i.e., function multiplier.

The esterification of terephthalic acid, A, with methanol, B, via the monoester, C, to give the diester, D, can be described by this mechanism (E denotes water). An analog computer study of this reaction has been carried out under much more complicated conditions employing a continuous reactor [22]. Another example of such a reversible reaction solved by an analog computer is the formation of a proline-chloranil complex, X, by the reaction of proline, A, and chloranil, B:

$$A + B \underset{\underline{k}_1}{\overset{k_1}{\rightleftarrows}} X. \tag{VII}$$

However, proline was present in excess so that the case reduced to pseudo-first-order kinetics [23]. Such reversible reactions are found regularly in enzyme kinetics.

2. Competitive and Consecutive Reactions

The anionic copolymerization of the monomers A and B can be described by the following reaction scheme [24, 25]:

$$\begin{aligned}
\sim A^- + A &\xrightarrow{k_{aa}} \sim A\text{-}A^- \\
\sim A^- + B &\xrightarrow{k_{ab}} \sim A\text{-}B^- \\
\sim B^- + B &\xrightarrow{k_{bb}} \sim B\text{-}B^- \\
\sim B^- + A &\xrightarrow{k_{ba}} \sim B\text{-}A^-.
\end{aligned} \tag{VIII}$$

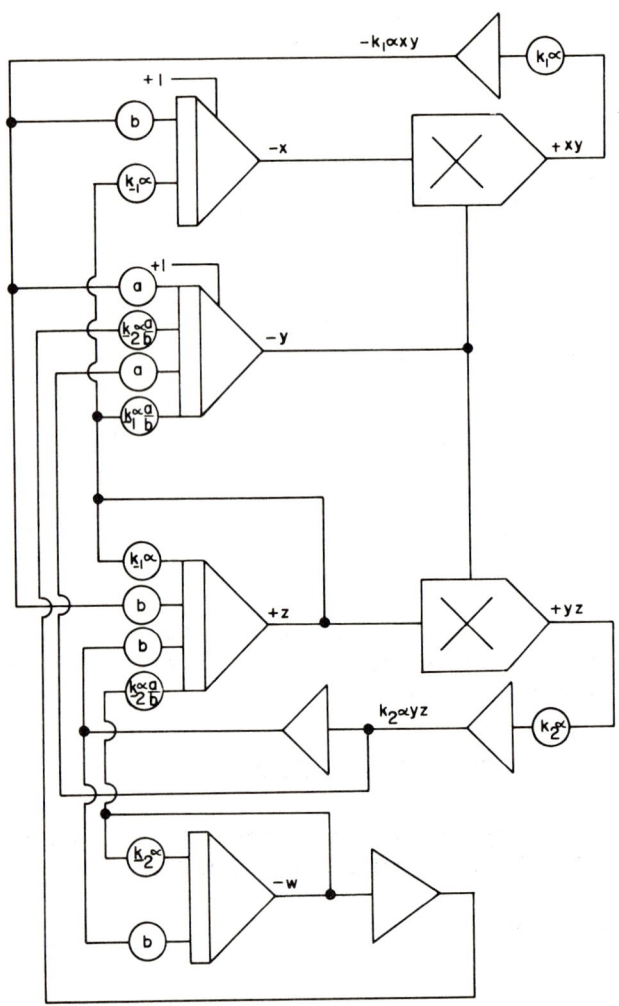

Fig. 6. Computer circuit diagram for second-order reversible reaction [type (VI)].

4. Simulation of Kinetic Models

Bearing in mind that the products on the right-hand side are again "living anions" ($\sim A^-$ or $\sim B^-$), we can write the following differential equations.

$$\frac{dA^-}{dt} = k_{ba} B^- A - k_{ab} A^- B \tag{12a}$$

$$\frac{dB^-}{dt} = k_{ab} A^- B - k_{ba} B^- A \tag{12b}$$

$$\frac{dA}{dt} = -k_{aa} A^- A - k_{ba} B^- A \tag{12c}$$

$$\frac{dB}{dt} = -k_{ab} A^- B - k_{bb} B^- B. \tag{12d}$$

The computer diagram for this case is shown in Fig. 7 and the concentration-time curves are shown in Fig. 8. The concentrations denoted as $A_M (B_M)$ and $A_E (B_E)$ are those obtained from Eqs. (12a)-(12d) by classical integration using the steady-state approximation

$$k_{ab} A^- B = k_{ba} B^- A \quad \text{(index M, Ref. [25])}$$

and

$$A^- \underset{k'_{ba}}{\overset{k'_{ab}}{\rightleftharpoons}} B^- \quad \text{(index E, Ref. [24])}$$

where $k_{ba} A_o = k'ba$ and $k_{ab} B_o = k'ab$.

The chlorination of triphenylsilylamines in solution with butylhypochlorite proceeds according to the overall reactions:

$$Ph_3SiNH_2 + (CH_3)_3COCl \rightarrow Ph_3SiNHCl + (CH_3)_3COH$$
$$Ph_3SiNHCl + (CH_3)_3COCl \rightarrow Ph_3SiNCl_2 + (CH_3)_3COH. \tag{IX}$$

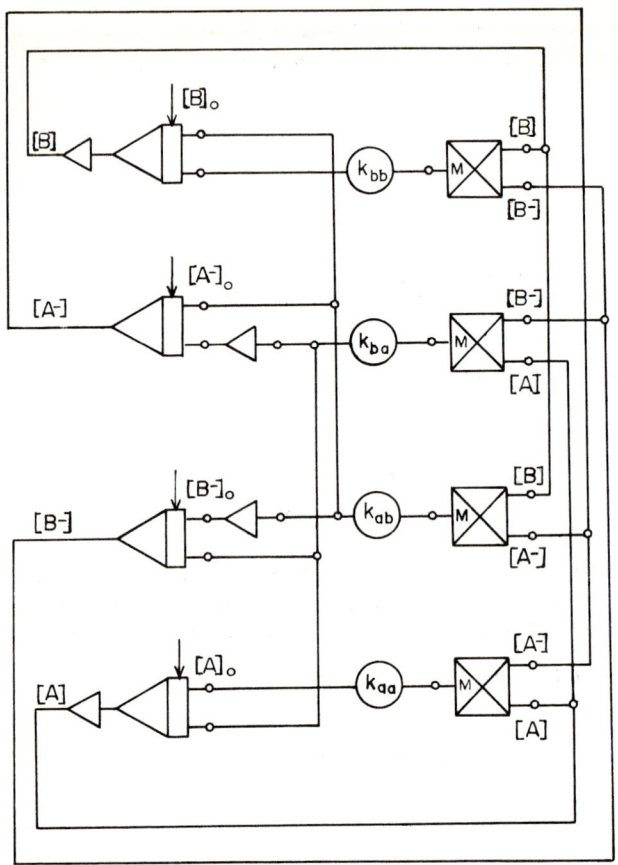

Fig. 7. Computer circuit diagram for second-order competitive consecutive reactions [reaction (VIII)]. Reprinted from Ref. [24], p. 333, by courtesy of VEB Deutscher Verlag für Grundstoffindustrie.

The mechanism of this reaction has been studied in detail [26] and the following reaction scheme, which fits the experimental data, has been proposed:

4. Simulation of Kinetic Models

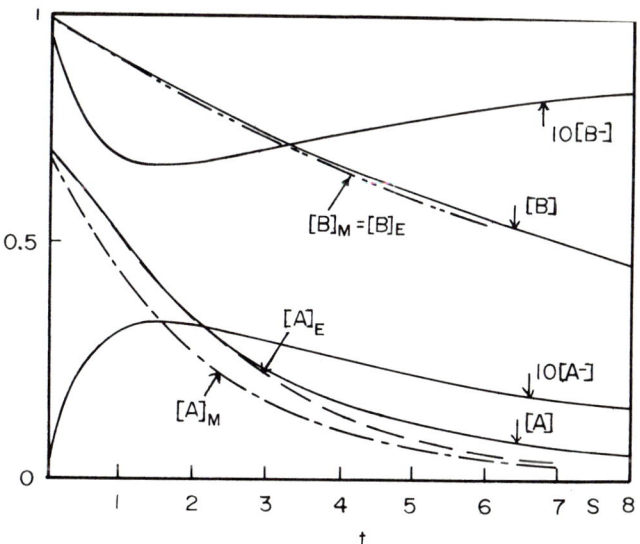

Fig. 8. Typical concentration-time curves for copolymerization reactions. Reaction parameters are as follows: $[A]_o = 0.7$; $[A^-]_o = 0$; $[B]_o = 1$; $[B^-]_o = 0.1$; $k_{aa} = 10$; $k_{ab} = k_{bb} = k_{ba} = 1$. Reprinted from Ref. [24], p. 334, by courtesy of VEB Deutscher Verlag für Grundstoffindustrie.

$$(CH_3)_3COCl \xrightarrow{k_1} (CH_3)_3CO\cdot + Cl\cdot$$

$$Ph_3SiNH_2 + Cl\cdot \xrightarrow{k_2} Ph_3SiNHCl + H\cdot$$

$$Ph_3SiNHCl + Cl\cdot \xrightarrow{k_3} E + H\cdot$$

$$(CH_3)_3COCl + H\cdot \xrightarrow{k_4} (CH_3)_3COH + Cl\cdot \qquad (X)$$

$$(CH_3)_3CO\cdot + H\cdot \xrightarrow{k_5} (CH_3)_3COH$$

$$E \xrightarrow{k_6} Ph_3SiNCl_2$$

$$E + Ph_3SiNH_2 \xrightarrow{k_7} 2Ph_3SiNHCl$$

The scaled differential equations for the analog evaluation are [27]

$$\frac{dM}{dt} = k_2'AX - k_3'MX + 2k_7'AE \tag{13a}$$

$$\frac{dD}{dt} = k_6 E \tag{13b}$$

$$\frac{dE}{dt} = k_3'MX - k_6 E - k_7'AE \tag{13c}$$

$$\frac{dX}{dt} = k_1 B - k_2'AX - k_3'MX + k_4'YB \tag{13d}$$

$$\frac{dY}{dt} = k_2'AX + k_3'MX - k_4'YB - k_5'YT \tag{13e}$$

$$\frac{dT}{dt} = k_1 B - k_5'YT. \tag{13f}$$

In addition, two mass-balance algebraic equations for the total concentrations of the amines, A, and chlorine, B, are used; thus

$$A = 1 - M - D - E \tag{13g}$$

and

$$B = \frac{a}{b} - M - 2D - 2E - X. \tag{13h}$$

The scaling procedure is slightly different from the one described earlier, which has been previously adopted. The variables A_1, B, ..., T are in fact computer variables obtained by the normalization of the respective chemical concentrations; thus

$$A = \frac{C_A}{a}, \quad B = \frac{C_B}{b}, \quad \ldots$$

where a is the initial concentration of triphenylsilylamine and b is the initial concentration of hypochlorite. The second-order rate constants were normalized in a similar way. Thus $k_2 = k_2'/a$, and so on. The analog computer diagram is shown in Fig. 9. The function divider employed in the circuit to produce B is in fact redundant as it performs division by a constant only. Excellent agreement between the computed and experimental curves is demonstrated in Fig. 10.

4. Simulation of Kinetic Models 231

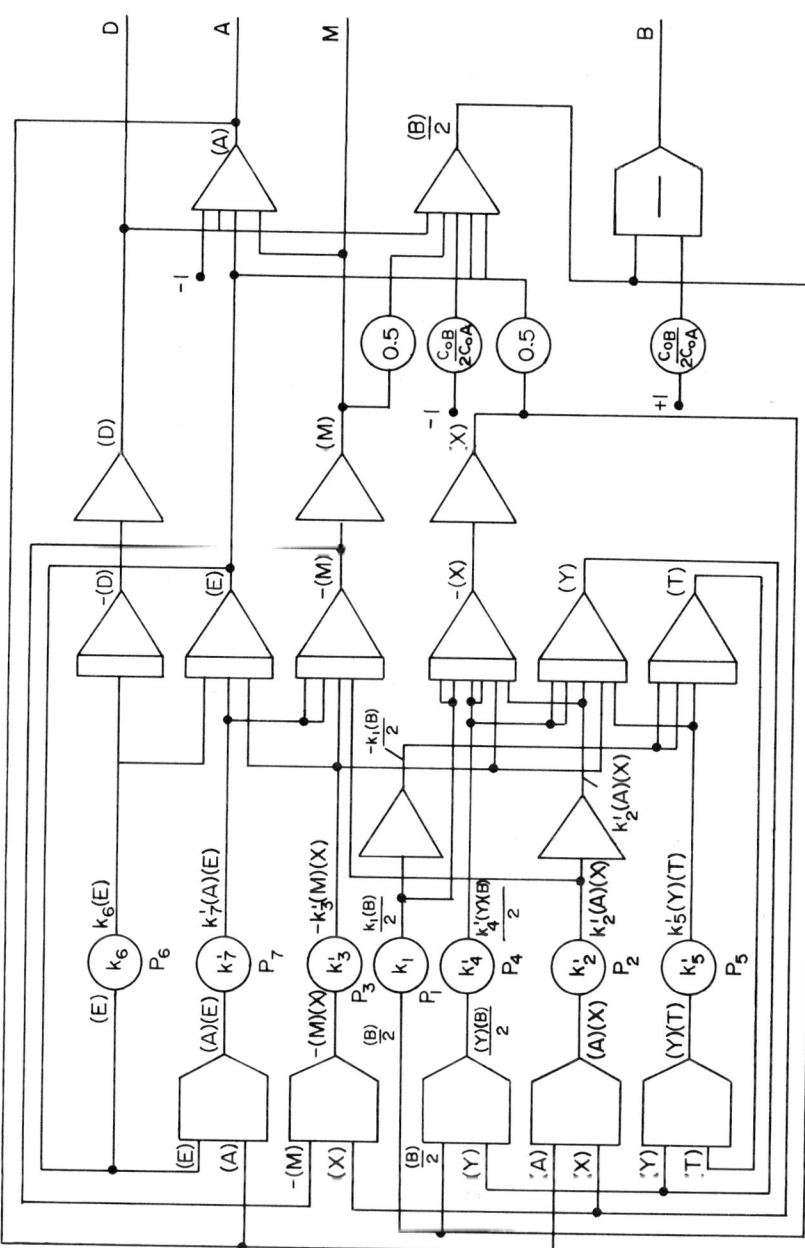

Fig. 9. Computer circuit diagram for second-order consecutive reactions [reaction (X)]. Reprinted from Ref. [27], p. 813, by courtesy of Chemie-Ingenieur-Technik. Note that parentheses are used in this figure instead of the usual square brackets notation for concentration.

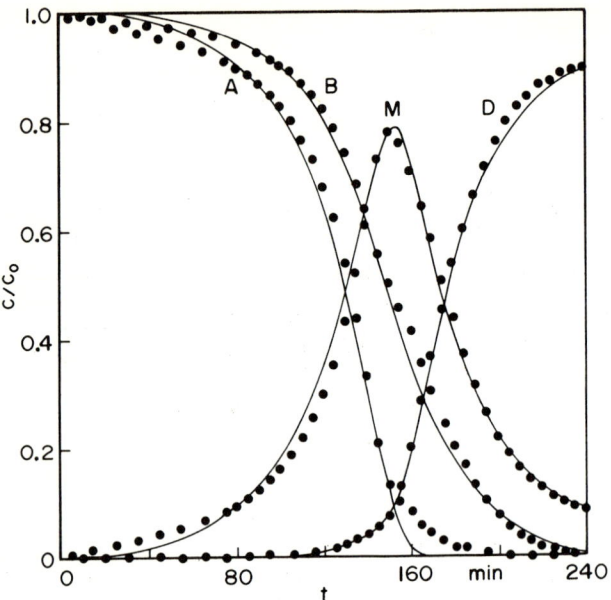

Fig. 10. Experimental and computed concentration-time curves for chlorination of triphenylsilylamines. Reprinted from Ref. [27], p. 815, by courtesy of Chemie-Ingenieur-Technik.

The thermal decomposition of 5-methyl-2-oxazolidinone (OM) to CO_2 and N-(2-hydroxypropyl)imidazolidinone (HPMI)

follows the kinetic scheme

$$2(OM) \xrightarrow{k_1} (HPMI) + CO_2$$
$$(OM) + (HPMI) \xrightarrow{k_2} (OM)(HPMI) \tag{XI}$$
$$2(OM)(HPMI) \xrightarrow{k_3} 3(HPMI) + CO_2.$$

4. Simulation of Kinetic Models

The agreement between experimental and computed concentration-time curves suggests that (HPMI) plays an autocatalytic role. This hypothesis has been supported by further experimental evidence [28]. Various degrees of autocatalysis were studied and a diagnostic procedure has been worked out for the estimation of the degree of autocatalysis (defined as the ratio k_3/k_1) from the peak concentrations of the intermediates.

Blackmore [29] has used both analog and digital computers in the kinetic study of reaction between oxygen and hydrogen. The time scaling of this particular reaction scheme presented problems, as some of the reactions were very much faster than others. To circumvent this, the author proposed either to split the reaction scheme and, hence, to study the first (faster) part in a different time scale or to treat the fast reactions algebraically and to use their steady-state concentrations in further calculations.

Other applications of the analog computer to competitive-consecutive second-order kinetics include the oxidation of p-amino-N-diethylaniline with persulfate [30] in solutions of various pH, consecutive chlorination of benzene [31], and study of the inhibition and promotion in branched-chain reactions of hydrocarbons [32].

III. OPEN KINETICS

A kinetic system is defined as open if during the reaction matter is being exchanged across boundaries [33]; such systems are common in chemical engineering as well as in biological kinetics [34]. Their mathematical treatment is inherently more complicated and analog computers have been used frequently for their solution. Many of the engineering applications are included in review articles of which the series by Matthews [35] is particularly recommended.

Another recent review of the analog computer treatment of open kinetics deals with the application of controlled current coulometry to reaction kinetics [36]. Here the reagent is generated electrolytically throughout the experiment, with a constant rate. It is a phenomenon well known to analytical chemists that some continuous titrations proceed relatively slowly, i.e., that there is a detectable concentration of both reagent and substrate present in the solution at the same time. As the concentrations of individual species depend on the velocity of their reaction, it is obvious that the concentration-time curves obtained in continuous titrations can be, in turn, used for the evaluation of rate constants. The main advantage of such an experimental approach is in the study of "medium" fast reactions. This is illustrated in Fig. 11 which

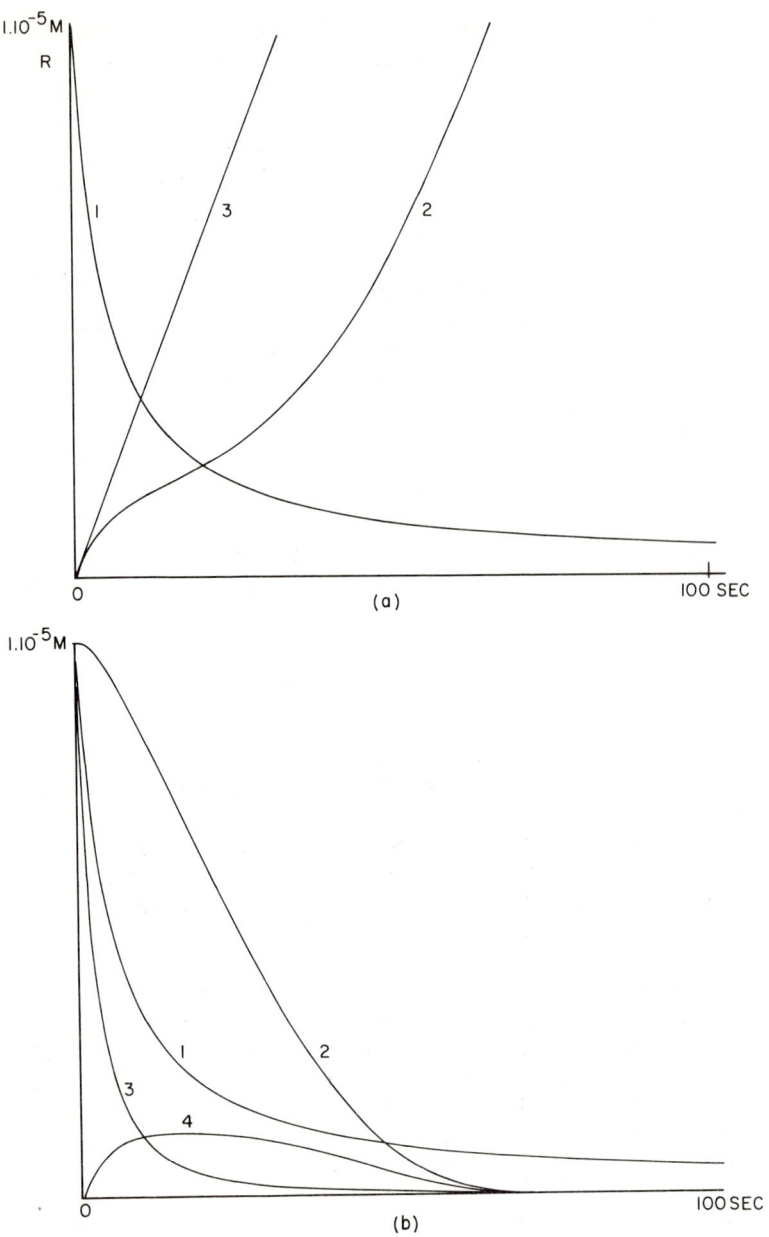

Fig. 11. Dependence of concentration of (a) R (b) A on time for reaction A + R → with continuously added reagent R. Rate of addition σ is 3×10^{-7} mole liter^{-1} sec^{-1}, $A_0 = 1 \times 10^{-5}$ M, $k = 2 \times 10^4$ liters mole^{-1} sec^{-1}.

4. Simulation of Kinetic Models

shows concentrations as a function of time substrate A (Fig. 11b, curves 1, 2) and a reagent R (Fig. 11a, curves 1, 2, 3) in a simple second-order reaction. In the classical arrangement where both A and R are initially present, the reaction proceeds at the beginning with the maximum velocity (Fig. 11b, curve 3). Therefore, the initial reaction period represents a major experimental uncertainty which is mainly a result of nonhomogeneity of the mixture (near the ampule, the syringe tip, etc.). For this reason the initial portion of the experimental curve is often extrapolated. This can be done, of course, only if the rate equation in the integrated form is known. In the case of continuously added reagent, the concentrations of both A and R change more slowly and the reaction velocity reaches its maximum about half-way through the experiment (Fig. 11b, curve 4). Furthermore, the initial slope is known from the blank (Fig. 11a, curve 3). This alone improves the accuracy and increases the margin of measurable rate constants by a factor of 5. Another advantage is that the same experimental setup can be used for the determination of concentrations. Further advantages arising from the electrolytical generation of the reagent have also been discussed [36]. They can be briefly summarized as follows:

1. High accuracy in preparation of very dilute solutions.

2. No need for standardization of added reagent.

3. No increase of volume.

4. Easily monitored on-off control (time base).

5. Possibility of generation of unstable and/or very reactive reagent.

There are, however, some electrochemical factors involved which have to be considered before this method can be applied [36].

In spite of their advantages open systems have found only a limited use in the kinetic studies. This has been largely due to their mathematical complexity. The evaluation of the experimental curves, however, presents no difficulties if an analog computer is used.

Let us define the rate of change of reagent concentration in the absence of any chemical reaction as

$$\rho = \frac{dR}{dt} \tag{14}$$

and in the case of electrolytically generated reagent

$$\rho = \frac{i}{nFV} \tag{15}$$

where i is the electrolytic current, n is the number of electrons, F is the faraday constant (96.494 C), and V is the volume. In the case of the solution of reagent being added continuously with a constant rate,

$$\rho = \frac{vC_R}{V}, \tag{16}$$

where v is the rate of addition in liters per second, V is the initial volume, and C_R is the concentration of the reagent solution. Equation (16) is, of course, valid only if total volume changes can be neglected [36]. The concentration-time curve is then represented in both cases by a straight line (Fig. 11a, curve 3). In the presence of A, concentration changes are depicted by curve 2 in both Fig. 11a and 11b and the corresponding differential equations are

$$\frac{dR}{dt} = -kAR + \rho \tag{17a}$$

$$\frac{dA}{dt} = -kAR \tag{17b}$$

where for $t = 0$, $A = A_0$ and $R = 0$. The introduction of time and variable scaling factors transforms the system of equations shown in (4) to

$$\frac{dx}{dT} = -k\alpha axy + \frac{\rho\alpha}{r} \tag{18a}$$

$$\frac{dy}{dT} = -k\alpha rxy \tag{18b}$$

where $a = A_{max} = A_0$ and $r = R_{max}$. The computer diagram is shown in Fig. 12. Note that this diagram is practically identical with the one for the equivalent closed system. The only difference is that the second input of the x integrator would be replaced by the initial condition on the same integrator for the closed system.

The reaction between diphenyl sulfide and electrolytically generated bromine [37] in acetic acid-water mixture has been found to obey Eqs. (18a) and (18b).

4. Simulation of Kinetic Models

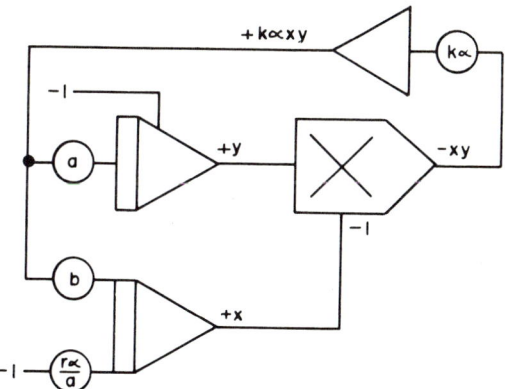

Fig. 12. Computer circuit diagram for second-order open-system reaction.

Cases in which the added reagent R reacts both with the substrate A and with the products of this primary reaction, B, can be schematically described as follows:

$$A + R \xrightarrow{k_1} B$$

$$B + R \xrightarrow{k_2} C. \qquad \text{(XII)}$$

The time changes of concentrations of the components of this system are given by the following rate equations:

$$\frac{dR}{dt} = -k_1 AR - k_2 BR + \rho \qquad (19a)$$

$$\frac{dA}{dt} = -k_1 AR \qquad (19b)$$

$$\frac{dB}{dt} = k_1 AR - k_2 BR \qquad (19c)$$

where for $t = 0$, $A = A_0$ and $B = R = 0$. The corresponding computer equations (in real time) are

$$\frac{dx}{dt} = -k_1 axy - k_2 axz + \frac{\rho}{r} \qquad (20a)$$

$$\frac{dy}{dt} = -k_1 rxy \tag{20b}$$

$$\frac{dz}{dt} = k_1 rxy - k_2 rxz \tag{20c}$$

where for $t = 0$, $y = 1$ and $x = z = 0$. The block circuit diagram representing this scheme is shown in Fig. 13. Typical curves showing the dependence of the concentration of reagent R on time are given in Fig. 14.

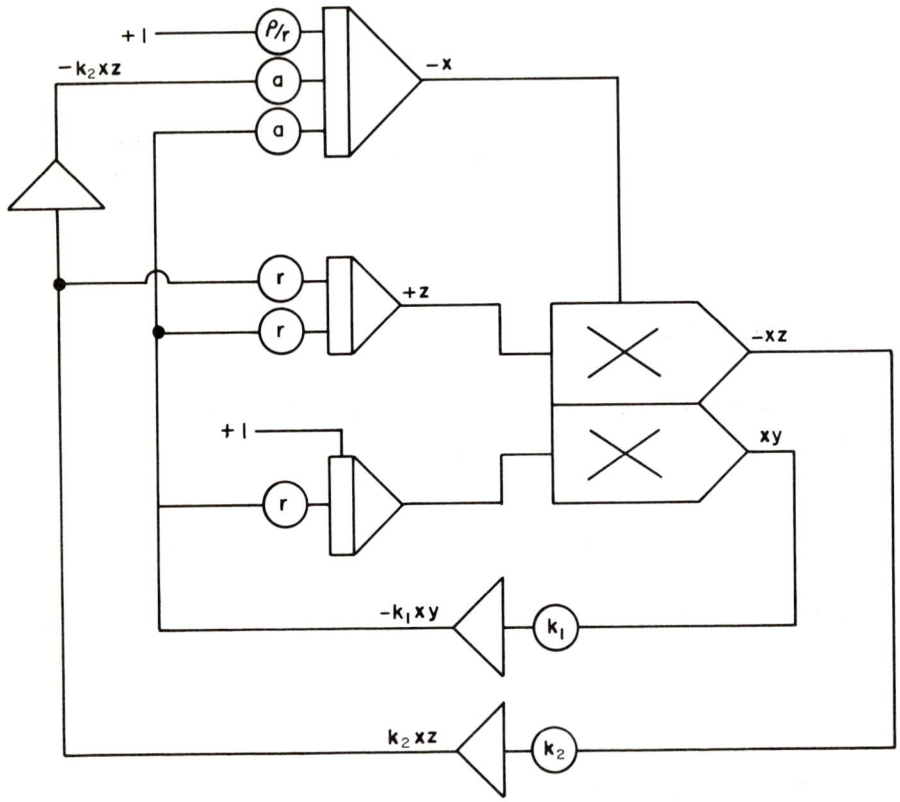

Fig. 13. Computer circuit diagram for two competitive consecutive reactions [type (XII)].

4. Simulation of Kinetic Models

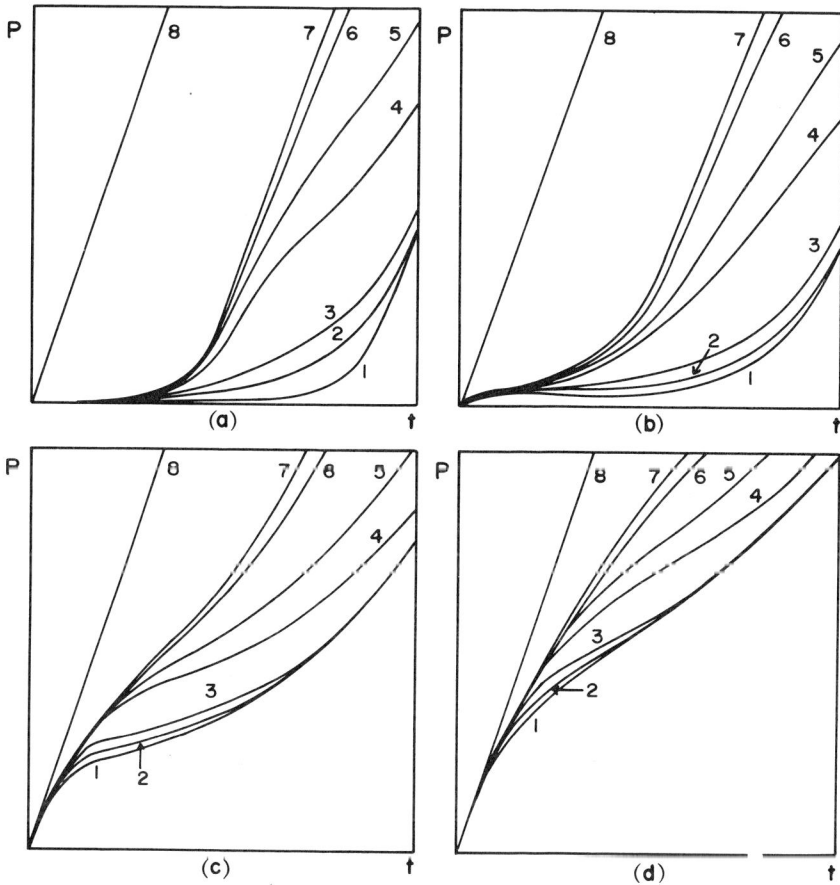

Fig. 14. Dependence of concentration of R on time for two competitive consecutive reactions; P is a parameter proportional to the concentration of R; $R_{max} = 7.79 \times 10^{-5}$ M; $A_0 = 9.34 \times 10^{-5}$ M. Values of k_1 (liters mole^{-1} sec^{-1}): (a) 1.18×10^4; (b) 2.35×10^3; (c) 2.35×10^2; (d) 1.18×10^2. Values of k_2 (liters mole^{-1} sec^{-1}) for each set: (1) 1.18×10^4; (2) 2.37×10^3; (3) 1.18×10^3; (4) 2.35×10^2; (5) 1.18×10^2; (6) 2.37×10^1; (7) 2.37; (8) blank.

The reaction between bromine and n-butyl benzyl sulfide was found to follow this scheme [38]. Another important feature of analog evaluation is illustrated in Fig. 15. No agreement between the experimental curve 1 and the curve computed for reaction $A + R \xrightarrow{k}$ could be obtained.

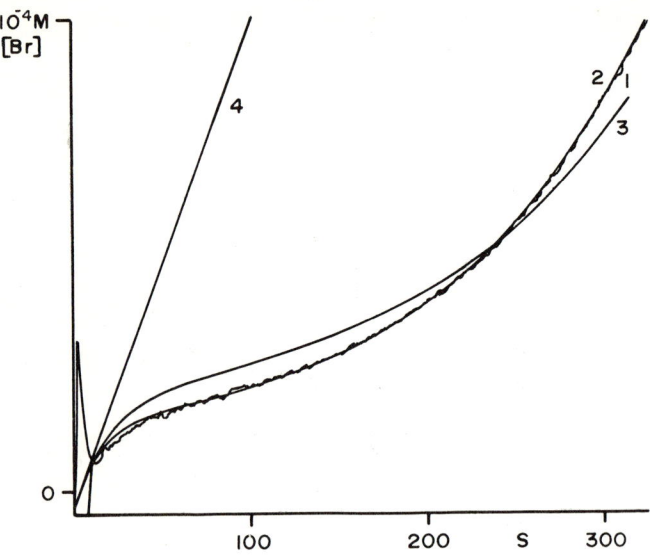

Fig. 15. Experimental and computed curves for reaction of electrolytically generated bromine with n-butyl benzyl sulfide in 95% acetic acid-5% water mixture: (1) concentration of bromine; (2) computed curve corresponding to type (XII); (3) best "fit" computed for reaction A + R → ; (4) experimental and computed blank.

However, the computed curve for the type (XII) matches exactly the experimental curve. This shows that it is possible to verify the proposed reaction scheme this way.

The overall reaction between bromine and various substituted 8-quinolinols which has been studied using an analog computer also fits this scheme [39]. Other hypothetical reaction schemes have been studied and their computer diagrams and concentration-time curves are given in Refs. [36], [38], and [40].

Another example of an open-system reaction is the radiolysis of p-xylene in liquid phase [41]:

4. Simulation of Kinetic Models

(XIII)

$$H_2O_2 \xrightarrow{k_2} \text{products}.$$

p-Xylene hydroperoxide is formed by a series of fast radical reactions. It subsequently reacts with p-toluic acid which itself is one of the products. The reaction is, therefore, autocatalytic. Hydrogen peroxide is also formed during this reaction and decomposes by a first-order reaction. The relevant rate equations are

$$\frac{dR}{dt} = -k_1 AR + \rho \tag{21a}$$

$$\frac{dA}{dt} = -\alpha k_1 AR \tag{21b}$$

$$\frac{dB}{dt} = \beta - k_2 B \tag{21c}$$

where A is p-toluic acid, B is hydrogen peroxide, and R is p-xylene hydroperoxide. There is a small amount of p-toluic acid present in the solution at the beginning of radiolysis. Thus, for $t = 0$: $A = A_0$, $R_0 = 0$, and $B_0 = 0$. Because the autocatalytic reactions (XIII) are slow, the concentrations of p-xylene and oxygen are constant and the rate of generation of p-xylene hydroperoxide, ρ, is also constant. The block diagram corresponding to the system of equations in (21) is shown in Fig. 16. Concentration–time curves for p-xylene hydroperoxide, p-toluic acid, and H_2O_2 are shown in Fig. 17.

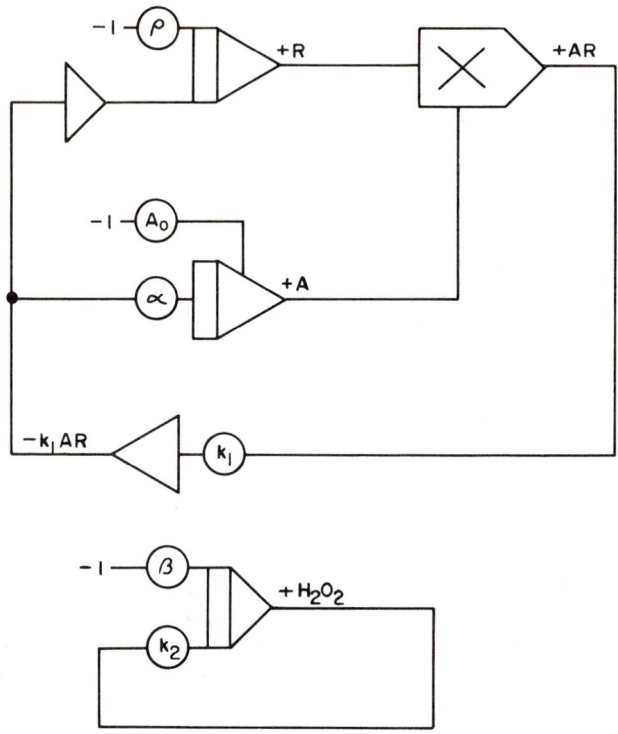

Fig. 16. Computer circuit diagram for radiolysis of p-xylene.

Strictly speaking this example does not fit exactly the definition of an open system as given above. It is the X-ray radiation (energy) which is causing the system to behave like an open one. From the point of view of the mathematical treatment any system which includes a zero-order reaction can be looked on as an open system.

It has been suggested [42] that it might be advantageous to choose the function ρ in such a way that the response signal, i.e., concentration, of reagent is a linear function of time. Because such a stressed ρ function would be rather complicated a special function generator would have to be used for its generation. This unique approach does not seem to offer any particular advantage as the reaction scheme would have to be known a priori to generate the appropriate ρ function.

4. Simulation of Kinetic Models

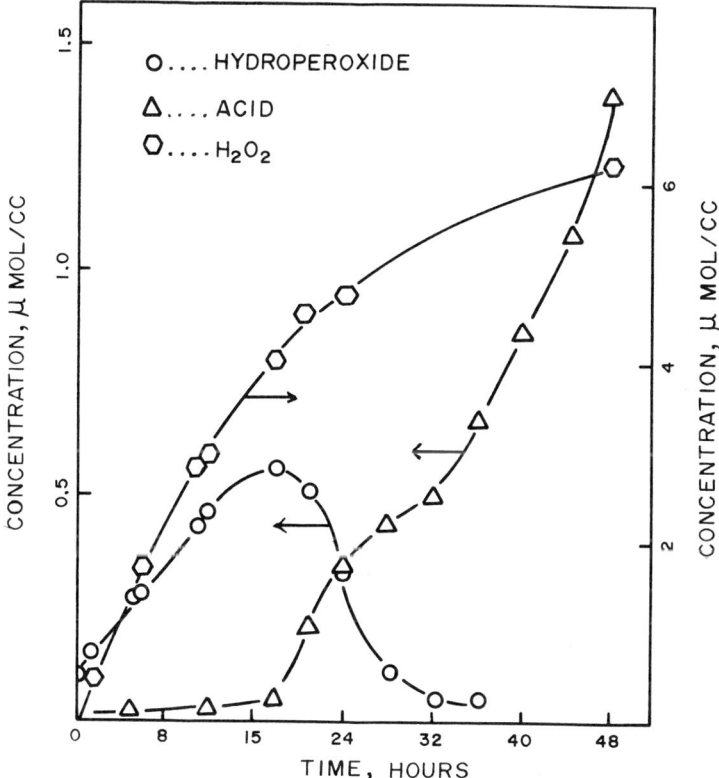

Fig. 17. Concentration-time curves for radiolysis of p-xylene. Reprinted from Ref. [41], p. 68, by courtesy of the American Institute of Chemical Engineers.

IV. BIOCHEMICAL APPLICATIONS

Many metabolic processes can be described by differential equations. It is not surprising that analog computers have found a wide application in this area and that the number of these applications is growing. They are divided about equally between pharmacokinetic and enzyme studies. There is also a smaller group of applications dealing with the metabolism of certain substrates not explicitly related to either of the two areas.

It is beyond the limits of this chapter to discuss all these applications. In order to do at least some justice to this subject, the titles of these applications for the period 1958-1970 are listed in the Appendix.

A. Pharmacokinetics

Reactions taking place in vivo can bear a close similarity to reactions in vitro even though the definition of the system boundaries is substantially different. In order to formulate the mathematical model of drug kinetics several assumptions must be made. The basic approach is to consider a series of compartments in the body which are linearily interconnected. It is assumed that there is a rapid equilibration between blood and other tissues, liquids, and body compartments. The rate of transfer between blood and the compartments is furthermore assumed to be independent of the amount of a drug. Under these conditions the whole system can be described by a series of first-order kinetic equations where integrators represent individual body compartments such as the stomach, blood, kidneys, and so on. Differential equations describing such a system closely resemble first-order kinetics of a conventional type. This idealized situation is valid only in a first approximation. Many other factors have to be taken into consideration in a real system, such as the mode of administration of the drug, nonlinear transport through membranes, saturation of compartments, and so on. To solve these complications by analytical mathematics is often impossible and an analog computer becomes very helpful.

Answers most often sought from a pharmacokinetic model are the distribution and levels of the drug throughout the body. Experimental data are usually obtained from monitoring excretion amounts of a drug and metabolites and/or their concentrations in blood. There are several reviews [13, 86, 94, 95, 99, 103, 111-113] on the application of analog computers into pharmacokinetics where the above-mentioned problems are discussed in detail.

The kinetics of absorption, metabolism, and excretion of (+)-amphetamine in man after administration of the drug in the form of a "free pellet" is described by the following scheme [43].

$$A \xrightarrow{k_a} \text{lag time} \longrightarrow M_1 \begin{array}{c} \xrightarrow{k_{m3}} M_2 \\ \xrightarrow{K_u} U \end{array} \qquad \text{(XIV)}$$

4. Simulation of Kinetic Models

where A is amphetamine in the gastrointestinal tract, M_1 is amphetamine in the body, M_2 is the amphetamine metabolite(s) in the body, and U is the amphetamine in the urine. The "lag time" is the time interval between ingestion of the dose and "zero time" when first-order loss from the gastrointestinal tract begins. The process has been followed under controlled acidic-urine conditions. The system is described by the following differential equations (post-lag-time period):

$$\frac{dA}{dt} = -k_a A \tag{22a}$$

$$\frac{dM_1}{dt} = k_a A - k_u M_1 - k_{m3} M_1 \tag{22b}$$

$$\frac{dM_2}{dt} = k_{m3} M_1 \tag{22c}$$

$$\frac{dV}{dt} = k_u M_1. \tag{22d}$$

The computer diagram is shown in Fig. 18. It is essentially a diagram for first-order consecutive reactions. In addition there is a parallel logic circuitry which enables simulation of lag time, τ. The time base, t, is compared with the lag time on a high-speed comparator. For $t - \tau > 0$ the comparator output becomes (logic) 1, the high-speed electronic switch (SW) becomes conducting, and the solution of Eqs. (22a)-(22d) commences. Very good agreement between computed and experimental data is illustrated in part (a) of Fig. 19. Part (b) of Fig. 19 shows the experimental and computed curves for multiple (3 x 5 mg dose) administration of the drug. Similar studies of drug kinetics can be found in references listed in the Appendix.

B. Enzyme Kinetics In Vivo

Pharmacokinetics and in vivo enzyme kinetics have many points in common. However, reversible reactions seem to occur more frequently in the latter. Analog computers have been used extensively in the investigation of the oscillatory behavior of regulating enzymic systems [44-50] which is associated with the question of the "biological clock." Such oscillations have been observed, for example, in the metabolism of glucose, and the following scheme has been proposed [49]:

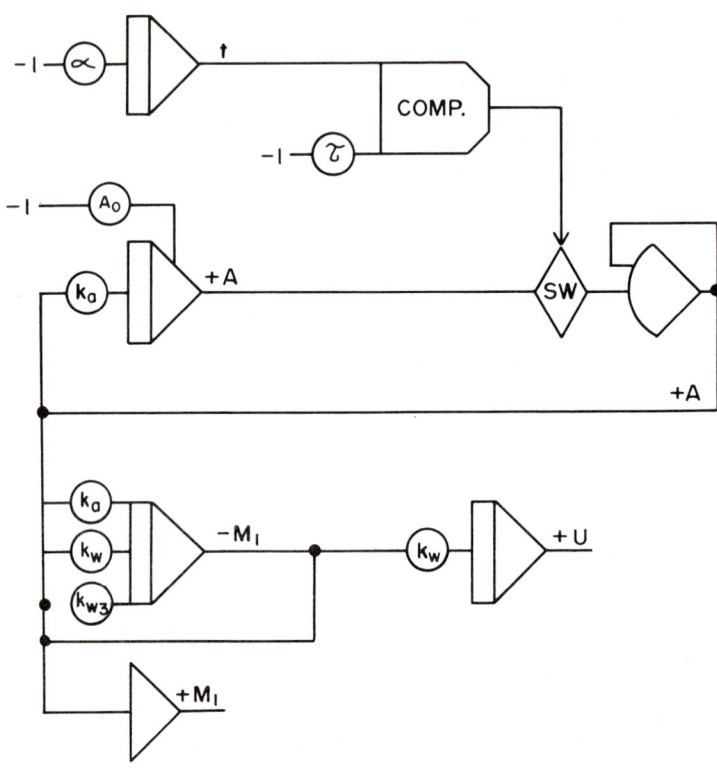

Fig. 18. Computer circuit diagram for metabolism of amphetamine drugs.

$$\text{glucose} \xrightarrow{k_1} \text{fructose-6-phosphate} \quad (F6P)$$

$$(F6P) + E_1^* \xrightarrow{k_2} E_1^* : (F6P)$$

$$E_1^* : (F6P) \xrightarrow{k_3} E_1^* + \text{fructose-1,6-diphosphate} \quad (FDP)$$

$$(FDP) + E_1 \underset{k_{-4}}{\overset{k_4}{\rightleftarrows}} E_1^* \tag{XV}$$

$$(FDP) + E_2 \xrightarrow{k_5} E_2 : (FDP)$$

$$E_2 : (FDP) \xrightarrow{k_6} E_2 + \text{dihydroxyacetone phosphate} + \text{glyceraldehyde-3-phosphate} \quad (GAP)$$

4. Simulation of Kinetic Models 247

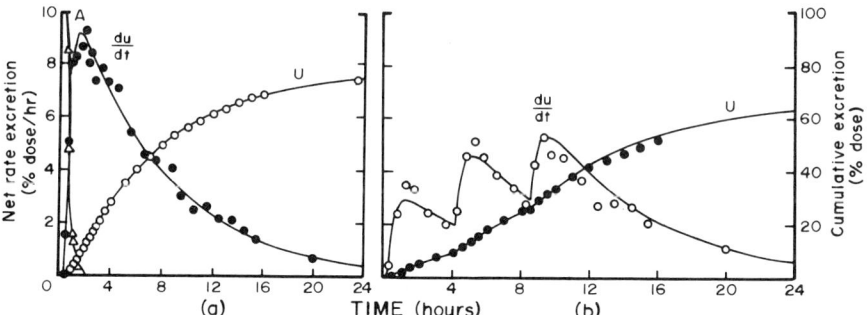

Fig. 19. Computer curves and experimental data points for the urinary excretion of amphetamine: (a) after oral administration of 15 mg (+)-amphetamine sulfate in "free pellet" form; (b) after a 3 x 5 mg dosage regimen of (+)-amphetamine sulfate orally in "free pellet" form. Reprinted from Ref. [43], p. 183, by courtesy of the Pharmaceutical Society of Great Britain.

where E_1^* represents an activated form of phosphofructokinase and E_2 is a combination of aldolase and triose phosphate isomerase. The reason for the oscillatory behavior of this system lies in the autocatalytic character of the reversible formation of $\underline{E}_1^*$. It slows down the production of (FDP) while decreasing the total concentration of (F6P) at the same time. A necessary condition for sustained oscillations to occur in a system is that it be steady state. Even then oscillations will take place only at a certain concentration of glucose. If the concentration of glucose decreases steadily during the experiment, only damped oscillations are observed (see Fig. 20). It is known that other glycolytic intermediates such as ATP and ADP are taking part in the simplified scheme above [44, 49]. This again causes the oscillations to change from a sustained to a damped mode. The detailed description of the analog computer program of this case has been published by Chance et al. [51].

Fukuda and Sugita [52-55] made extensive use of analog computers controlled by parallel logic in their analysis of various metabolisms. They have postulated a general scheme where the rate of change of metabolite m in compartment is given by the differential equation

$$\frac{dm_i}{dt} = \sum_j J_{ji} - \sum_k J_{ik}$$

where J_{ji} is the flux of m from compartment i to k, and J_{ji} and J_{ik} are functions of the metabolites m_i. Thus

$$J_{ji} = R_{ji} f_{ji}(\ldots, m_{i-1}, m_i, \ldots)$$

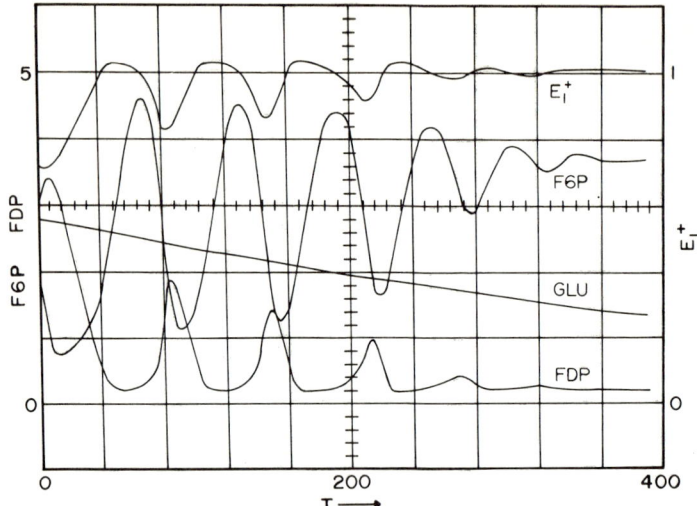

Fig. 20. Damped oscillations arising through depletion of glucose. Both time and concentration scales are in normalized units. Reprinted from Ref. [49], p. 992, by courtesy of the New York Academy of Sciences.

where f_{ji} is the ordinary kinetic function of the metabolites m_i which can be simulated on an analog computer. On the other hand, R_{ji} is the logic function of a particular metabolite m_i; it can assume only two values, i.e., logic 1 (existence) or 0 (nonexistence). In the biological system this corresponds to a repressor which operates only above a certain threshold concentration. This digital type of information is thought to be transmitted mainly on the genetic level. At the same time a kinetic (analog) type of information is working in the cytoplasm. The above-mentioned authors used this concept in their analog computer study of the enzymatic system of E. coli, in the study of regulation of tryphtophan synthetase, and in the study of metabolism of iodine in the thyroid gland.

V. MISCELLANEOUS APPLICATIONS

A. Kinetic Analysis

Let us assume two first-order competitive reactions having one of the products in common

$$A \xrightarrow{k_1} P + P_1 + \cdots$$
$$B \xrightarrow{k_2} P + P_2 + \cdots$$

(XVI)

4. Simulation of Kinetic Models

where for $t = 0$, $A = A_0$ and $B = B_0$. Corresponding differential equations are

$$\frac{dA}{dt} = -k_1 A \tag{23a}$$

$$\frac{dB}{dt} = -k_2 B \tag{23b}$$

$$\frac{dP}{dt} = k_1 A + k_2 B \tag{23c}$$

and the computer diagram is shown in Fig. 21. It is obvious that the shape of the P vs. t curves will depend not only on the values of rate constants k_1 and k_2 but also on initial concentrations A_0 and B_0. In other words, by knowing values of k_1 and k_2 potentiometers representing A_0 and B_0 can be varied systematically until agreement between the computed and experimental curves is obtained. Initial concentrations of species A and B can then be calculated from the potentiometer settings. Values of k_1 and k_2 can be determined beforehand in the same way using known concentrations of A_0 and B_0.

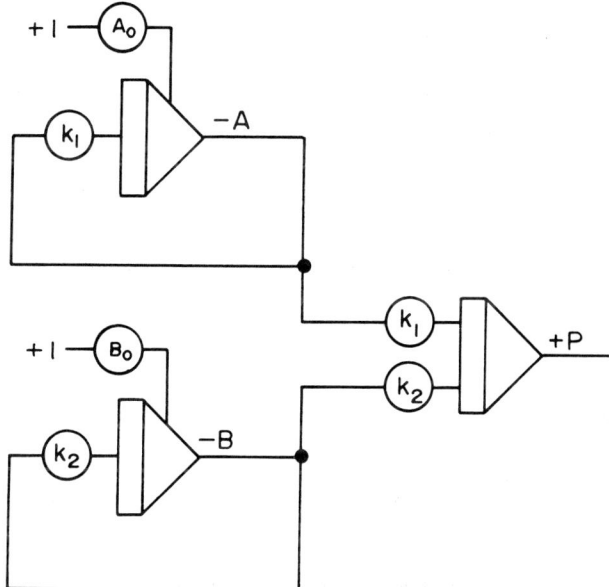

Fig. 21. Computer circuit diagram for two competitive reactions [type (XVI)].

The analysis of a mixture of ketones based on their reaction with hydroxylamine

$$R_1R_2CO + NH_3OH^+ \xrightarrow{k} R_1R_2CNOH + H^+ \qquad (XVII)$$

follows this principle [56]. The reaction is made pseudo-first-order by employing an excess of hydroxylamine hydrochloride. An electronic pH-stat holds the pH of the solution constant throughout the reaction by adding a small amount of sodium hydroxide solution whenever the pH drops. The output signal of the pH-stat which is proportional to the total amount of the ketones reacted is monitored and used for the evaluation. A special-purpose analog computer designed to solve the above-mentioned equations has been used [57]. An application of commercial analog computers to second-order kinetic analysis has been considered also by Mark et al. [58] and Janata [36]. Special-purpose analog computers have been constructed [59, 60] to solve simultaneous differential equations and operate on real-time basis.

It is fair to say that the trend now is away from analog toward digital instrumentation. This topic has been reviewed recently by Pardue [61].

B. Simulation of Log (K)-pH Profile

The hydrolytic degradation of both aspirin [62] and phenethicillin [63] can be represented by Eq. (24)

$$k_o = k_1H^+ + [(k_2 - k_1)H^+ + k_3 + k_4OH^-]\frac{K_a}{K_a + H^+} \qquad (24)$$

where k_o is the observed pseudo-first-order rate constant measured at a constant pH and K_a is the dissociation constant of respective acids. In this mechanism both the Bronsted acid and base forms undergo specific catalysis (rate constants k_1 and k_2). Only the base form is characterized by a spontaneous degradation (k_3) and by a specific base catalysis (k_4). An analog computer can be used to simulate the log(K)-pH profile.

Since the pH profile is a plot of log k_o as a function of pH (i.e., log H^+) it was found convenient to program Eq. (24) in the logarithmic form [64] (see Fig. 22)

$$\log(k_o) = \log\ \text{antilog}(F_1) + \text{antilog}(F_2)$$

$$+ \text{antilog}(F_3) + \text{antilog}(F_4) \qquad (25)$$

4. Simulation of Kinetic Models

where

$$F_1 = \log k_1 - pH = a - pH \tag{26a}$$

$$F_2 = \log(k_2 - k_1) - pH + \log \frac{K_a}{K_a + H^+} = b - pH + V_1 \tag{26b}$$

$$F_3 = \log k_3 + \log \frac{K_a}{K_a + H^+} = C + V_1 \tag{26c}$$

$$F_4 = \log k_4 - pOH + \log \frac{K_a}{K_a + H^+} = d - pOH + V_1. \tag{26d}$$

The independent variable, pH, is generated by integrating a constant. The function $K_a/(K_a + H^+)$ has been produced from the function generator V_1. In the original paper [64], the transducers LOG, ANTILOG, and the function generator V_1 are based on operational amplifiers with transistors in the feedback loop or at the input, respectively. The scaling of Eqs. (25) and (26) is done with respect to the type of transistors used; the original paper should be consulted for details. The characteristic log(K)-pH profile for the degradation of aspirin together with the experimental values is shown in Fig. 23.

It is convenient to use the computer in the repetitive operation mode and to display the results on an oscilloscope screen. The shape of the profile can then be varied continuously by adjusting potentiometers corresponding to individual rate constants. This exercise has been found particularly useful for lecture demonstration of drug decomposition kinetics.

C. Kinetics of Chemisorption

Although the question of its theoretical significance still remains open, the Elovich equation is often used in chemisorption studies [65]. According to the Elovich equation the rate of adsorption is governed by the differential equation

$$\frac{dq}{dt} = a \exp(-\alpha q) \tag{27}$$

where q is the amount of material absorbed at time t, and a and α are constants at a given temperature. The integrated forms of Eq. (27) for

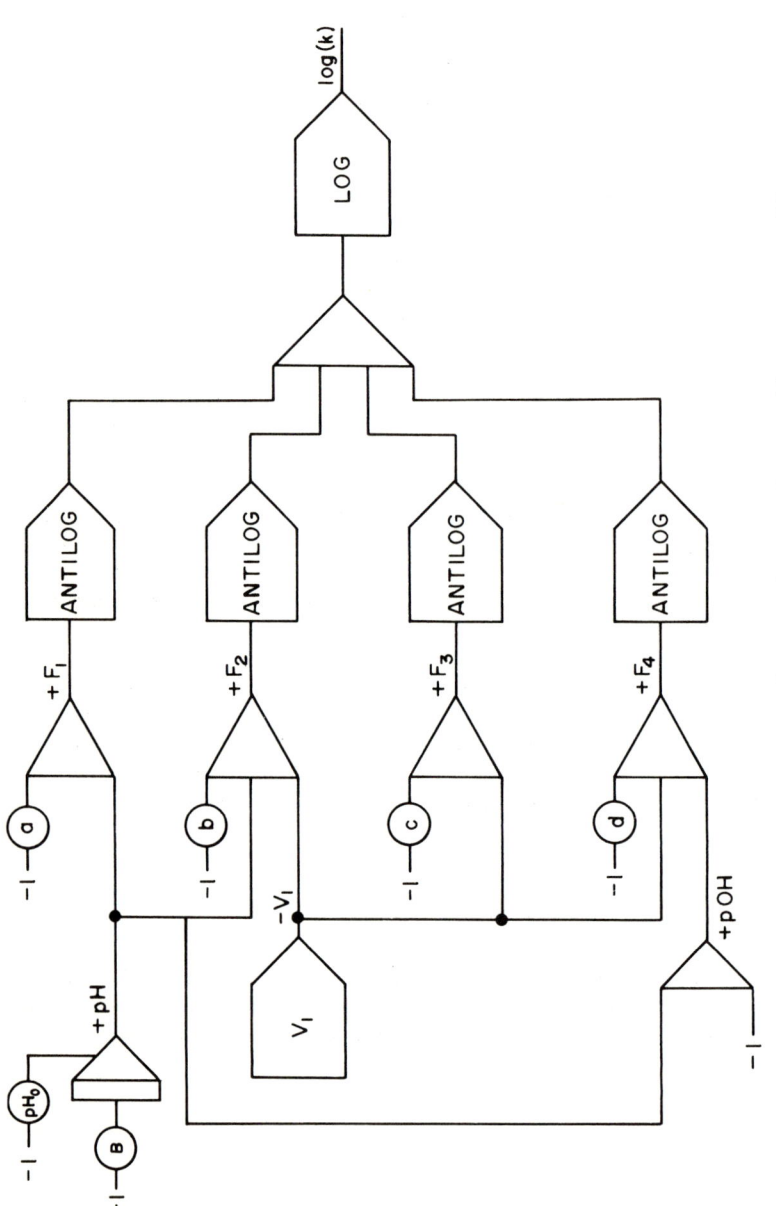

Fig. 22. Computer circuits diagram for simulation of log(K)-pH profile.

4. Simulation of Kinetic Models

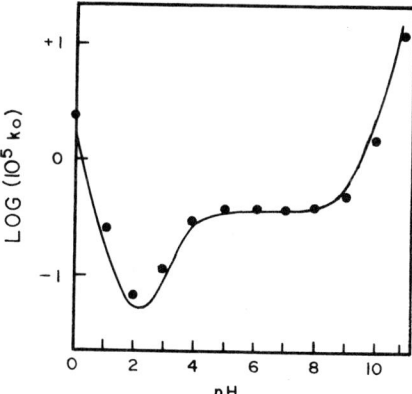

Fig. 23. Log(K)-pH profile characterizing aspirin degradation in aqueous solution at 25°C. Reprinted from Ref. [64], p. 863, by courtesy of the Division of Chemical Education, American Chemical Society.

various initial conditions are known; however, their application for evaluation of experimental data is rather laborious. It has been suggested that an analog computer can greatly facilitate this task.

The term $\exp(-\alpha q)$ can be generated from a function generator or by using standard components. To illustrate the flexibility in the programming, the latter approach will be demonstrated here. The derivative of the term $\exp(-\alpha q)$ is

$$\frac{d(\exp(-\alpha q))}{dt} = -\alpha \exp(-\alpha q) \frac{dq}{dt}, \qquad (28)$$

thus

$$\exp(-\alpha q) = -\alpha \int \exp(-\alpha q) \frac{dq}{dt} \cdot dt \qquad (29)$$

The term $\exp(-\alpha q)$ will therefore appear at the output of integrator 1 (see Fig. 24). This quantity, multiplied by the constant a, yields the value dq/dt according to Eq. (27), which in turn is then used to produce the right-hand side of Eq. (29). The output of amplifier 4 is integrated to give q, which is plotted against time. There are three possible initial conditions: Either there is no material adsorbed at $t = 0$ ($q_0 = 0$), or the adsorbed amount ($q_0 > 0$) does not or does comply with the exponential law. This

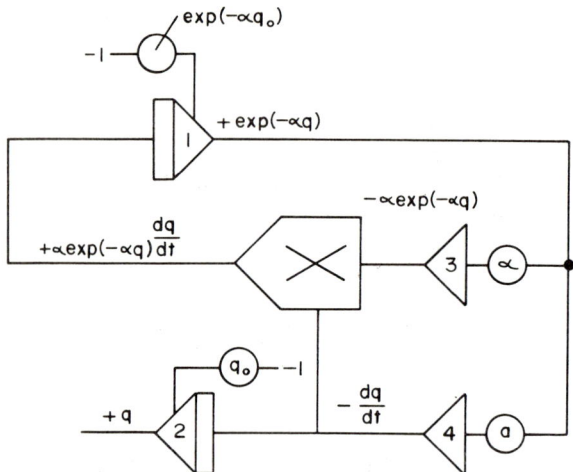

Fig. 24. Computer circuit diagram for solution of Elovich equation.

situation, which can make the classical evaluation so cumbersome, presents no problems in the analog computer evaluation as the initial conditions can be set easily on the appropriate integrators.

The adsorption of hydrogen on $2MnO \cdot Cr_2O_3$ was compared with calculated curves and the results are shown in Fig. 25. Analog computer studies of kinetics of adsorption controlled by diffusion on plane, spherical, and dropping electrodes have been carried out by Holub and Nemec [67, 68].

D. Nonisothermal Kinetics

It has been assumed in all examples discussed up to now that the reaction temperature remains constant. There are, however, many cases, particularly in engineering applications, when the reaction takes place under nonisothermal conditions. This problem has been discussed by Williams [69, 70], from the point of view of process control.

Let us assume reaction scheme

$$A + B \xrightarrow{k_1} C + D$$
$$A + C \xrightarrow{k_2} E$$

(XVIII)

4. Simulation of Kinetic Models

where for $t = 0$: $A = A_0$, $B = B_0$, and $C = D = E = 0$. Reaction velocities for the two reactions are

$$V_1 = -\left(\frac{dA}{dt}\right)_1 = \frac{dB}{dt} = \left(\frac{dC}{dt}\right)_1 = k_1 A^{a_1} B^{b_1} \tag{30a}$$

$$V_2 = -\left(\frac{dA}{dt}\right)_2 = -\left(\frac{dC}{dt}\right)_2 = k_2 A^{a_2} C^{c_2} \tag{30b}$$

and the Arrhenius equations for the respective rate constants are

$$k_1 = k_1^\circ \exp\left(-\frac{E_1}{RT}\right) \tag{31a}$$

$$k_2 = k_2^\circ \exp\left(-\frac{E_2}{RT}\right). \tag{31b}$$

For the purpose of programming it is convenient to rewrite Eqs. (30) and (31) in logarithmic form. Thus

$$\ln v_1 = \ln k + a_1 \ln A + b_1 \ln B \tag{32a}$$

$$\ln v_2 = \ln k_2 + a_2 \ln A + c_2 \ln C \tag{32b}$$

$$\ln k_1 = \ln k_1^\circ - \frac{E}{R} \cdot \frac{1}{T} \tag{32c}$$

$$\ln k_2 = \ln k_2^\circ - \frac{E_2}{R} \cdot \frac{1}{T} \tag{32d}$$

where a_1, a_2, b, and c_2 are the (fractional) reaction orders [71]. Total concentration changes of any species are then derived from Eqs. (30a) and (30b). Therefore

$$\frac{dA}{dt} = \left(\frac{dA}{dt}\right)_1 + \left(\frac{dA}{dt}\right)_2 = -v_1 - v_2 \tag{33a}$$

$$\frac{dB}{dt} = -v_1 \tag{33b}$$

$$\frac{dC}{dt} = \left(\frac{dC}{dt}\right)_1 + \left(\frac{dC}{dt}\right)_2 = v_1 - v_2. \tag{33c}$$

The actual form of the temperature change depends on the experimental arrangement. It can vary in discrete steps (e.g., variable isothermal) or

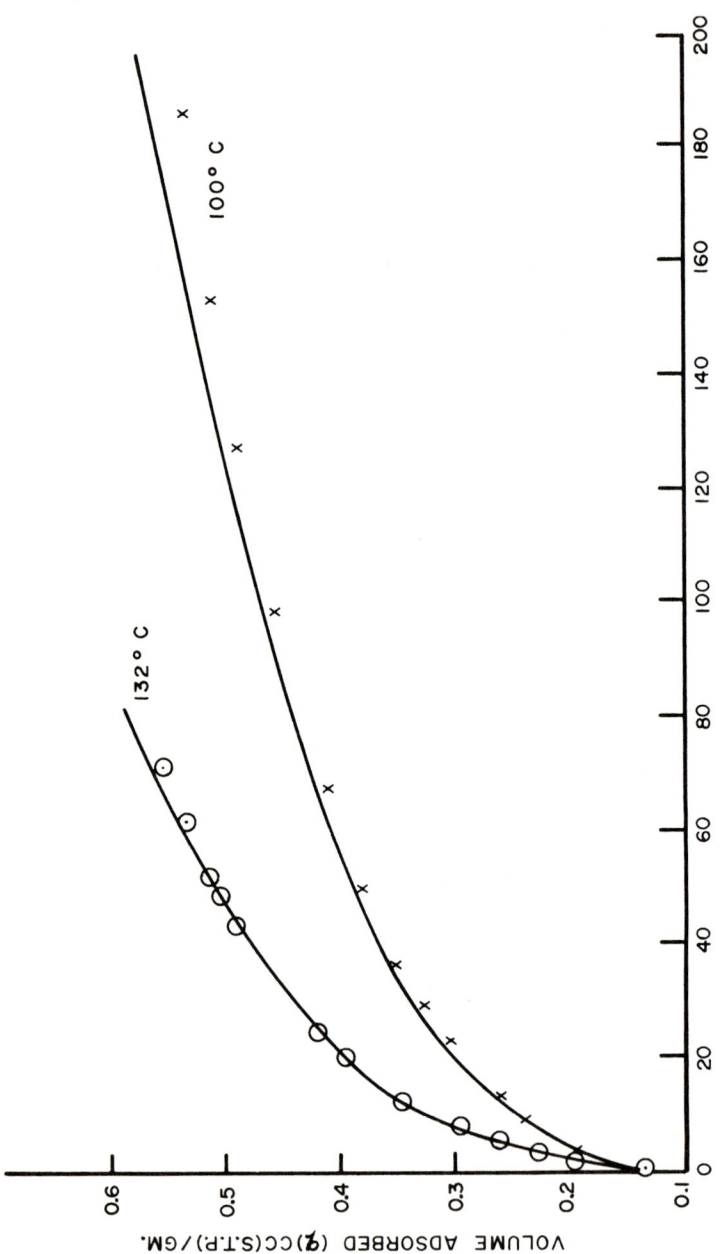

Fig. 25. Experimental and computed curves for chemisorption of H_2 on $2MnO \cdot Cr_2O_3$. The equation parameters are as follows: for $t = 132°C$: a (ml g^{-1} min^{-1}) = 0.0364, α (g ml^{-1}) = 6.92, $\circ$ (ml g^{-1}) = 0.13; for $t = 100°C$: a (ml g^{-1} min^{-1}) = 0.0166, α (g ml^{-1}) = 8.37, $\circ$ (ml g^{-1}) = 0.14. Reprinted from Ref. [66], p. 138, by courtesy of the Instruments and Control Systems.

4. Simulation of Kinetic Models

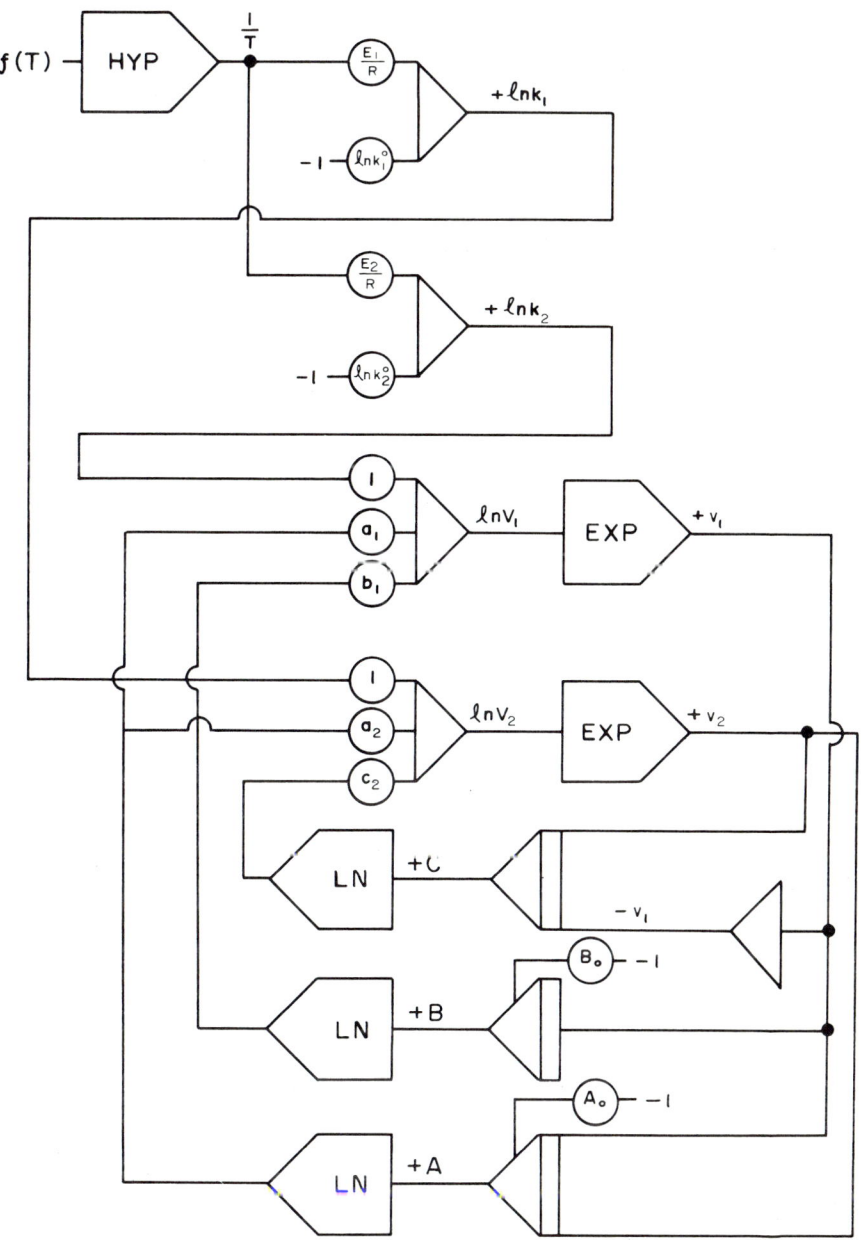

Fig. 26. Computer circuit diagram for simulation of nonisothermal kinetics [type (XVIII)].

it can assume a complex function if a heat transfer is involved and/or if the reaction is adiabatic. Nevertheless, in all cases a hyperbolic function E/RT must be generated. The output of the hyperbolic function generator (see Fig. 26) is then summed with constants representing $\ln k_1^°$ and $\ln k_2^°$ to give $\ln k_1$ and $\ln k_1$ and $\ln k_2$, respectively. Equations 4 are then realized by summing logarithms of concentrations of A, B, and C with $\ln k_1$ and $\ln k_2$. Exponentiation and integration which follow this step provide the concentrations of A, B, and C.

It is obvious that this exercise requires a number of nonlinear elements (two exponential, three logarithmic, and one hyperbolic function generator in our example). It is cases such as this where a digitally simulated analog computer is particularly useful.

Other applications of analog computers into simulation of kinetic systems include the study of ion-pair recombination rates in low dielectric constant liquids [72], simulation of a reaction of zero order taking place in a laminar catalyst under nonisothermal conditions [73], and the study of energy partitioning in a linear collision model for the reaction of Cs with alkyl iodides [74].

REFERENCES

[1]. R. G. E. Franks and W. E. Schiesser, Chem. Eng. Prog., 63: No. 4, 68 (1967).

[2]. R. D. Brennan and T. Z. Fahidy, Instrum. Control Systems, 39:133 (1966).

[3]. G. J. Farris and L. E. Burkhart, Chem. Eng. Prog. Symp. Ser., 61:16 (1965).

[4]. N. Markin, Computer Design, 7:38 (1968).

[5]. J. R. Hurley, Proceedings of the 1963 Spring Joint Computer Conference, 69, 000.

[6]. R. A. Gaskill, J. W. Harris, and A. L. McKnight, Proceedings of the 1963 Spring Joint Computer Conference, 83, 000.

[7]. G. A. Korn and T. M. Korn, Electronic Analog and Hybrid Computers, McGraw-Hill, New York, 1964.

4. Simulation of Kinetic Models

[8]. A. S. Jackson, Analog Computation, McGraw-Hill, New York, 1964.

[9]. J. R. Ashley, Introduction to Analog Computation, Wiley, New York, 1963.

[10]. EA1 Handbook of Analog Computation, Vol. I, Electron Associates Inc., Princeton, N.J., 1967.

[11]. M. Yano and T. Imoto, Kagaku, 17:728 (1962).

[12]. L. Boydzhirev, Khim. Ind. (Sofia), 408 (1968); through CA, 70: 84167v (1968).

[13]. H. Roepke and J. Riemann, Analog Computer in Chemie und Biologie, Springer, Berlin, 1969.

[14]. H. Langemann, Dechema Monograph, 53:161 (1964).

[15]. T. R. Crossley and M. A. Slifkin, Progress in Reaction Kinetics, 5:409 (G. Porter, ed.), Pergamon, New York, 1970.

[16]. R. Repges and W. Boguth, Ber., 69:638 (1965).

[17]. V. V. Biryukov and V. V. Kafarov, Zh. Fiz. Khim, 38:2023 (1964).

[18]. S. J. Wajc, A. H. Mansour, and R. Jottrand, Rev. Inst. Franc. Petrole Ann. Combust. Liquides, 20 (5):849 (1965).

[19]. T. Suzuki and Y. Tanimura, Chem. Pharm. Bull., 15:1055 (1967).

[20]. B. Bonath, B. Foertsch, and R. Saemann, Chem.-Ing.-Tech., 38:739 (1966).

[21]. R. O. Butterfield, E. D. Bitner, C. R. Scholfield, and H. J. Dutton, J. Am. Oil Chem. Soc., 41:29 (1964).

[22]. J. E. R. Rijnsdorp, R. Vichnevetsky, and J. G. van de Vusse, Analog Comp. Appl. Study Chem. Process., Int. Sem. Proc. Brussels, 136 (1960); through CA, 57:14475c (1960).

[23]. T. R. Crossley and M. A. Slifkin, Educ. Chem., 4:280 (1967).

[24]. A. Schmidt, Plaste Kaut., 16:332 (1969).

[25]. F. R. Mayo and F. M. Lewis, J. Am. Chem. Soc., 66:1594 (1944).

[26]. P. Bekiaroglou, Z. Phys. Chem., N. F. 64:263 (1969).

[27]. P. Bekiaroglou, Chem.-Ing.-Tech., 40:811 (1968).

[28]. W. E. Walles and A. E. Platt, Ind. Eng. Chem., 59:41 (1967).

[29]. D. Blackmore, Trans. Faraday Soc., 62:415 (1966).

[30]. J. Eggers, Z. Elektrochem., 61:1210 (1957).

[31]. A. Carlson, Instrum. Control Systems, 38:147 (1965).

[32]. M. Lucquin, J. Monastier, F. Langrand, A. Perez, and A. Perche, J. Chim. Phys. Physicochim. Biol., 66:1714 (1969).

[33]. A. A. Frost and R. G. Pearson, Kinetics and Mechanism, 2nd ed., Wiley, New York, 1961.

[34]. C. Walter, Enzyme Kinetics, Ronald, New York, 1966.

[35]. T. Matthews, Chem. Eng., 71:77, 93, 137 (1964); 72:79 (1965).

[36]. J. Janata and H. B. Mark Jr., Electroanalytical Chemistry, 3:1 Dekker, New York, 1969.

[37]. J. Janata and O. Schmidt, J. Electroanal. Chem., 11:224 (1966).

[38]. J. Janata, O. Schmidt, and P. Zuman, Collect. Czech. Chem. Commun., 31:2344 (1966).

[39]. G. O'Dom and Q. Fernando, Anal. Chem., 38:844 (1966).

[40]. J. Janata, Thesis, Charles University, Prague, 1965.

[41]. R. F. Bradley and J. A. Roth, Chem. Eng. Prog., 63(4):67 (1967).

[42]. R. E. Cover, J. Phys. Chem., 71:1990 (1967).

[43]. A. H. Beckett and G. T. Tucker, J. Pharm. Pharmacol., 20:174 (1968).

4. Simulation of Kinetic Models

[44]. C. Walter, Ref. [34].

[45]. B. C. Goodwin, Adv. Enzymes Reg., 3:425 (1965).

[46]. F. Heinmets, Math. Biosci., 3:175 (1968).

[47]. R. A. Spangler and F. M. Snell, Nature, 191:457 (1961).

[48]. C. Walter, Nature, 209:404 (1966).

[49]. J. Higgins, Proc. Natl. Acad. Sci, 51:989 (1964).

[50]. K. Yokota, Kagaku To Seibutsu, 8:259 (1970); through CA, 73:94937t (1970).

[51]. B. Chance, D. S. Greenstein, J. Higgins, and C. C. Yang, Arch. Biochem. Biophys., 37:322 (1957).

[52]. N. Fukuda, J. Theor. Biol., 1:440 (1961).

[53]. M. Sugita, J. Theor. Biol., 4:179 (1963).

[54]. M. Sugita and N. Fukuda, J. Theor. Biol., 5:412 (1963).

[55]. M. Sugita, J. Theor. Biol., 13:330 (1966).

[56]. E. C. Torren and M. K. Gnuse, Anal. Letters, 1:295 (1968).

[57]. E. C. Torren and J. E. Davis, Anal. Letters, 1:289 (1968).

[58]. H. B. Mark, Jr., L. J. Papa, and C. N. Reilley, Advances in Analytical Chemistry and Instrumentation (C. N. Reilley, ed.), Vol. 2:255, Wiley-Interscience, New York, 1963, p. 255.

[59]. D. Pinkel and H. B. Mark, Jr., Talanta, 12:491 (1965).

[60]. T. E. Weichselbaum, W. H. Plumpe, R. E. Adams, J. C. Hagerthy, and H. B. Mark, Jr., Anal. Chem., 41:725 (1969).

[61]. H. L. Pardue, Advances in Analytical Chemistry and Instrumentation (C. N. Reilley, ed.), 7, 141 Wiley-Interscience, New York, 1969.

[62]. E. R. Garrott, J. Am. Chem Soc., 79:3401 (1957).

[63]. M. A. Schwartz, A. P. Granatek, and F. H. Buckwalter, J. Pharm. Sci., 51:523 (1962).

[64]. N. G. Lordi, J. Chem. Educ., 12:861 (1969).

[65]. J. N. Sarmousakis and M. J. D. Low, J. Chem. Phys., 25:178 (1956).

[66]. J. T. Sommerfeld, Instrum. Control Systems, 41(4):137 (1968).

[67]. K. Holub and L. Nemec, J. Electroanal. Chem., 11:1 (1966).

[68]. K. Holub and L. Nemec, J. Electroanal. Chem., 18:209 (1968).

[69]. T. J. Williams, Control Eng., 5(7):100 (1958).

[70]. T. J. Williams, System Engineering for Process Industries, McGraw-Hill, New York, 1961.

[71]. Ref. [17].

[72]. P. K. Ludwig, J. Chem. Phys., 50:1787 (1969).

[73]. K. Oestergaard, Dan. Kemi, 44:27 (1963).

[74]. C. Ottinger, J. Chem. Phys., 51:1170 (1969).

APPENDIX

A. Reviews of Engineering Applications

[75]. T. Ankel and P. Wolf, Chem.-Ing.-Tech., 41:951 (1969). Application of analog computers to problems of control and process engineering.

[76]. P. C. Distefano and W. Richards, Chem. Eng., 75:195 (1968). Hybrid computers in the control and process industry.

[77]. A. Frank and L. Lapidus, Chem. Eng. Prog., 62:66 (1966). Hybrid simulation of chemical engineering systems.

[78]. B. Hedstroem, Svensk. Kem. Tidskr., 78:500 (1966). The use of digital and analog computers in chemical engineering.

[79]. J. Malenge, Ind. Chem. Belge, 30:1319 (1965). Hybrid computation, a promising technique for chemical engineering.

4. Simulation of Kinetic Models   263

[80]. T. Matthews, Ref. [35].

[81]. T. J. Williams, Ind. Eng. Chem., 55 (11):43 (1963). Computers, automation, and process control.

[82]. T. J. Williams, Met. Soc. Conf., 32:1 (1964). Computer simulation — its place and accomplishments in process control.

[83]. T. J. Williams, Ind. Eng. Chem., 57 (12):33 (1965). Process control.

[84]. T. J. Williams, Ind. Eng. Chem., 61(1):76 (1969). Computers and process control.

[85]. J. W. Womack, Chem. Process Eng., 45:72 (1964). The analog computer as a chemical engineering tool.

B. Biochemical Applications

[86]. E. Ackerman, E. Strickland Jr., J. B. Hazelrig, and L. G. Gatewood, Clin. Pharmacol. Ther., 8:170 (1967). Computers in biomathematical applications.

[87]. B. E. Ballard and J. E. Goyan, Med. Biol. Eng., 4:483 (1966). Application of analog computer techniques to in vivo drug kinetics.

[88]. A. H. Beckett, R. N. Boyes, and P. J. Appleton, J. Pharm. Pharmacol. Suppl., 18:76 (1966). The metabolism and excretion of lignocaine in man.

[89]. E. Cerasi, Acta Endocrinol., 55:163 (1967). An analog computer model for the insulin response to glucose infusion.

[90]. B. Chance, J. Biol. Chem., 235:2440 (1960). Analog and digital representation of enzyme kinetics.

[91]. J. T. Dolukio, W. G. Crouthamel, G. H. Tan, J. V. Swintosky, and L. W. Dittert, J. Pharm. Sci., 59:72 (1970). Drug absorption. III, Effect of membrane storage on the kinetics of drug adsorption.

[92]. B. R. Fish, Health Phys., 1:276 (1958). Application of an analog computer to analysis of distribution and excretion data.

[93]. E. R. Garrett, A. J. Aglen, and A. J. Lambert, Int. Z. Klin. Pharmakol. Theor. Toxikol 1-14 (1967); through CA, 69: 94722e (1967). Pharmacokinetic analysis of receptor-site models in multicompartmental systems.

[94]. E. R. Garrett and C. D. Alway, Int. Congr. Chemotherapy Proc. 3rd, Stuttgart, 2:1666 (1963). Drug distribution and dosage. Complex pharmacokinetic models and the analog computer.

[95]. E. R. Garrett and H. J. Lambert, J. Pharm. Sci., 55:626 (1966). Analog computers in drug dosage and formulation design.

[96]. E. R. Garrett, P. C. Thomas, D. P. Wallock, and C. D. Alway, J. Pharmacol. Exp. Ther., 130:106 (1960). Psicufuranine, kinetics, and mechanism in vivo with the application of the analog computer.

[97]. R. G. Janes and J. O. Osburn, J. Physiol., 181:59 (1965). Analysis of glucose measurements by computer simulation.

[98]. V. L. Kaler and A. A. Sergeev, Dokl. Akad. Nauk Belorus. SSR, 11:551 (1967); through CA, 68:503v (1967). Simulation of the simplest enzymic reaction in an open system on analog computer.

[99]. W. Kaplan, EAI Applications Reference Library 4.4.10a (1968). Pharmacokinetics.

[100]. G. Knowles, A. L. Downing, and M. J. Barnett, J. Gen. Microbiol., 38:263 (1965). Determination of kinetic constants for nitrifying bacteria in mixed culture with the aid of an electronic computer.

[101]. J. A. Luetscher, C. A. Camargo, R. A. Cheville, E. W. Hancock, A. J. Dowdy, and G. W. Nobes, Steroid Dyn. Proc. Symp. Tokyo, 341 (1965); through CA, 68:19260k (1965). Conjugation and excretion of aldosterone; testing of models with an analog computer.

[102]. W. W. Mapleson, J. Appl. Physiol., 18:197 (1963). An electronic analog for uptake and exchange of inert gases and other agents.

[103]. P. Pejovic, Automatika (Ljubljana), 6:1069 (1965); through CA, 63:18505a (1965). Application of general-purpose analog computers in the modeling of biochemical experiments.

[104]. E. B. Reeve and L. J. Kulhanek, Phys. Bases Circ. Transp. Regul. Exch. Proc. Conf., Denver, Colorado, 1966, 151; through CA, 68:56588w (1966). Regulation of body water content.

4. Simulation of Kinetic Models

[105]. J. G. Reich, Eur. J. Bioch., 6:395 (1968). Analog computer analysis of tracer flow patterns throughout the glycolytic and related pathways in erythrocytes and other intact metabolic systems.

[106]. J. Riemann, H. Roepke, E. Gerhardt, K. H. Kolb, and H. Gibian, Arzheimforsch., 18:1443 (1968). Simulated biodynamics of glycodiazine using an analog computer.

[107]. J. Riemann, H. Roepke, K. Kieslich, H. J. Koch, and H. Gibian, Eur. J. Bioch., 6:60 (1968). Simulation of a microbiological hydroxylation with an analog computer.

[108]. J. R. Robinson and S. P. Eriksen, J. Pharm. Sci., 55:1254 (1966). Theoretical formulation of sustained-release dosage forms.

[109]. K. Rommel, Verk. Dtsch. Ges. Pathol., 53:58 (1969). Analog computer analysis of intestinal absorption using galactose absorption as an example.

[110]. K. Rommel and K. Grimmel, Klin. Wochenschr., 47:541 (1969). Analog computer analysis of galactose conversion in persons with and without hepatic diseases.

[111]. U. M. Stauffer, Clin. Pharmacol. Ther., 8:194 (1967). Computers in physiological analysis and simulation.

[112]. T. Suzuki, Kagaku No Ryoiki, 19:297 (1965); through CA, 63:16125f (1965). The analog computer and its applications to pharmacology.

[113]. J. G. Wagner, Clin. Pharmacol. Ther., 8:201 (1967). Use of computers in pharmacokinetics.

[114]. G. R. Wilkinson and A. H. Beckett, J. Pharm. Sci., 57:1933 (1968). Absorption, metabolism, and excretion of the ephedrines in man. II, Pharmacokinetics.

[115]. W. G. Yeisley and E. C. Pollard, J. Theor. Biol., 7:485 (1964). An analog computer study of differential equations concerned with bacterial cell synthesis.

Chapter 5

THE APPLICATION OF THE MONTE CARLO METHOD
TO CHEMICAL KINETICS

John J. Manock

Chemistry Department
Western Carolina University
Cullowhee, North Carolina 28723

I. INTRODUCTION 267

II. SIMULATION OF CHEMICAL REACTIONS 269

    A. First-Order Rate Equations 269
    B. Consecutive First-Order Reactions 271
    C. Reversible Reactions 277
    D. Second-Order Rate Equations 278
    E. Complex Reactions 283

III. CONCLUSION 288

    A. Chemical Kinetics 288
    B. Statistical Considerations 290
    C. Other Uses 290

    REFERENCES 291

I. INTRODUCTION

In recent years there have been numerous articles and books written on the Monte Carlo method. For the individual who desires a detailed description, the author refers him to a few of these works [1-6] for a general treatment on its mathematical justification and limitations. The purpose of this chapter is not to develop the general mathematical foundation of the Monte Carlo method, but rather to apply this method of

Copyright © 1973 by Marcel Dekker, Inc. All Rights Reserved. Neither this work nor any part may be reproduced or transmitted in any form or by any means, electronic or mechanical, including photocopying, microfilming, and recording, or by any information storage and retrieval system, without permission in writing from the publisher.

simulation specifically to the area of chemical kinetics and illustrate its justification and limitations with regard to reaction simulation.

Basically, the Monte Carlo method (or method of statistical trials) can be used to simulate the final results of a series of random events, where each particular event has a certain probability of occurring. These events can occur simultaneously, consecutively, or repeatedly. The result obtained from the Monte Carlo method is the number of "successful" events found compared to the number of random events considered. To obtain a statistically accurate result a large number of random events must be considered. This requirement implies the use of a digital computer, and indeed, a digital computer is necessary to obtain a statistically accurate simulation by the Monte Carlo method.

The Monte Carlo method can be applied to two specific types of problems.

1. Problems involving analytical expressions that are converted to an equivalent statistical problem by randomly choosing samples and then obtaining the probability of a particular event occurring. An example of this type of problem is the use of the Monte Carlo method to integrate an analytical function [7].

2. The simulation of a physical system by a random process such as the propagation of neutrons through matter [8]. Previously, a random process was considered solved when it could be reduced to an analytical expression, but with the use of the Monte Carlo method this restriction is not required. The simulation of chemical reactions by the Monte Carlo method belongs to this classification.

There have been several articles recently which have applied the Monte Carlo method of simulation to chemical kinetics [9-12] and although this method is not considered new it has become more useful to the chemist as a tool to simulate chemical reactions with the increasing number of digital computers now available. In applying this method to chemical reactions, it is not necessary to solve differential equations and one does not need to incorporate the restraints of differential equations nor to accept the inherent errors of steady-state approximations in order to solve difficult differential equations in the closed form [13].

Let us now consider some examples of simulating chemical reactions with the Monte Carlo method. In each example comparison will be shown between the integrated results and the Monte Carlo results.

5. Application of the Monte Carlo Method

II. SIMULATION OF CHEMICAL REACTIONS

A. First-Order Rate Equations

In developing the application of the Monte Carlo method to the area of chemical kinetics, consider first the simplest example of a monomolecular reaction

$$A \xrightarrow{k_1} B$$

where the reactant molecule A is decomposing by first-order kinetics to form the product molecule B.

To simulate this reaction by the method of statistical trials, the reaction vessel is represented as a storage array of numbers in the computer memory bank with digits representing the corresponding molecules. In the example chosen, 1's may be placed in the storage array to represent the molecule A. The initial molar concentration of A is proportional to the initial number of 1's present in the storage array, i.e., $(1)_0 \propto (A)_0$ or $(1)_0 = \alpha (A)_0$. Nonreactive molecules, such as solvent molecules or inhibitors, may be represented by the digit 0. When these digits are randomly chosen, no reaction will result. To be completely representative of a reaction vessel consisting of reactant and solvent molecules, the digits representing the reactant molecules should be placed randomly in the storage array. However, this is not necessary since the final selection of the digits will be random.

As the reaction is started, an area in the storage array is randomly selected and tested for the digit 1. If the digit 1 is found, then a "fruitful" event has occurred, which corresponds to the molecule A reacting. The digit 1 is replaced with the digit 2, representing the product molecule B, and in addition, an element of time is noted. If the randomly chosen digit is not the digit 1, then a "fruitful" event has not occurred, and the passage of an element of time is simply noted. As discussed in the Introduction this is the result desired from the Monte Carlo method, which is the number of successful events found relative to the number of random events considered.

To be more quantitative, assume there are 1000 storage areas in the storage array representing the reaction vessel and a digit 1 is initially placed in each storage area. Random numbers between 1 and 1000 are generated. Suppose that the first random number generated is 324. The 324th storage area is tested for the digit 1, a "fruitful" event is noted, and this storage area is filled with the digit 2. Later, the number 324 is again randomly generated, but this time the digit 2 is found and in this case no reaction occurs and an element of time is recorded.

Evaluating this model statistically, one would state that the probability of a particular digit being randomly chosen per unit of time would be a constant value which would depend on the size of the storage array. In the example cited with 1000 areas in the storage array and random numbers between 1 and 1000 being generated, the probability of a particular storage area being selected would be 1/1000.

The rate equation for this model, which would be the rate of the digit 1 being replaced by the digit 2, would be represented mathematically as

$$\frac{-d(1)}{dN_1} = \beta (1) \qquad (1)$$

where β is the probability constant for a particular molecule being chosen, and N_1 is the number of attempts per unit of time to find digit 1. Obviously, as the digit 1 becomes replaced by the digit 2, the rate of the digit 1 being replaced by the digit 2, or the rate of the reaction, will decrease.

Integration of Eq. (1) gives the result

$$\ln \frac{(1)_0}{(1)} = \beta N_1 \qquad (2)$$

where $(1)_0$ is the initial "concentration" of digits representing the molecule A, and (1) is the "concentration" of digits representing the molecule A at time t. Since $(1)_0 = \alpha (A)_0$ and $(1) = \alpha (A)$, Eq. (2) can be written in the usual form for first-order rate equations

$$\ln \frac{(A)_0}{(A)} = \beta N_1 \qquad (3)$$

with the proportionality constants α canceling.

Equating this result from the Monte Carlo model to the result obtained from the first-order rate equation

$$\ln \frac{(A)_0}{(A)} = k_1 t \qquad (4)$$

5. Application of the Monte Carlo Method

yields the identity

$$\beta N_1 = k_1 t. \tag{5}$$

Equation (5) shows that the rate constant for a first-order reaction times a unit of time is equal to the probability of a given molecule being randomly chosen times the number of attempts to find a molecule per unit of time. Thus, a first-order reaction having a certain specific rate constant can be simulated such that the time scale for the Monte Carlo model coincides with the real time scale.

This Monte Carlo model is totally analogous to a monomolecular reaction, such as illustrated by a radioactive decay process where a given molecule has a constant probability of decaying per unit of time. The rate of the reaction for a radioactive decay process is equal to the number of molecules present times the probability of a particular molecule decaying. Basically, this is the physical picture represented by the Monte Carlo model.

In developing systems which consist of combinations of first-order and pseudo-first-order reactions, such as parallel, consecutive, and/or reversible, or any combination of these, the model already formulated has simply to be expanded. Since the mathematics required to solve the differential equations for these systems in the closed form can become prohibitive, the Monte Carlo model becomes extremely powerful and useful.

Two examples of common first-order reactions, i.e., first-order consecutive reactions and first-order reversible reactions, are now analyzed along with the computer logic applied to simulate these reactions. In addition, comparisons between the integrated results and the Monte Carlo results are also shown. After consideration of these two examples, the Monte Carlo model can readily be extended to more complex first-order systems.

B. Consecutive First-Order Reactions

In the case of the first-order consecutive reaction

$$A \xrightarrow{k_1} B \qquad \text{Branch 1}$$

$$B \xrightarrow{k_2} C \qquad \text{Branch 2}$$

the rate equations for this reaction are

$$\frac{-d(A)}{dt} = k_1(A) \tag{6}$$

$$\frac{d(B)}{dt} = k_1(A) - k_2(B) \tag{7}$$

and

$$\frac{d(C)}{dt} = k_2(B). \tag{8}$$

With $(A) = (A)_0$, $(B) = 0$, and $(C) = 0$ when $t = 0$, these rate equations can be integrated to give the following equations for the concentrations of A, B, and C as functions of time

$$(A) = (A)_0 \exp(-k_1 t) \tag{9}$$

$$(B) = \frac{k_1 (A)_0}{k_2 - k_1} \left[\exp(-k_1 t) - \exp(-k_2 t) \right] \tag{10}$$

and

$$(C) = (A)_0 \left[1 + \frac{1}{k_1 - k_2} \left[k_2 \exp(-k_1 t) - k_1 \exp(-k_2 t) \right] \right]. \tag{11}$$

Although the mathematics required to obtain these solutions is not necessarily prohibitive, the substitution required to solve the rate equation for (B) may be somewhat difficult for junior-senior chemistry students. In addition, it is difficult for the student to visualize (C) as a function of time.

With the Monte Carlo model this first-order consecutive reaction is represented by the rate equations

$$\frac{-d(1)}{dN_1} = \beta_1 (1) \tag{12}$$

and

$$\frac{-d(2)}{dN_2} = \beta_2 (2) \tag{13}$$

5. Application of the Monte Carlo Method

where N_1 is the number of attempts to find molecule A per unit of time and N_2 is the number of attempts to find molecule B per unit of time. If the molecules (or digits) are stored in the same sample array, the probability factors β_1 and β_2 are equal since the individual probability factor of a specific molecule being randomly selected is dependent only on the size of the sample array.

Integration of these two rate equations gives the same results as previously obtained,

$$\beta N_1 = k_1 t \tag{14}$$

and

$$\beta N_2 = k_2 t. \tag{15}$$

The ratio of these two equations produces the result

$$\frac{N_1}{N_2} = \frac{k_1}{k_2}. \tag{16}$$

Thus, when considering a system which consists of several first-order reactions, the branches need to be tested only in the ratio of the corresponding relative rate constants.

A flow chart which illustrates a method to solve this particular problem with a digital computer is shown in Fig. 1. The logic developed in the flow chart can be applied to simulate any combination of the general first-order reaction

$$A \rightleftharpoons B \rightleftharpoons C \rightleftharpoons D, \ldots$$

The computer program, written in FORTRAN IV, corresponding to this flow chart is in current use, and a copy can be obtained from the author upon request.

The logic of the flow chart can be illustrated best by considering a specific reaction, such as the first-order consecutive reaction previously discussed

$$A \xrightarrow{1.0} B \xrightarrow{2.0} C.$$

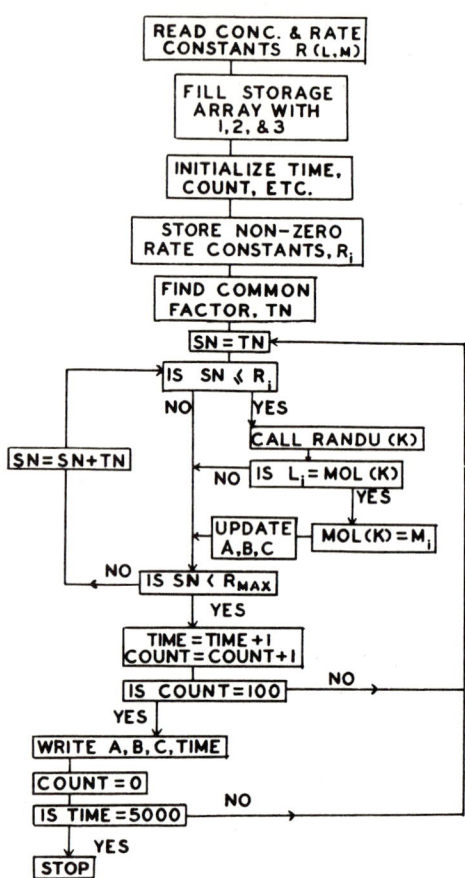

Fig. 1. Flow chart for the general first-order reaction simulation by the Monte Carlo method.

In the following discussion digit 1 will correspond to molecule A, digit 2 will correspond to molecule B, and digit 3 will correspond to molecule C. The storage array is initially filled with 1's and the time is initialized to zero. The nonzero rate constants R(L, M), where L corresponds to the reactant molecule and M corresponds to the product molecule, are provided by the investigator. In the example chosen, R(1, 2) = 1.0 and R(2, 3) = 2.0. The largest common factor among the nonzero rate constants is determined, and is referred to as the variable TN in the flow

5. Application of the Monte Carlo Method

chart. Initially, the variable SN, which will be referred to as the incremental test factor, is set equal to the largest common factor TN.

Each specific nonzero rate constant is then tested in succession and compared to the variable SN, and if a specific rate constant is larger than or equal to this variable, that particular molecule can react. A random number within the range of the limits of the storage array is generated from the subroutine, RANDU, which generates pseudorandom numbers. RANDU is part of the IBM scientific subpackage. The digit corresponding to this storage area is tested to see if it is the molecule desired. If it is the desired molecule, a "fruitful" event has occurred and this reactant is replaced with the corresponding product. If the randomly chosen molecule is not the molecule desired, no "fruitful" event occurs and the next nonzero rate constant is considered. When all nonzero rate constants have been tested, SN is incremented by the incremental test factor TN, and the procedure is repeated. When SN becomes larger than a particular nonzero rate constant, this branch is no longer tested during this unit of time. When SN becomes larger than the maximum nonzero rate constant, an element of time is noted, after which SN is initialized to TN and the entire process is repeated.

In the specific example cited, the largest common factor TN is 1.0. Referring to the flow chart, in the first cycle the rate constant R(1, 2) is equal to SN; thus, molecule A can react, and then a molecule is randomly chosen from the storage area and tested to see if it is molecule A. If the randomly chosen digit is a 1, then a "fruitful" event has occurred and this storage area is filled with the digit 2. The concentrations of A, B, and C are updated. If the randomly chosen digit is not 1, then the reaction is simply continued. Also, in the first cycle, R(2, 3) is greater than SN and digit 2 is considered in the same manner. However, in the second cycle, SN is equal to 2.0, and thus R(1, 2) is less than SN, and only R(2, 3) is considered. At this point SN is made equal to 3.0, none of the branches are considered, and an increment of time is added. SN is then initialized to TN and the entire procedure is repeated. This requires that during an element of time the branches are tested in the ratio of their respective rate constants and thus fulfills the requirement for the model that

$$\frac{N_1}{N_2} = \frac{k_1}{k_2} . \tag{17}$$

After the designated periods of time shown in the flow chart as COUNT = 100, the concentrations of the components and the time are tabulated and recorded. This entire procedure is repeated until the reaction shows completion. The results for the first-order consecutive reaction are shown in Fig. 2. Using 5000 molecules as the sample array and N_1 equal to 1, the value for the rate constant of the reaction simulated would be 2.0×10^{-4} min^{-1}. The integrated values with $R(1, 2) = 2.0 \times 10^{-4}$ min^{-1} and the Monte Carlo results are shown in Fig. 2, and the agreement is within 0.2% standard deviation. The chemical reaction for any specific rate constant can be made to coincide with the Monte Carlo time scale by using the equation

$$\beta N_1 = k_1 t \tag{18}$$

and adjusting either the size of the sample array or the number of times that molecule A is searched for per unit of time.

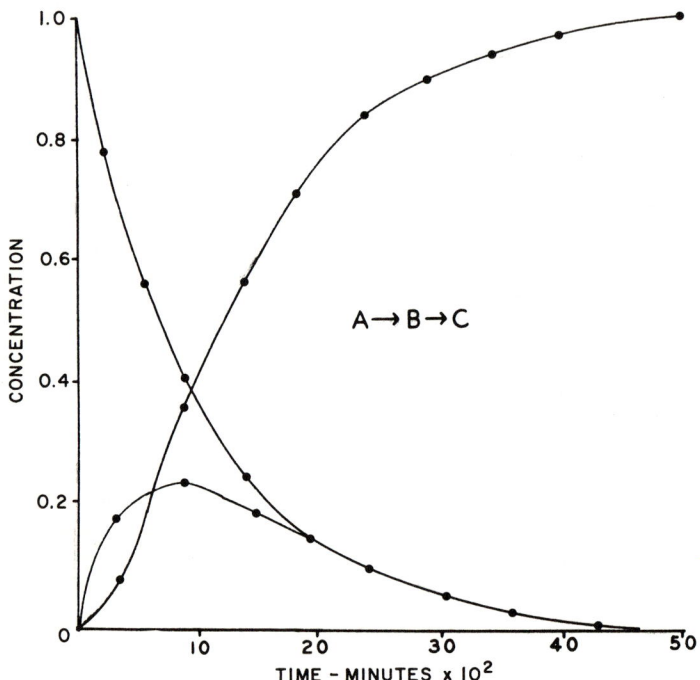

Fig. 2. An illustration of a first-order consecutive reaction with $k_2/k_1 = 2$. Comparison is shown between the integrated results (———) and the Monte Carlo results (· · ·).

5. Application of the Monte Carlo Method

The statistical accuracy of this model is dependent on the size of the sample array and the number of events considered, and is discussed later in the chapter (Section III. B).

C. Reversible Reactions

In discussing reversible reactions, consider the example of the first-order reversible reaction

$$A \underset{k_2}{\overset{k_1}{\rightleftarrows}} B.$$

The logic of the flow chart (Fig. 1) which applies to this reaction is the same as previously discussed for the first-order consecutive reaction except in this example k_1 [R(1, 2)] = 2 and k_2 [R(2, 1)] = 5, and therefore, the forward and reverse reactions are tested in the ratio of the frequency of occurrence of the numbers 2 and 5. All other rate constants in the general first-order model are zero.

The rate equations for this reaction are

$$\frac{d(A)}{dt} = -k_1(A) + k_2(B) \tag{19}$$

and

$$\frac{d(B)}{dt} = \frac{-d(A)}{dt}. \tag{20}$$

With the conditions $(A) = (A)_0$ and $(B) = (B)_0$ when $t = 0$, the above rate equations integrate to give the equation

$$\ln\left[\frac{k_1 A - k_2 B}{k_1(A)_0 - k_2(B)_0}\right] = -(k_1 + k_2)t \tag{21}$$

expressing (A) and (B) as functions of time. Rewriting Eq. (21) for (A) as a function of time gives

$$(A) = \frac{k_2[(A)_0 + (B)_0] + [k_1(A)_0 - k_2(B)_0]\exp[-(k_1 + k_2)t]}{k_1 + k_2} \tag{22}$$

with

$$(B) = (A)_0 + (B)_0 - (A).$$

As previously discussed, for the Monte Carlo model with 5000 molecules in the sample array and $N_1 = 2$, the calculated values for the rate constants of the reaction simulated would be $k_1 = 4.00 \times 10^{-4}$ min^{-1} and $k_2 = 1.00 \times 10^{-3}$ min^{-1}. Figure 3 shows the comparison between the integrated results and the results obtained from the Monte Carlo model with $(A)_0 = 1.0$ moles/liter and $(B)_0 = 0$. The results for the two methods agreed to better than 0.1%.

The results illustrate quite clearly the concept of mass action, where at equilibrium the rates of the forward and reverse reactions are equal and

$$k_1 (A)_{eq} = k_2 (B)_{eq} \tag{23}$$

or

$$K_{eq} = \frac{(B)_{eq}}{(A)_{eq}} = \frac{k_1}{k_2} .$$

This reversible first-order reaction can be considered as an illustration of how the Monte Carlo model can readily be adapted to explain the concept of equilibrium. An example of this would be liquid-vapor equilibrium where the liquid molecules would have a constant probability of evaporating and the gaseous molecules would have a given probability of combining to form the liquid state. The rates of these respective processes would be equal to the number of molecules present in each state times the probability of that particular event occurring. Equilibrium is established when the rates become equal.

D. Second-Order Rate Equations

The Monte Carlo model can readily be extended to include the consideration of second- and higher-order rate equations. The only additional restriction necessary to include the consideration of second-order reactions is that two molecules, whether they be of similar or different types, must be found in order for a reaction to occur. In addition the concept of collisions in bimolecular reactions can be illustrated very effectively by varying the number of digits corresponding to each reactant molecule, thus showing the effect of the relative concentrations of the reacting species on the rate of the reaction. After applying the Monte Carlo model to second-order reactions, the extension to higher-order reactions will be obvious.

5. Application of the Monte Carlo Method

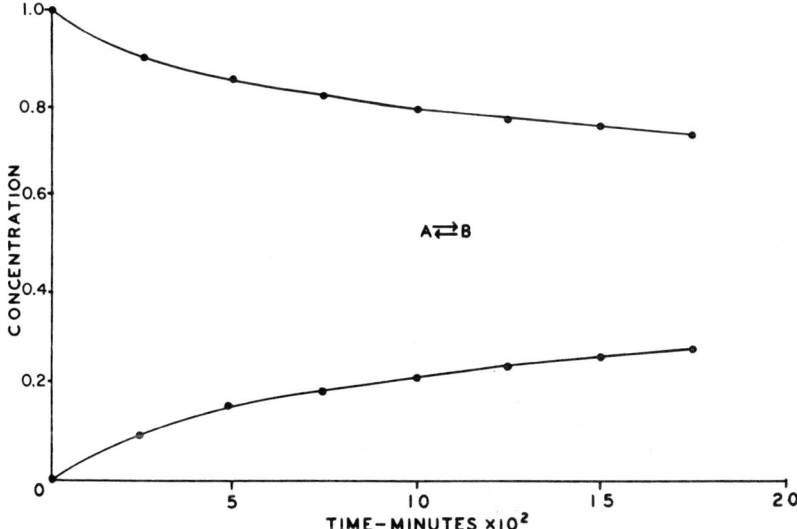

Fig. 3. An illustration of a reversible first-order reaction with $k_1/k_2 = 2/5$. Comparison is shown between the integrated results (——) and the Monte Carlo results (···).

Consider the case of the second-order reaction

$$A + B \xrightarrow{k_2} C.$$

In the following discussion molecule A will be designated by the digit 1, molecule B by the digit 2, and molecule C by the digit 3.

The rate equation of the Monte Carlo model for this reaction would be formulated as

$$\frac{-d(1)}{dN} = \beta_1(1)\beta_2(2). \tag{24}$$

This states that the rate for the decrease of digit 1 (molecule A) per number of attempts to find digits 1 and 2 per unit of time is equal to the number of 1's present times the number of 2's present times the probabilities of finding these particular digits (β_1 and β_2); β_1 will equal β_2 if the molecules are stored in the same sample array or arrays of equal size. Writing Eq. (24) in terms of the rate of increase in digit 3 gives the equation

$$\frac{d(3)}{dN} = \beta\left[(1)_0 - (3)\right]\beta\left[(2)_0 - (3)\right] \tag{25}$$

with $(1)_0$ and $(2)_0$ equal to the initial number of 1's and 2's present. Integrating Eq. (25) gives the result

$$\beta^2 N = \frac{1}{(1)_0 - (2)_0} \ln\left[\frac{(1)(2)_0}{(2)(1)_0}\right]. \tag{26}$$

As in the case of the examples discussed for the first-order rate equations, $(1)_0 = \alpha(A)_0$ and $(2)_0 = \alpha(B)_0$. However when considering second-order rate equations it is important that the proportionality constant α be considered, since it does not cancel as it did for the first-order rate equations. The same proportionality constant α must be used for both A and B. If this condition is satisfied, Eq. (26) can be reduced to the familiar form of the second-order rate equation

$$\alpha\beta^2 N = \frac{1}{(A)_0 - (B)_0} \ln\frac{(A)(B)_0}{(A)_0(B)}. \tag{27}$$

Equating Eq. (27) and the result for the second-order rate equation when the concentrations of the reactant components are not equal,

$$k_2 t = \frac{1}{(A)_0 - (B)_0} \ln\frac{(A)(B)_0}{(A)_0(B)} \tag{28}$$

yields the result,

$$\alpha\beta^2 N = k_2 t. \tag{29}$$

Thus, if it is desired to simulate a reaction for a specific rate constant (liter mole^{-1} time^{-1}) in order for the simulated concentration curve and experimental curve to become superimposed, it is necessary to consider the size of the sample array, $1/\beta$, the proportionality constant α, and the number of times per unit of time required to search for the reactant molecules N. It is important when simulating experimental results for a given rate constant that the investigator understand the relation [Eq. (29)] between the rate constant and the Monte Carlo model.

Figure 4 shows the comparison between the integrated results and the Monte Carlo results for the second-order reaction where $(A)_0 = 3.0$ moles/liter, $(B)_0 = 2.0$ moles/liter, and $k_2 = 8.0 \times 10^{-4}$ liter

5. Application of the Monte Carlo Method

mole^{-1} min^{-1}. The calculated rate constant ($\alpha \beta^2 N$) and the value found from the least-squares treatment on the simulated concentrations agreed to within 1% deviation.

When dealing with systems of second order which consist of more than one branch, such as

$$A + B \xrightarrow{k_1} C \qquad \text{Branch 1}$$

and

$$B + C \xrightarrow{k_2} D \qquad \text{Branch 2}$$

the same result is obtained from the Monte Carlo model as in the case of the first-order reactions which consisted of more than one branch; i.e.,

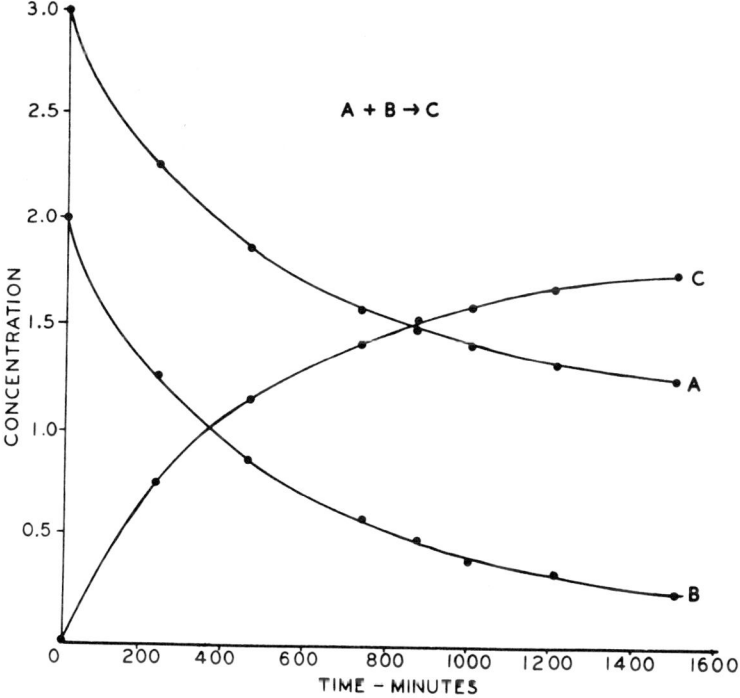

Fig. 4. An illustration of a second-order reaction with the comparison of the integrated results (——) and the Monte Carlo results (...).

the individual branches are tested in the ratio of their respective rate constants. From the Monte Carlo model the two branches in the example cited would reduce mathematically to

$$\alpha \beta^2 N_1 = k_1 t \qquad (30)$$

and

$$\alpha \beta^2 N_2 = k_2 t \qquad (31)$$

and thus,

$$\frac{N_1}{N_2} = \frac{k_1}{k_2} . \qquad (32)$$

As before, N_1 is equal to the number of times branch 1 is considered per unit of time, and N_2 is equal to the number of times branch 2 is considered per unit of time. In addition the same restrictions as previously discussed are imposed on the proportionality constants β and α so that they will cancel in the final ratio equation. The flow chart shown in Fig. 1 for the general first-order rate equations can be applied to a general second-order equation with the additional modification that two specific molecules be found in order for a "fruitful" event to occur. It should be noted that one product digit (molecule C) is replacing two reactive digits (molecules A and B) and thus one of the reactive digits is replaced with the digit 3 (molecule C) and the other digit is replaced in the array with a "nonreactive" digit, such as 0.

The rate equations for the example above are

$$\frac{-d(B)}{dt} = k_1 (A)(B) + k_2 (B)(C) \qquad (33)$$

$$\frac{d(C)}{dt} = k_1 (A)(B) - k_2 (C)(B) \qquad (34)$$

$$\frac{-d(A)}{dt} = k_1 (A)(B) \qquad (35)$$

$$\frac{d(D)}{dt} = k_2 (B)(C). \qquad (36)$$

Dividing the rate equation (34) by Eq. (35) yields the equation

$$\frac{d(C)}{d(A)} = -1 + \frac{k_2 (C)}{k_1 (A)} , \qquad (37)$$

5. Application of the Monte Carlo Method

thereby removing time as a variable [14]. Substituting $(C) = (A)x$ and $K = k_2/k_1$ gives the equation

$$\frac{dx}{d \ln(A)} = -1 + (K - 1)x. \tag{38}$$

With $(C) = 0$ and $(A) = (A)_o$ when $t = 0$, Eq. (38) integrates to give the solution

$$(C) = \frac{(A)}{K - 1} \left[1 - \left(\frac{(A)}{(A)_o}\right)^{K-1} \right]. \tag{39}$$

Likewise, Eqs. (33) and (35) can be combined to remove the time dependence, yielding an expression for (B)

$$(B)_o - (B) = \frac{2K - 1}{K - 1} \left[(A) - (A)_o\right] - \frac{(A)_o}{K - 1} \left[1 - \left(\frac{(A)}{(A)_o}\right)^K\right]. \tag{40}$$

To enable these solutions to be solved in closed form, let $K = k_2/k_1 = 1/2$, and $(B)_o = 2(A)_o$, thus the expression for (B) reduces to

$$(B) = 2(A)_o \left[\frac{(A)}{(A)_o}\right]^K. \tag{41}$$

Using Eq. (35) the expression for (A) is found to be

$$(A) = \frac{(A)_o}{(1 + (A)_o k_1 t)^2} \tag{42}$$

With this example the usefulness and power of the Monte Carlo method becomes increasingly evident and the restrictions on k_2/k_1 and $(B)_o = 2(A)_o$ for the closed-form solution are not required. Comparison of the results obtained for this example with $k_1 = 1.33 \times 10^{-4}$ liter mole^{-1} min^{-1}, $k_1 = 2k_2$, $(A)_o = 1.0$ moles/liter, and $(B)_o = 2.0$ moles/liter is shown in Fig. 5. A least-squares treatment of $1/(A)$ versus time showed that the experimental rate constant (k_1) and the calculated rate constant ($\alpha \beta^2 N_1$) agreed to better than 1%. Again, this agreement can be improved by increasing the size of the sample array.

E. Complex Reactions

The Monte Carlo model derived from the consideration of suitable models for first- and second-order rate equations can be used to enable simulation of any system which consists of any combination of these

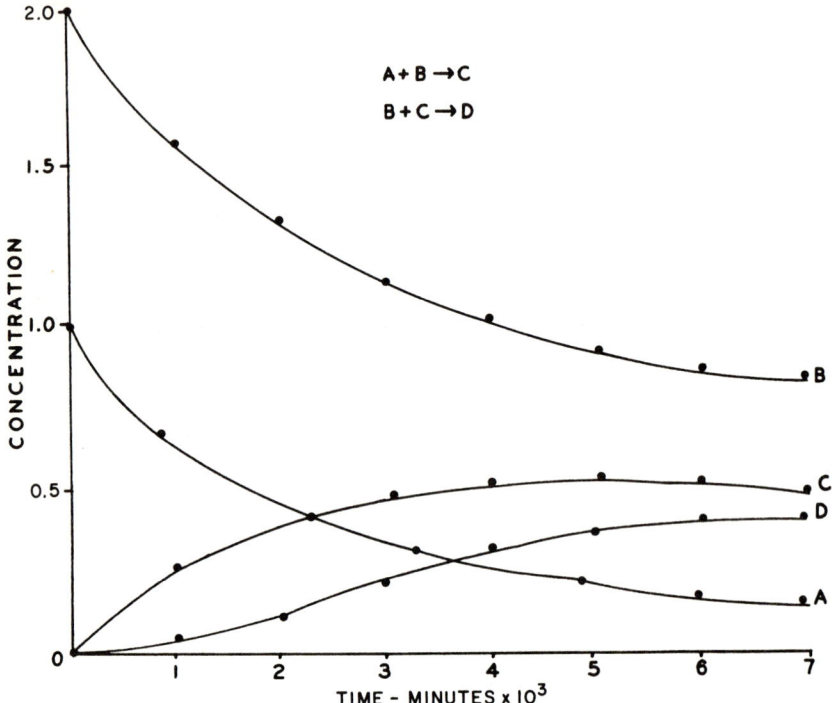

Fig. 5. An illustration of a system with two second-order reactions with $k_1/k_2 = 2$ and $(B)_0 = 2(A)_0$. Comparison is shown between the integrated results (——) and the Monte Carlo results ($\cdots$).

reactions. As the system under consideration becomes more complex and as the ability to obtain a mathematical solution becomes increasingly difficult or perhaps impossible, the value of the Monte Carlo method of simulation becomes more evident. Even in the cases where there is a mathematical solution in the closed form, the physical result is often lost in the complexity of the mathematical solution.

To give an illustration of a system which consists of both a first- and a second-order reaction, consider the following case

$$A \xrightarrow{k_1} B \qquad \text{Branch 1}$$

$$A + B \xrightarrow{k_2} C \qquad \text{Branch 2}$$

5. Application of the Monte Carlo Method

where branch 1 is a first-order reaction and branch 2 is a second-order reaction. The rate equations for this system are

$$\frac{-d(A)}{dt} = k_1(A) + k_2(A)(B) \tag{43}$$

$$\frac{d(B)}{dt} = k_1(A) - k_2(A)(B) \tag{44}$$

and

$$\frac{d(C)}{dt} = k_2(A)(B). \tag{45}$$

As in the previous example, these equations can be solved by removing time as the independent variable. Dividing Eq. (44) by Eq. (43) gives the equation

$$\frac{d(B)}{d(A)} = \frac{(B-K)}{(B+K)} \tag{46}$$

where K is equal to k_1/k_2. With the condition $(B) = 0$ when $(A) = (A)_0$, Eq. (46) is integrated to give

$$(B) + 2K \ln\left[\frac{K-(B)}{K}\right] = (A) - (A)_0. \tag{47}$$

In developing the Monte Carlo model, the branches 1 and 2 are simply considered as first- and second-order reactions, respectively, thus obtaining the equations

$$\beta_1 N_1 = k_1 t \tag{48}$$

and

$$\alpha \beta_1 \beta_2 N_2 = k_2 t, \tag{49}$$

with all terms having the same meanings as discussed in the previous sections. As before, if the molecules A and B are represented in the same storage array or arrays of the same size, then $\beta_1 = \beta_2$ since each particular molecule will have the same probability of being selected. Therefore, the number of times branch 1 is tested per unit of time relative to the number of times branch 2 is considered per unit of time is

$$\frac{N_1}{N_2} = \frac{\alpha \beta k_1}{k_2}. \tag{50}$$

The results for the Monte Carlo model and the results from the analytical expression are shown in Fig. 6 where $(A)_0 = 2.0$ moles/liter, $(B)_0 = 0$, and $k_1/k_2 = 2$. These results are compared for a particular concentration of (B) by evaluating the concentrations for (A) and (C) from the analytical expression and then superimposing the curves for the Monte Carlo model and the analytical expression. It should be noted that there is a significant tailing at the end of the curves for (A) and (C). However, this is to be expected since the concentration curves for (B) are forced to fit; the statistical inaccuracy is manifested in the curves for (A) and (C).

The last example of a complex reaction to be considered is

$$A \xrightarrow{k_1} B + C \xrightarrow{k_2} D$$
$$A \xrightarrow{k_3} D$$

for which the rate equations are

$$\frac{d(A)}{dt} = -k_1(A) - k_3(A) \tag{51}$$

$$\frac{d(B)}{dt} = \frac{d(C)}{dt} = k_1(A) - k_2(B)(C) \tag{52}$$

and

$$\frac{d(D)}{dt} = k_3(A) + k_2(B)(C). \tag{53}$$

These rate equations can be integrated [15] to obtain the expression for (B) as a function of time, which is

$$(B) = (A)_0 \frac{k_1}{k_1 + k_3} \left(\frac{\gamma}{K}\right)^{1/2} \frac{iJ_1(2i(\gamma K)^{1/2}) - \delta H_1(2i(\gamma K)^{1/2})}{J_0(2i(\gamma K)^{1/2}) + i\delta H_0(2i(\gamma K)^{1/2})} \tag{54}$$

where

$$\gamma = \exp(-(k_1 + k_3)t), \quad K = \frac{k_1 k_2 (A)_0}{(k_1 + k_3)^2}$$

5. Application of the Monte Carlo Method

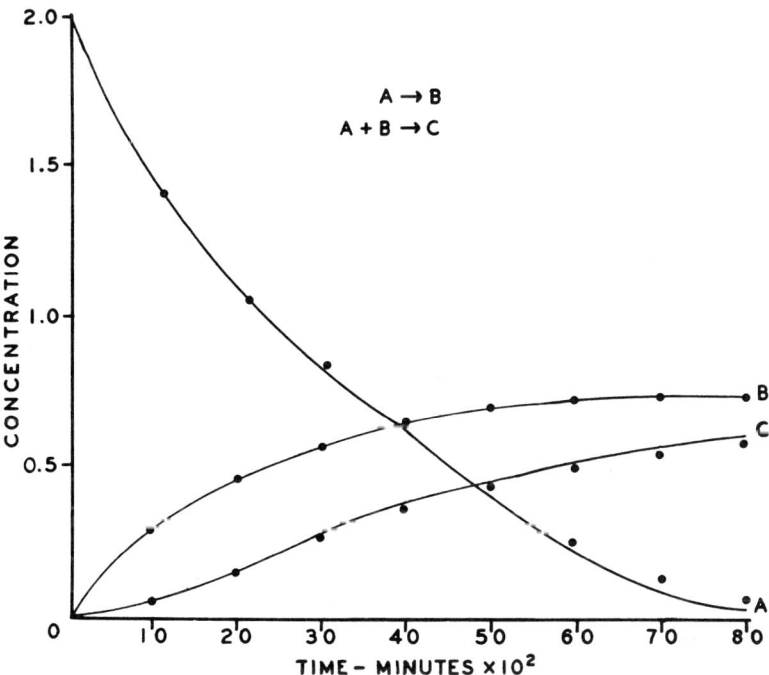

Fig. 6. An illustration of a system consisting of both a first- and a second-order reaction with $k_1/k_2 = 2$. Comparison is shown between the integrated results (——) and the Monte Carlo results ($\cdots$).

and

$$\delta = \frac{iJ_1(2i\sqrt{K})}{H_1(2i\sqrt{K})} .$$

The terms J and H are Bessel functions which can be found in tables [16]. This reaction has been chosen since it has been discussed previously by Schaad [9] as an application of the Monte Carlo method for simulating chemical reactions. Schaad developed a method for testing random number overflow in determining the ratio to which the individual branches were considered. However, using the method developed in this chapter the branches are tested in the ratios

$$\frac{N_1}{N_2} = \frac{k_1}{k_2} \qquad (55)$$

$$\frac{N_1}{N_2} = \frac{\alpha \beta k_1}{k_2} \qquad (56)$$

and

$$\frac{N_3}{N_2} = \frac{\alpha \beta k_3}{k_2} . \qquad (57)$$

There is an additional difficulty in simulating this particular system, similar to the problem discussed in connection with the second-order reaction, where the number of digits change during the course of the reaction. In the previous example, this was not a problem since the excess digit could simply be replaced by a "nonreactive" digit. In this case, however, when a "fruitful" event is noted for branch 1, k_1, two digits (molecules B and C) are required to replace molecule A.

This problem can be solved with the initial filling of the storage array. A section of the storage array is filled with "nonreactive" digits, such as 0. When a "fruitful" event is found for branch 1, one of these digits is replaced with the corresponding product digit (either molecule B or C). The other product digit is placed in the storage area previously occupied by molecule A.

Figure 7 shows the integrated results for (B) and the values obtained from the Monte Carlo model. The integrated values were obtained for $(A)_0 = 2.0$ moles/liter, $k_1 = 1.333 \times 10^{-3}$ min^{-1}, $k_2 = 3.33 \times 10^{-4}$ liter mole^{-1} min^{-1}, and $k_3 = 3.33 \times 10^{-4}$ min^{-1}. The average deviation between the integrated results and the Monte Carlo results was less than 2%.

III. CONCLUSION

A. Chemical Kinetics

In the area of chemical kinetics, the Monte Carlo method of simulation provides the chemist with several extremely powerful methods for attacking kinetic problems. First, the chemist has available a tool to evaluate readily a proposed mechanism for given empirical data. Often a proposed mechanism can be accepted or rejected simply by visual inspection. In order to do this the only requirement is a proposed relative ratio of the

5. Application of the Monte Carlo Method

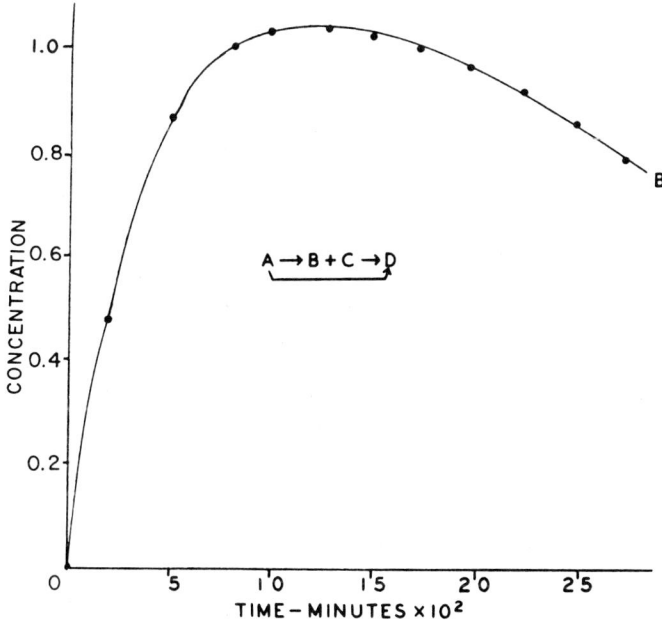

Fig. 7. A plot of (B) as a function of time for the complex reaction discussed in the text. Comparison is shown between the integrated results (——) and the Monte Carlo results (...).

rate constants. The empirical curve and the calculated curve can be compared by superimposing the two curves and noting any discrepancies.

In addition, by evaluating the differences between the two curves, the relative rate constants can be changed by a corrective iteration process and the reaction simulated again. This provides the investigator with a method to force his proposed mechanism upon the empirical data in order to obtain a predetermined precision. Once the curves have been superimposed, the absolute value for the rate constants can be determined from the scaling of the Monte Carlo time scale and the utilization of the equations developed in this chapter.

Another powerful tool provided by the Monte Carlo method that has been developed in detail in this chapter is the ability to simulate precisely a system where the rate constants are known. In developing this model, the investigator must understand his model and determine how many times he wishes to search for a particular molecule or test a particular branch per unit of time, and thus remove the arbitrary units on the time scale. In all the examples discussed in this chapter, where the integrated values were compared to the Monte Carlo values, the rate constants were determined prior to the simulation.

B. Statistical Considerations

The error involved from using the Monte Carlo method to simulate a particular reaction cannot be estimated prior to the reaction simulation. The statistical agreement is usually determined by evaluating the standard deviation of the quantities simulated, as illustrated in this work. In the case of reaction simulation the statistical agreement between the calculated values from the Monte Carlo method and those values obtained from the integration of the differential equations will be a function of the size of the sample array and the number of branches involved in the particular reaction. Obviously, the more decisions or branches under consideration in the particular system, the more statistically interdependent these decisions will become. McCracken [17] states that the accuracy will increase proportionally to the square of the sample array. Thus to double the accuracy of the simulation, the sample size must be quadrupled.

The statistical considerations are the most serious limitation of the Monte Carlo method, although for chemists a standard deviation of less than 5% is acceptable in most cases. The examples discussed in this work were developed with a sample array of 5000 molecules and the ratio of the rate constants were restricted to small whole-number ratios. In these cases usually 1% standard deviation or better was obtained. If the reaction to be simulated involves rate constants which do not reduce to small whole-number ratios, then a large sample array and a large number of random events must be considered to produce the statistical accuracy.

In the examples discussed above using a sample size of 5000 molecules, the execution time on an IBM 360 Model 30 computer was approximately 2 min. By increasing the sample size to 45,000 molecules, the execution time increased to over 30 min. Thus, in order to use a larger sample array and to increase the accuracy of the simulation, the computer execution time will increase proportionally. The accuracy one can obtain from the Monte Carlo method is dependent on the amount of effort one is willing to spend.

C. Other Uses

As discussed in the Introduction there are many areas in which the Monte Carlo method of simulation is currently being applied. For the chemist the Monte Carlo method has potential in explaining and simulating

5. Application of the Monte Carlo Method

phenomena such as equilibrium processes, mechanisms of gaseous reactions, catalytic mechanisms, and so on. The method can readily be adapted to any problem which can be simulated by random events where each event has a given probability of occurring.

Although it is doubtful that the Monte Carlo method of simulation should be presented to junior-senior undergraduate students in order to aid the teaching of chemical kinetics, any advanced course on chemical kinetics designed for graduate students should incorporate this method to illustrate a useful tool for simulation of chemical reactions. The general model provides a method to simulate any combination of first-order and/or second-order reactions, and thus, enables the student to observe reactions which could not be evaluated in the closed form. In addition, the model offers a sophisticated use of the digital computer and illustrates its usefulness for problems which are impossible for the slide rule or desk calculator. In making the student aware of methods of simulation, the usefulness of this approach could be applied to other areas such as experimental design. This could possibly lead to results previously considered impossible due to the mathematical restrictions resulting from the differential equation approach.

REFERENCES

[1]. R. Lattes, Methods of Resolution for Selected Boundary Problems in Mathematical Physics, Documents on Modern Physics, Gordon & Breach, New York, 1969, p. 125.

[2]. A. S. Householder, G. E. Forsythe, and H. H. Germond, Monte Carlo Method, National Bureau of Standards Applied Mathematics Series 12, Washington, D.C., 1951.

[3]. J. V. Brown, Modern Mathematics for the Engineer (E. Reckenbach, ed.) McGraw-Hill, New York, 1961.

[4]. Symposium on Monte Carlo Methods (H. A. Meyer, ed.), Wiley, New York, 1956.

[5]. J. M. Hammersley, J. Assoc. Comput. Mach., 3:No. 2, 73 (1956).

[6]. A. S. Householder, Principles of Numerical Analysis, McGraw-Hill, New York, 1953.

[7]. T. R. Dickson, The Computer and Chemistry, Freeman, San Francisco, Calif., 1968, p. 129.

[8]. Y. A. Schreider (ed.), The Monte Carlo Method, Pergamon, New York, 1966, Chap. III.

[9]. L. J. Schaad, J. Am. Chem. Soc., 85:3588 (1963).

[10]. B. Rabinovitch, J. Chem. Educ., 46:262 (1969).

[11]. J. J. Manock and D. L. Hooper, J. Chem. Educ., 48:530 (1971).

[12]. J. J. Manock and D. L. Hooper, Proceedings of the Conference on Computers in Chemical Education and Research, DeKalb, Illinois, 1971.

[13]. C. W. Pyun, J. Chem. Educ., 48:194 (1971).

[14]. S. W. Benson, The Foundations of Chemical Kinetics, McGraw-Hill, New York, 1960.

[15]. R. G. Pearson, L. C. King, and S. H. Langer, J. Am. Chem. Soc., 23:4149 (1951).

[16]. E. Jahnke and F. Emde, Tables of Functions, Dover, New York, 1943, pp. 224-229, 236-249.

[17]. D. McCracken, Sci. Am., 192:No. 5, 90 (1955).

Chapter 6

INTEGRATION OF COMPLEX RATE EQUATIONS
USING INFINITE SERIES

Norman C. Peterson

Kirk Chemical Laboratory
Polytechnic Institute of Brooklyn
Brooklyn, New York

and

Howard J. Butcher

Nadir Metallurgical Co.
Elmont, New York

I.	INTRODUCTION	293
II.	NUMERICAL INTEGRATION	295
III.	ANALYTICAL FUNCTIONS SOLUTIONS	297
IV.	INFINITE SERIES SOLUTIONS	299
	REFERENCES	310

I. INTRODUCTION

Because electronic digital computers are commonly available to chemists, chemical kineticists have begun to rethink their approaches to problems. They have had to examine ways of solving problems which were dismissed as too difficult prior to the advent of machine computation. One such area is that of reaction mechanisms. A description of all the elementary steps through which a reaction proceeds is known as the mechanism, and it specifies the time course from reactants to products. A collection of steps postulated for a given reaction is a possible mechanism which implies a set of differential equations for the time rate of change of concentrations. Solutions to such systems of differential equations are integrated rate laws* describing concentrations as functions of time. Those mechanisms can be distinguished which correspond to different systems of differential equations and therefore to different integrated rate laws.

A sensitive test of mechanism, favored by the authors, is to examine the fit of experimental concentration-time data to the integrated rate law by iterative nonlinear least squares. That integrated rate law in best agreement with the experimental data during the entire history of the reaction corresponds to the mechanism best describing the system among those considered. The computational problem generated by this technique has two parts: first, solving the differential equations, and second, applying some statistical criterion for judging the fit of the data to the resulting integrated rate equations. The latter question has been examined elsewhere [1-5].

Testing a mechanism requires that predicted concentrations be compared with measured ones at a number of selected times. The rate constants in the mechanism are initially unknown. The comparison procedure is repeated, iteratively searching for rate constants giving the best fit for that mechanism. In this type of comparison a table of numerical values of the concentrations for various times is required; thus the integrated rate law can be obtained from a numerical solution of the rate equations.

Several approaches available for obtaining the integrated rate law are surveyed below before giving detailed examples of solution by infinite series.

The mathematical problem to be solved for a complex mechanism, excluding diffusion, consists of M independent ordinary differential equations together with the initial conditions of the form:

---

*The term rate law is used instead of rate equation to emphasize that it may not be possible to express the solution in terms of an equation.

6. Integration of Complex Rate Equations

$$\frac{dc_1}{dt} = k_1 R_{11} + k_2 R_{12} + \cdots + k_N R_{1N}$$

$$\vdots \qquad (1)$$

$$\frac{dc_M}{dt} = k_1 R_{M1} + k_2 R_{M2} + \cdots + k_N R_{MN}$$

where the k's are rate constants and the R's are products of some of the concentrations, $c_1, c_2, \ldots, c_M$, to integral powers. Initial conditions specify each of the c's at $t = 0$. This set of complex equations is described mathematically as first order, i.e., containing only first derivatives. Each equation is linear only if each of the R's has a single concentration raised to a power at most 1. The independent variable, time, does not appear explicitly in these equations.

Approaches to the solution of the rate equations (1) include the following three categories:

    I.   Numerical integration.

    II.   Analytical functions solutions.

    III. Infinite series solutions.

II. NUMERICAL INTEGRATION

The problem of finding a numerical solution of simultaneous differential equations has been thoroughly treated in the literature on numerical analysis [6]. There are several algorithms to choose from, including those described here. Programs containing these algorithms are often found in mathematical subroutine libraries as part of the software available for a computer. To use these packaged programs it is only necessary to encode the differential equations to be solved, that is, the computation of the rate in each equation, and to specify the initial conditions. Numerical methods seem attractive when confronted with nonlinear differential equations. The mechanism problem has been discussed by Detar [8-10], who favors a simpler numerical integration scheme than those which follow.

One numerical analysis procedure often used in solving rate problems [11-13] is known as fourth-order Runge-Kutta integration. The procedure is documented in various mathematical texts [14, 15], and is outlined below without quoting the equations.

First, a step size is chosen. The rate is evaluated four times for each step: once at the beginning of the step, twice in the center, and once at the end of the step. The concentration at the end of the step is computed using a weighted average of the four rates. This procedure is repeated for each differential equation to be solved simultaneously. The entire time-concentration curve is traced out by repeating the same computation. The initial conditions for each subsequent step are taken as the result of the step just finished.

The step size is chosen to give reasonable accuracy, demonstrated by comparing the entire solution computed for steps of half-size with that computed using full-size steps. An estimate of the error of the result of the former computation is less than the difference between the two computations [15]. As the steps are made smaller, increased machine time will be required, and round-off errors may degrade the solution.

To ensure that one has computed an accurate solution it is important that the concentration-time law does not change as the step size is reduced. A step size which is much too large can lead to apparently negative concentrations, but positive concentrations do not guarantee accuracy. In summary, the Runge-Kutta algorithm can be easily reduced to a computer program for an arbitrary mechanism, but the procedure must be tested with care to establish accuracy.

The Runge-Kutta procedure has the disadvantage that each step in the integration is treated as a fresh problem, without benefit of prior information. In a predictor-corrector algorithm, concentrations evaluated earlier in the solution are used to estimate the concentration after the next step, and that result is corrected for the derivative at that point. This procedure works well when the solution can be approximated accurately as a polynomial over a short interval of time. A solution developed in this way has the advantages that the derivative need be evaluated but once for each step and the step size can be adjusted during the calculation for minimizing computing time.

Several algorithms of this type are in common use. A major disadvantage is the possibility of unstable, oscillating solutions resulting from the effects of noise, from say round off, in each step being amplified in subsequent steps. The programming of these algorithms is more complex than that of the Runge-Kutta algorithm. As they are not self-starting, the Runge-Kutta algorithm is used to produce enough of the solution required for the predictor formula. Here too the results must be tested carefully to ensure that they are correct. We recommend comparing the numerical results for at least one case for which the exact solution is known. Agreement in this test is necessary to ensure that the numerical integration program can be used with some confidence.

Examples of predictor-corrector procedures include Adams-Bashforth and Milne's methods. Hamming has reviewed them critically [6, 16]. They are described in the standard works on numerical analysis.

The Monte Carlo numerical integration of the rate equations is described in detail in Chapter 5 of this volume. The reaction mixture is represented by a finite collection of digits, each representing individual molecules. The fate of these individuals is determined by chance encounters selected by random numbers, and the rules for the encounters are in correspondence with those given by the rate equations. Published accounts of this procedure [17, 18] suggest that its main use is pedagogical, and practical application seems restricted by the requirements for computer storage and time. If only modest accuracy is required, the computer time might be as short as that needed for some other approach. Nonlinear problems arising from complex mechanisms are sufficiently difficult to recommend solution by several approaches to test the quality of one of them.

III. ANALYTICAL FUNCTIONS SOLUTIONS

For a small number of rate equations the solutions are known in terms of elementary functions. These can be found in various textbooks [19-22]. Any complex mechanism which consists of kinetic first-order steps coupled in arbitrary ways has solutions expressible exactly in terms of the exponential function [21, 23].

Solutions to a small number of other mechanism schemes have been discovered which are expressible exactly in terms of tabulated functions. The various mathematical procedures for discovering these solutions are sufficiently elusive to preclude any general outline of an attack on a new problem. The treatise by Murphy [24] can be consulted for a systematic approach to finding closed-form solutions.

We have summarized analytical solutions known to us for consecutive reactions in Table 1. Although these are exact solutions, the programming involved in their use may be difficult. Those planning to use the results cited in the table should be prepared to confront the problem of possible typographical errors in the original publications.

The use of these exact solutions for computer evaluation of concentrations requires that computer routines be available to evaluate the special functions. Frequently it is necessary to generate these programs for the specific problem at hand.

TABLE 1

Some Consecutive Mechanisms with Exact Solutions

Mechanism	Functions in solution	Ref.
A ⟶ B B ⟶ C and other first-order coupled systems, reversible and irreversible	Exponential functions	Rodiguin and Rodiguina [23]
A ⟶ B B + B ⟶ C	Bessel functions of the first and third kinds	Chien [25]
A ⟶ B B + B ⟶ C A ⟶ C	Bessel functions of the first and third kinds	Pearson et al. [26]
A ⟶ B B + D ⟶ C	Incomplete gamma function and the exponential integral	Chien [25]
A + A ⟶ B B ⟶ C	Exponential integral	Balandin and Leibenson [27]
A + A ⟶ B B + B ⟶ C	Elementary functions	Chien [25]
A + A ⟶ B B + D ⟶ C	Incomplete gamma function	Chien [25]
A + B ⟶ C C + D ⟶ E	Hypergeometric function	Kelen [28]
A + B ⟶ C C ⟶ E	Incomplete beta function	Kelen [29]
A + B ⟶ C C + C ⟶ E	Legendre functions of the first and second kinds	Kelen [29]
A + A ⟶ C A + C ⟶ E	Incomplete beta function and the exponential integral	Kelen [30]
A ⟶ C A + C ⟶ E	Incomplete gamma function	Kelen [31]
A + B ⟶ C A + C ⟶ E	Incomplete beta function	Kelen [31]

6. Integration of Complex Rate Equations

IV. INFINITE SERIES SOLUTIONS

A common computer approximation for many of these functions is a truncated infinite series. The following section explores the procedure for finding infinite series solutions directly, avoiding the step of trying to find the solution in terms of tabulated functions.

In our experience, solutions based on expansions in an infinite series can be found more readily than analytic solutions. Such infinite series expansions are of a convenient form for generating a computer program. We have not been able to find other examples of series solution in the literature of integrated rate laws. It is hoped that examples of this procedure will help to acquaint chemical kineticists with some classical tools for solving differential equations.

Solutions in infinite series are useful when confronted with an ordinary inhomogeneous linear differential equation in a single concentration which cannot be integrated in terms of elementary functions. Rate equations with constant coefficients* are readily solved in terms of exponential functions. A nonlinear equation of the Riccati type can be converted to a linear equation having a convenient series solution. First we outline the latter case and then give an example of two types in which the details of the series solution are given.

The set of M rate equations (1) are equivalent to M-1 independent first-order differential equations relating the concentrations

$$\frac{dc_1}{\sum_{i=1}^{N} k_i R_{1i}} = \frac{dc_2}{\sum_{i=1}^{N} k_i R_{2i}} = \cdots = \frac{dc_M}{\sum_{i=1}^{N} k_i R_{Mi}}. \quad (2)$$

This system can sometimes be solved more easily than the original equations.

The original set (1) is also equivalent to a set of differential equations of higher order in each of the concentration variables. For example, consider the mechanism

$$A \underset{k_2}{\overset{k_1}{\rightleftarrows}} B$$

---

*Constant coefficients refer to the coefficients of the independent variable and its derivatives.

$$B \xrightarrow{k_3} 2C$$

$$2C \xrightarrow{k_4} E$$

for which the rate equations are

$$\frac{dA}{dt} = -k_1 A + k_2 B \tag{3a}$$

$$\frac{dB}{dt} = k_1 A - (k_2 + k_3) B \tag{3b}$$

$$\frac{dC}{dt} = 2k_3 B - k_4 C^2. \tag{3c}$$

Upon differentiating Eqs. (3a)-(3c) with respect to time and using the relations in (3), two equations of second order can be found each containing a single concentration variable

$$\frac{d^2 A}{dt^2} + (k_1 + k_2 + k_3) \frac{dA}{dt} + k_1 k_3 A = 0$$

$$\frac{d^2 B}{dt^2} + (k_1 + k_2 + k_3) \frac{dB}{dt} + k_1 k_3 B = 0. \tag{4}$$

These are homogeneous linear equations with constant coefficients. Repeated differentiation of the third equation and elimination of A and B give a nonlinear equation in C of fourth order.

Because the equations shown in (4) have constant coefficients, they can be solved by standard methods for A and B in terms of exponentials. Equation (3c) is nonlinear but, because B can be obtained as a known function of time, $B(t)$, it is a type of nonlinear differential equation known as a Riccati equation which is convertible to a linear equation of second order. The general Riccati equation [15]

$$\frac{dy}{dt} + Q(t)y + R(t)y^2 = P(t) \tag{5}$$

is convertible to a linear equation by the transformation

6. Integration of Complex Rate Equations

$$y = \frac{1}{Ru} \frac{du}{dt}$$

which transforms it to the second-order linear equation (6):

$$R \frac{d^2u}{dt^2} - \left(\frac{dR}{dt} - QR\right) \frac{du}{dt} - PR^2 u = 0. \tag{6}$$

The prospects for solution of a second-order linear equation are much happier than for the fourth-order nonlinear equation referred to previously. If we let $R(t) = k_4$, $P(t) = 2k_3 B$, and $Q = 0$, then we can make the transformation

$$C = \frac{1}{k_4 u} \frac{du}{dt}$$

resulting in a linear equation for u:

$$\frac{d^2u}{dt^2} - 2k_3 k_4 B(t) u = 0. \tag{7}$$

The Frobenius solution of second-order linear differential equations in terms of infinite series is well known in the mathematical theory of differential equations. This could be accomplished easily for Eq. (7) if $B(t)$ were a polynominal in t. If a change of independent variable can be found to transform the equation to one of this form, then a series solution of the resulting equation is readily found [14].

This example seems to suggest that a nontrivial number of mechanism problems may be amenable to solutions in terms of infinite series. The solutions of two consecutive-parallel mechanism schemes are given below. The development of the series solutions hopefully contains sufficient detail to provide the reader with clues for attacking similar problems. The differential equations solved below also have been solved analytically independently by Kelen [29], but the results were not known to us when these solutions were developed in infinite series.

The irreversible mixed second-order, first-order sequence with a parallel path to the end product can be symbolized

$$C + B \xrightarrow{k_1} D$$

$$D \xrightarrow{k_2} E$$

$$C + B \xrightarrow{k_3} E.$$

The corresponding rate equations are

$$\frac{-dB}{dt} = \frac{-dC}{dt} = (k_1 + k_3)BC \tag{8}$$

$$\frac{dD}{dt} = k_1 BC - k_2 D \tag{9}$$

$$\frac{dE}{dt} = k_3 BC + k_2 D. \tag{10}$$

Taking B_o and C_o as the initial values of B and C, the instantaneous concentrations of the reactants, Eq. (8) is easily integrated to give the well-known result:

$$\frac{1}{B_o - C_o} \ln\left[\frac{C_o B}{B_o C}\right] = (k_1 + k_3)t. \tag{11}$$

After making the substitution:

$$G = B_o - C_o,$$

and noting the stoichiometric requirement that $B - C = B_o - C_o$, solving for B and C gives:

$$B = \frac{GB_o}{B_o - C_o \exp[-G(k_1 + k_3)t]}$$

and

$$C = B(C_o/B_o) \exp[-G(k_1 + k_3)t].$$

The time rate of change of the intermediate concentration D is given by Eq. (9). After eliminating B and C, we obtain the following linear inhomogeneous equation

$$\frac{dD}{dt} + k_2 D = \frac{k_1 G^2 (C_o/B_o) \exp[-G(k_1 + k_3)t]}{\{1 - (C_o/B_o) \exp[-G(k_1 + k_3)t]\}^2}. \tag{12}$$

Now let $Z = (C_o/B_o) \exp[-G(k_1 + k_3)t]$ and let $\mu = k_2/[G]$ and $\gamma = Gk_1/(k_1 + k_3)$. The differential equation (12) can then be transformed to one with polynomial coefficients in Z:

$$\frac{dD}{dZ} - \frac{\mu D}{Z} = \frac{\gamma}{(1-Z)^2}.$$

6. Integration of Complex Rate Equations

This equation is multiplied by the integrating factor $Z^{-\mu}$ to give

$$Z^{-\mu} \frac{dD}{dZ} - \mu D Z^{-\mu-1} = \frac{-\gamma}{Z^{\mu}(1-Z)^2}$$

or

$$d(Z^{-\mu} D) = \frac{-\gamma \, dZ}{Z^{\mu}(1-Z)^2}$$

The solution to this linear inhomogeneous equation is obtained as an infinite series [32]. The series arises as a convenient way of evaluating the integral

$$R_{\mu}(Z) = \int \frac{dZ}{Z^{\mu}(1-Z)^2}.$$

To aid in evaluating the integral, the binomial expansion [33] is used to expand the integrand:

$$(1-Z)^{-2} = \sum_{j=0}^{\infty} (j+1) Z^j.$$

This series is convergent for $Z^2 < 1$.

Now by choosing B as the reagent in excess, it can be assumed that $B_0 > C_0$ and $G > 0$ because of the symmetry of the equations with respect to interchange of B and C. Thus, it is evident that $Z^2 < 1$ for all t. The series is therefore uniformly convergent. After being expanded, $R_{\mu}(Z)$ becomes

$$R_{\mu}(Z) = \int \sum_{j=0}^{\infty} (j+1) Z^{j-\mu} \, dZ.$$

This series is readily integrated term by term for noninteger μ, obtaining $R_{\mu}(Z)$ as a convergent series:

$$R_{\mu}(Z) = Z^{-\mu} \sum_{j=1}^{\infty} \frac{j Z^j}{j-\mu}.$$

The function R_{μ} can be integrated in closed form for $\mu = 1$:

$$R_1(Z) = \frac{1}{(1-Z)} - \ln\left[\frac{(1-Z)}{Z}\right]. \tag{13}$$

For integer $\mu > 1$, the recursion formula (14) gives R_m in terms of R_{m-1} [33]

$$R_m(Z) = \frac{-1}{(m-1)Z^{m-1}(1-Z)} + \frac{m}{m-1}R_{m-1}(Z). \tag{14}$$

The solution is for $D = 0$ at $t = 0$:

$$D = \gamma Z^\mu \left[R_\mu(Z_0) - R_\mu(Z)\right]. \tag{15}$$

When computing $R_\mu(Z)$ by machine, one approximates the infinite series by an nth partial sum. Thus there is a truncation error. For example, when $Z = 0.5$, $\mu = 0.90909...$, the series was evaluated until the nth term was less than 10^{-6} of the nth partial sum. This required 22 terms and led to a calculated truncation error $\Delta_{22} < 10^{-6.4}$ where the sum itself is 6.29286. For larger t, Z is smaller, and fewer terms are needed to obtain an accurate value of the sum. The mechanism described here has a pathway in addition to the mechanism described in [32]. This more general solution collapses to that for the mechanism $B + C \rightarrow D$; $D \rightarrow E$, when $k_3 = 0$. The quantity γ then becomes G and the equations retain their form.

Calculated curves of the intermediate concentration vs. time are shown in Fig. 1 in which the rate constant k_1 is varied near 1.0. The curve C corresponds to the case in which the closed-form solution applies, i.e., μ is an integer.

A second mechanism is the irreversible, mixed second-order, second-order sequence with an alternate pathway.

The solution of the differential equations partly resembles the one developed above. Some of the equations are repeated for clarity. The mechanism is

$$A + B \xrightarrow{k_1} D$$

$$2D \xrightarrow{k_2} 2E$$

$$A + B \xrightarrow{k_3} E.$$

6. Integration of Complex Rate Equations

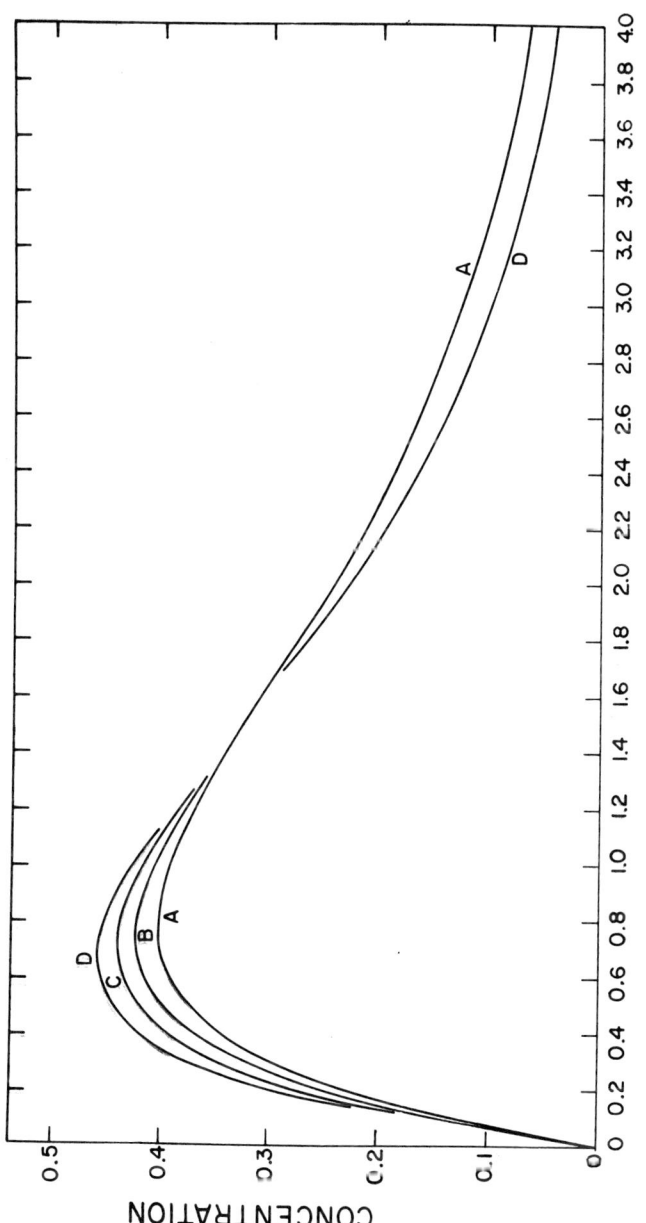

Fig. 1. Concentration of intermediate, D, vs. time, for the mixed second-order, first-order mechanism (p. 301); $C_0 = 1$, $B_0 = 2$, $k_2 = 1$, $k_3 = 0$. Curve A, $k_1 = 0.8$; curve B, $k_1 = 0.9$; curve C, $k_1 = 1$; curve D, $k_1 = 1.25$.

The rate equations are

$$\frac{-dA}{dt} = \frac{-dB}{dt} = (k_1 + k_3)AB \tag{16}$$

$$\frac{dD}{dt} = k_1 AB - k_2 D^2 \tag{17}$$

$$\frac{dE}{dt} = k_2 D^2 + k_3 AB. \tag{18}$$

Taking A_0 and B_0 as the initial values of A and B, the instantaneous concentrations of the reactants, Eq. (16) is first integrated to give the previous result:

$$\frac{1}{A_0 - B_0} \ln\left[\frac{B_0 A}{A_0 B}\right] = (k_1 + k_3)t. \tag{19}$$

Again making the substitution

$$G = A_0 - B_0,$$

and noting the stoichiometric requirement that $A - B = A_0 - B_0$, gives the following for A and B, as before:

$$A = \frac{GA_0}{A_0 - B_0 \exp[-G(k_1 + k_3)t]}$$

and

$$B = A \frac{B_0}{A_0} \exp[-G(k_1 + k_3)t].$$

The time rate of change of the concentration of intermediate D is given by Eq. (17) or

$$\frac{dD}{dt} + k_2 D^2 = \frac{k_1 G^2 (B_0/A_0) \exp[-G(k_1 + k_3)t]}{\{1 - (B_0/A_0) \exp[-G(k_1 + k_3)t]\}^2}. \tag{20}$$

Now let $Z = (B_0/A_0) \exp[-G(k_1 + k_3)t]$, $a = (k_1 + k_3)G/k_2$, and $b = k_1 G/(k_1 + k_3)$. The differential equation (20) can then be transformed to

6. Integration of Complex Rate Equations

$$\frac{dD}{dZ} - \frac{D^2}{aZ} + \frac{b}{(1-Z)^2} = 0. \tag{21}$$

This is seen to be a Riccati equation, and may be transformed to a second-order linear equation by the change of variable

$$D = \frac{-aZ}{u} \frac{du}{dZ} \tag{22}$$

resulting in

$$Z(1-Z)^2 \frac{d^2u}{dZ^2} + (1-Z)^2 \frac{du}{dZ} - \eta u = 0 \tag{23}$$

where $\eta = b/a$. This equation can be solved directly by the method of Frobenius, as follows: First assume a solution of the form

$$u = Z^r \sum_{j=0}^{\infty} a_j Z^j. \tag{24}$$

Substituting (24) into the differential equation and equating the coefficient of the first term to zero results in the indicial equation $r^2 = 0$. One solution can then be found from the series expansion

$$u_1 = \sum_{j=0}^{\infty} a_j Z^j. \tag{25}$$

Substituting the expansion (25) into the differential equation (23), and setting the coefficients of Z^j equal to zero gives the following relations among the a_j:

$$a_1 = a_0 \eta \tag{26}$$

$$a_{j+1} = \left[(2j^2 + \eta) a_j - (j-1)^2 a_{j-1} \right] / (j+1)^2. \tag{27}$$

The coefficient a_0 can be chosen arbitrarily and others can be obtained using the recursion formulas (26) and (27).

A second-order differential equation has two linearly independent solutions. The second solution for this equation has the form

$$u_2 = u_1 \ln Z + \sum_{j=0}^{\infty} b_j Z^j. \tag{28}$$

By substituting Eq. (28) for u in the differential equation and equating coefficients of Z^j to zero, it is found that b_0 is arbitrary and the other coefficients are related by the recursion formulas

$$b_1 = \eta b_0 - 2a_1 \tag{29}$$

$$b_{j+1} = \left[-2(j+1)a_{j+1} + 4ja_j - 2(j-1)a_{j-1} + (2j^2 + \eta)b_j - (j-1)^2 b_{j-1} \right] / (j+1)^2. \tag{30}$$

The concentration of D is obtained from Eq. (22)

$$D = -a \left\{ \frac{Z(du_1/dZ) + \lambda Z(du_2/dZ)}{u_1 + \lambda u_2} \right\} \tag{31}$$

where λ is a constant of integration, and the terms in the numerator are

$$Z \frac{du_1}{dZ} = \sum_{j=0}^{\infty} a_j j Z^j \tag{32}$$

and

$$Z \frac{du_2}{dZ} = Z \frac{du_1}{dZ} \ln Z + \sum_{j=0}^{\infty} b_j j Z^j + u_1. \tag{33}$$

The integration constant λ can be obtained from the initial condition, say at $t=0$, $D=0$, and $Z = Z_0 = B_0/A_0$. The condition that the numerator in (31) must be zero at $t=0$ gives:

$$\lambda = - \frac{(du_1/dZ)_{Z=Z_0}}{(du_2/dZ)_{Z=Z_0}}.$$

The concentration of product E is obtained by difference, viz., $E = E_0 + A_0 - A - D$.

6. Integration of Complex Rate Equations

The differential equation (23) has regular singularities at $Z=0$ and $Z=1$. It can be proved [14, 34] that an expansion about $Z=0$ has a region of convergence $0 < Z < 1$. Now it can be assumed as before that $B_0 < C_0$ and $G < 0$ because of the symmetry of the equation with respect to interchange of A and B. Thus, it is evident that $Z < 1$ for all t. The series is therefore uniformly convergent.

Sample calculations were performed in which the infinite series were truncated when the last term for du_2/dZ was less than ε, and it was found that the number of terms required decreased with increasing values of t. For example, when $C_0 = 1.0$, B_0 was 2.0, $k_1 = k_2 = 1.0$, $k_3 = 0$, and $\varepsilon = 10^{-9}$; 28 terms were required at $t = 0.1$, and 18 terms at $t = 0.6$. Computed concentrations of intermediate were 0.17271636 and 0.53396965, respectively. Repeating the calculations with $\varepsilon = 10^{-6}$ required 19 and 13 terms, obtaining concentrations of 0.17271640 and 0.53396835, respectively.

Calculated curves of the intermediate concentration vs. time are shown in Figs. 2 and 3 in which B_0 and the rate constant k_2 are varied.

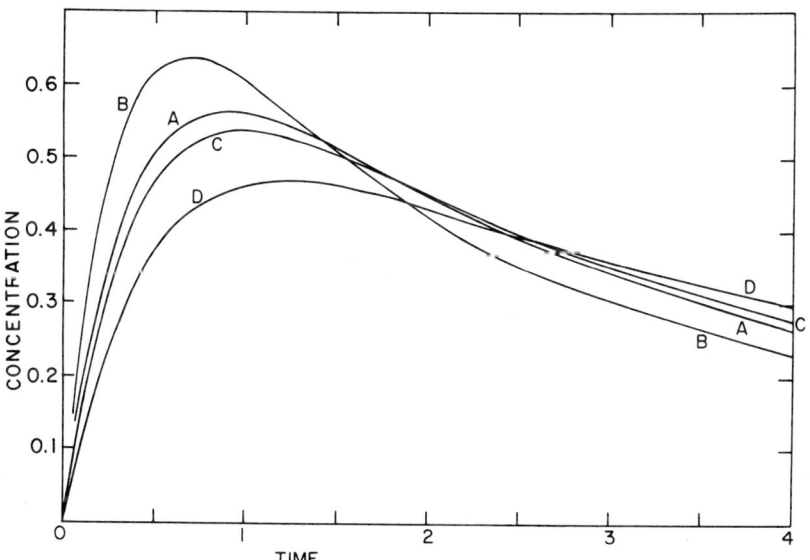

Fig. 2. Concentration of intermediate vs. time for the mixed second order, second-order sequence; $C_0 = 1$, $k_1 = 1$, $k_3 = 0$. Curve A, $B_0 = 2.0$; curve B, $B_0 = 3.0$; curve C, $B_0 = 1.75$; curve D, $B_0 = 1.25$.

Inspection shows that

$$\eta = \frac{b}{a} = \frac{k_1 k_2}{(k_1 + k_3)^2} .$$

The alternate path is removed for $k_3 = 0$ and $\eta = k_2/k_1$. The solution remains valid for the simpler mechanism [32].

It was convenient to choose examples of mechanisms where the rate equations are linear or reducible to linear. A system of first-order nonlinear rate equations such as (1) has a convergent series solution [35]. Series expansions for solving systems of nonlinear rate equations potentially are a useful alternate to strictly numerical methods of integration. We are unable to provide an example of a direct series solution of coupled nonlinear equations, but see the text by Hochstat [35] and the references cited therein for an introduction.

We hope that this introduction will have made the reader aware of the utility of series solutions for rate equations. Those wanting to pursue this

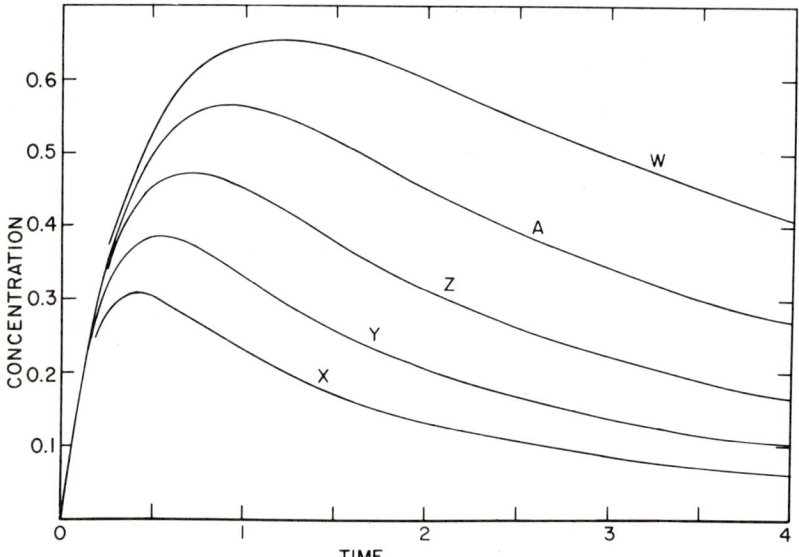

Fig. 3. Concentration of intermediate vs. time as in Fig. 2 but with $B_0 = 2.0$, $k_1 = 1.0$, $k_3 = 0$. Curve A, $k_2 = 1$; curve X, $k_2 = 8.0$; curve Y, $k_2 = 4.0$; curve Z, $k_2 = 2.0$; curve W, $k_2 = 0.5$.

approach for other mechanisms are encouraged to refer to and use the extensive mathematical literature on differential equations. We hope that the suitability of series solutions to machine computation is evident. Compared to particular analytic solutions, series solutions seem more readily derived. Compared to numerical solutions, series solutions are inherently more accurate. A combination of these methods may be most useful in attacking and taming difficult problems.

It has occurred to us that chemists would have use for compendia of tested computing techniques. We note the lack of a forum in the literature for the dissemination of computational methods in chemistry, especially those that are machine oriented. We suggest that a journal or a continuing series of volumes could fulfill this need. Such an addition to the literature explosion clearly could have the positive effect of reducing much duplication of effort.

REFERENCES

[1]. J. R. Kittrell, R. Mezaki, and C. C. Watson, Ind. Eng. Chem., 58:51 (1966).

[2]. T. I. Peterson, Chem. Eng. Process/Symp. Ser., 56:111 (1960).

[3]. T. I. Peterson, Chem. Eng. Sci., 17:203 (1962).

[4]. D. M. Himmelblau, C. R. Jones, and K. B. Bischoff, Ind. Eng. Chem. Fundamentals, 6:539 (1967).

[5]. D. W. Marquardt, J. Soc. Ind. Appl. Math., 11:431 (1963).

[6]. R. W. Hamming, Numerical Methods for Scientists and Engineers, McGraw-Hill, New York, 1962.

[7]. D. F. Detar and C. E. Detar, J. Phys. Chem., 70:3842 (1966).

[8]. D. F. Detar, Use of Computer Techniques in the Study of Reaction Mechanism, Institute of Molecular Biophysics and Department of Chemistry, Florida State University, Bulletin No. 27.

[9]. D. F. Detar, J. Chem. Educ., 44:191 (1967).

[10]. D. F. Detar, J. Chem. Educ., 44:193 (1967).

[11]. P. P. Sorokin, J. R. Lankard, V. L. Moruzzi, and F. C. Hammond, J. Chem. Phys., 48:4726 (1968).

[12]. M. J. Kurylo, N. C. Peterson, and W. Braun, J. Chem. Phys., 53:2776 (1970).

[13]. R. A. Keller, E. Zalewski, and N. C. Peterson, J. Opt. Soc., 62:319 (1972).

[14]. E. L. Ince, Ordinary Differential Equations, Dover, New York, 1956, p. 398.

[15]. H. T. Davis, Introduction to Nonlinear Differential and Integral Equations, Dover, New York, 1962.

[16]. R. W. Hamming, J. Assoc. Comput. Mach., 6:37 (1959).

[17]. L. J. Schaad, J. Am. Chem. Soc., 85:3588 (1963).

[18]. B. Rabinovitch, J. Chem. Educ., 46:262 (1969).

[19]. E. A. Moelwyn-Hughes, Physical Chemistry, 2nd rev. ed., Pergamon, New York, 1961, Appendix 9.

[20]. S. W. Benson, The Foundations of Chemical Kinetics, McGraw-Hill, New York, 1960.

[21]. A. A. Frost and R. G. Pearson, Kinetics and Mechanism, Wiley, New York, 1961.

[22]. K. J. Laidler, Chemical Kinetics, McGraw-Hill, New York, 1965.

[23]. N. M. Rodiguin and E. N. Rodiguina, Consecutive Chemical Reactions, Van Nostrand, Princeton, N. J., 1964.

[24]. G. M. Murphy, Ordinary Differential Equations and Their Solutions, Van Nostrand, Princeton, N. J., 1960.

[25]. J. Y. Chien, J. Am. Chem. Soc., 70:2256 (1948).

[26]. R. G. Pearson, L. C. King, and S. H. Langer, J. Am. Chem. Soc., 73:4149 (1951).

[27]. A. S. Baladin and L. S. Leibenson, Dokl. Akad. Nauk SSSR, 39:22 (1943).

[28]. T. Kelen, Z. Phys. Chem., N.F. 58:268 (1968).

[29]. T. Kelen, Z. Phys. Chem., N.F. 60:191 (1968).

6. Integration of Complex Rate Equations

[30]. T. Kelen, Acta Chim. Acad. Sci. Hungaricae, 59:323 (1969).

[31]. T. Kelen, Acta Chim. Acad. Sci. Hungaricae, 60:87 (1969).

[32]. H. J. Butcher and N. C. Peterson, Proceedings of the Conference on Computers in Chemical Education and Research, 3-39, Northern Illinois University, DeKalb, Illinois, 1971.

[33]. I. S. Gradshteyn and I. M. Ryzhik, Tables of Integrals, Series, and Products, 4th ed., Academic Press, New York, 1965.

[34]. E. T. Whittaker and G. N. Watson, Modern Analysis, 4th ed., Cambridge Univ. Press, London, 1935.

[35]. H. Hochstat, Differential Equations, Holt, New York, 1964.

AUTHOR INDEX

Numbers in parentheses are reference numbers and indicate that an author's work is referred to although his name is not cited in the text. Underlined numbers give the page on which the complete reference is listed.

A

Adams, R. E., 143, 146(19), 147, 148(26), 149(26), 155, 156(19), 205, 250(60), 261
Archard, J. F., 46(24), 102
Ashley, J. R., 211(9), 259

B

Baladin, A. S., 298(27), 312
Bauder, A., 25(4), 41
Beckett, A. H., 244(43), 247(43), 260
Bekiaroglou, P., 228(26), 230(27)-232(27), 260
Bennett, H. E., 62(43), 63(43), 66(43), 103
Bennett, J. M., 62(43), 63(45), 66(43), 103
Benson, S. W., 283(14), 292, 297(20), 312
Beveridge, G. S. G., 70(53), 103
Biryukov, V. V., 211(17), 214(17), 255(71), 259, 262(17)
Bischoff, K. B., 294(4), 311
Bitner, Ed., 219, 220(21), 221(21), 259
Bittikofer, J. A., 195(56), 196(56), 207
Blackmore, D., 233, 260
Blaedel, W. J., 111(11), 117(11), 152(11), 159(11), 205
Blount, H. N., 44(9, 11), 101, 102

Boguth, W., 211(16), 259
Bonath, B., 217, 259
Born, M., 48(32), 102
Boydzhirev, L., 211(12), 259
Braaf, Th., 65, 103
Bradley, R. F., 240(41), 243(41), 260
Braun, W., 295(12), 311
Brennan, R. D., 211(2), 258
Brown, J. V., 267(3), 291
Brown, K. M., 51(56), 70(55), 72, 73, 104
Buckwalter, F. H., 250(63), 261
Burke, M. F., 173(49), 174(49), 207
Burkhart, L. E., 211(3), 258
Bush, L. W., 6(1), 41
Butcher, H. J., 303(32), 304(32), 310(32), 313
Butterfield, R. O., 219, 220(21), 221(21), 259

C

Cahan, B. D., 65(48), 103
Carlson, A., 233(31), 260
Chance, B., 247, 261
Chien, J. Y., 298, 312
Clegg, P. L., 44(16), 46(24), 102
Conte, S. D., 70(55), 73, 104
Cordos, E. M., 160, 162(44), 164(44)-166(44), 194, 206
Cover, R. E., 242(42), 260
Crook, A. W., 44(16), 102

Crossley, T.R., 211(15), 225(23), 259
Crouch, S.R., 127(13), 130(13), 146(24), 151(13), 153(36), 157(41), 159(41-43), 160(41), 161(41), 162(44), 164(44), 165(36, 44), 166(44), 167(47)-172(47), 179(52), 180(52), 182(52), 183(52), 194, 205, 207

D

Dahl, W.E., 156, 157(40), 158(40), 206
Davis, H.T., 295(15), 296(15), 300(15), 312
Davis, J.E., 190, 207, 250(57), 261
Davis, W.S., 36(6), 41
Delaney, C.J., 173(48), 175(48), 207
Deming, S.N., 143, 145, 153, 201, 202(37), 203, 205, 206
Den Boef, G., 65, 103
Detar, C.E., 311
Detar, D.F., 295(8-10), 311
Dickson, T.R., 268(7), 292
Ditchburn, R.W., 44(1), 101
Dow, J.D., 57(39), 103
Dutton, H.J., 219, 220(21), 221(21), 259

E

Eggers, J., 233(30), 260
Eggert, A.A., 167(45, 46), 190(45), 194(46), 195(46), 200, 201(45), 206, 207
Emde, F., 287(16), 292
Enke, C.G., 139(17, 18), 140(17), 151(31), 205, 206

F

Fahidy, T.Z., 211(2), 258

Fahrenfort, J., 46(26, 27), 49(26), 102
Farris, G.J., 211(3), 258
Feil, P.D., 148, 205
Fernando, Q., 240(39), 260
Feynman, R.P., 48(31), 102
Fochs, P.D., 45(21), 102
Foertsch, B., 217, 259
Forsythe, G.E., 267(2), 291
Franklin, M.L., 146(25), 205
Franks, R.G.E., 211(1), 258
Frings, C.S., 173(48), 175(48), 207
Frost, A.A., 233(33), 260, 297(21), 312
Frysinger, J.R., 197(59)-199(59), 207
Fukuda, N., 247(52, 54), 261

G

Garrett, E.R., 250(62), 261
Gaskill, R.A., 211(6), 223(6), 258
Germond, H.H., 267(2), 291
Gnuse, M.K., 190, 207, 250(56), 261
Goodwin, B.C., 245(45), 261
Gradshteyn, I.S., 303(33), 304(33), 313
Granatek, A.P., 250(63), 261
Greenstein, D.S., 247, 261
Gucker, F.T., 70(54), 103
Guilbault, G.G., 109(4-6), 148(28a), 204, 206

H

Hagerty, J.C., 143, 146(19), 147, 155, 156(19), 205, 250(60), 261
Hammersley, J.M., 267(5), 291
Hamming, R.W., 295(6), 297(6, 16), 311, 312
Hammond, E.C., 295(11), 311

AUTHOR INDEX

Hansen, W. N., 44(7, 8, 15), 46(28, 30), 48, 49(28), 51(28), 52(7), 54, 55(28), 56, 60, 62(46), 65(47), 88, 101-103
Harrick, N. J., 46(29), 50(29), 52(29), 57(35), 60(29, 42), 102, 103
Harris, J. W., 211(6), 223(6), 258
Hass, G., 44(6), 101
Heavens, O. S., 44(3, 5), 46(25), 58, 101-103
Heinmets, F., 245(46), 261
Hicks, G. P., 111(11), 117(11), 152(11), 153(32, 33), 159(11), 167(45, 46), 172(33), 190(45), 194(46), 195(46), 200, 201(45), 205-207
Higgins, J., 245(49), 247(49, 51), 248(49), 261
Himmelblau, D. M., 294(4), 311
Hochstat, H., 310, 313
Holub, K., 254(67, 68), 262
Hooper, D. L., 268(11, 12), 292
Horkans, J., 65(48), 103
Horlick, G., 146(25), 205
Horton, J. A., 46(30), 62(46), 102, 103
Householder, A. S., 267(2, 6), 291
Hurley, J. R., 211(5), 223(5), 258

I

Imoto, T., 211(11), 259
Ince, E. L., 295(14), 301(14), 309(14), 312
Ingle, Jr., J. D., 127(13), 130(13), 146(24), 151(13), 167(47)-172(47), 205, 207

J

Jackson, A. S., 211(8), 223(8), 259
Jahnke, E., 287(16), 292
James, G. E., 131(14), 177, 178(51), 179, 180(51), 191, 192(14), 193(14), 205, 207
Janata, J., 233, 235, 236(37, 38), 240(40), 250, 260
Javier, A. C., 153(36), 165(36), 206
Johnson, D. E., 151(31), 206
Jones, C. R., 294(4), 311
Jones, D. O., 173(49), 174(49), 207
Jones, R. N., 39(7), 40(8, 10), 41
Jottrand, R., 214(18), 259

K

Kafarov, V. V., 211(17), 214(17), 255(71), 262(17), 259
Kelen, T., 298(28-31), 301, 312, 313
Keller, R. A., 295(13), 312
Kinbara, A., 57(36), 103
King, L. C., 286(15), 292, 298(26), 312
Kingdon, K. H., 45(18), 102
Kittrell, J. R., 294(1), 311
Kolb, D. M., 65(49), 103
Korn, G. A., 211(7), 223(7), 258
Korn, T. M., 211(7), 223(7), 258
Kubler, D. G., 148, 205
Kurylo, M. J., 295(12), 311
Kuwana, T., 44(7, 9-11), 52(7), 101, 102

L

Laidlor, K. J., 297(22), 312
Laitinen, H. A., 135(15), 205

Langemann, H., 211(14), 259
Langer, S.H., 286(15), 298(26), 292, 312
Langmuir, I., 45(18), 102
Langrand, F., 233(32), 260
Lankard, J.R., 295(11), 311
Lao, B.Y., 57(39), 103
Latterell, J.J., 175, 176(50), 207
Lattes, R., 267(1), 291
Leighton, R.B., 48(31), 102
Levinstein, H., 58(41), 103
Lewis, F.M., 225(25), 227(25), 260
Liebenson, L.S., 298(27), 312
Loach, P.A., 143, 144(20), 205
Lordi, N.G., 250(64), 251(64), 253(64), 262
Low, M.J.D., 251(65), 262
Loyd, R.J., 15(2), 41, 143, 144(20), 205
Lucquin, M., 233(32), 260
Ludwig, P.K., 257(72), 262

M

Maki, A.H., 26(5), 41
Male', D., 45, 102
Malmstadt, H.V., 139(18), 146(25), 150(29), 153(32-36), 157(41), 159(41, 43), 161(41), 162(44), 166(44), 165(36, 44), 166(44), 172(29, 33, 35), 194, 205
Manock, J.J., 268(11, 12), 292
Mansour, A.H., 214(18), 259
Margerum, D.W., 175, 176(50), 195(56), 196(56-58), 197(59), 198(58, 59), 199(59), 207
Mark, Jr., H.B., 44(8, 12-15), 52(33), 58(12), 101-103, 109(7), 110(7, 9), 111(7, 9), 112(12), 114(12), 117(7),
133(7, 9), 135(7, 9), 136(7, 12), 137(16), 138(7), 143, 146(19), 147, 148(26), 149(26), 155, 156(19), 188, 189(16), 197(7), 204, 205, 233, 235, 236, 240, 250(36, 56-60), 260, 261
Markin, N., 211(4), 258
Marquardt, D.W., 294(5), 311
Matthews, T., 233, 260
Mattson, J.S., 44(12, 14), 52(33), 58(12), 102, 103
Mayo, F.R., 225(25), 227(25), 260
McCracken, D., 290, 292
McIntyre, J.D.E., 65(49), 103
McKnight, A.L., 211(6), 223(6), 258
Mezaki, R., 294(1), 311
Moelwyn-Hughes, E.A., 297(19), 312
Monastier, J., 233(32), 260
Moruzzi, V.L., 295(11), 311
Müller, R.H., 146(23), 205
Murphy, G.M., 297, 312
Myers, R.J., 25(4), 41

N

Nemec, L., 254(67, 68), 262
Newman, L.G., 22(3), 36(6), 41
Nyssen, G.A., 196(57), 207

O

O'Dom, G., 240(39), 260
Oestergaard, K., 257(73), 262
Osteryoung, R.A., 44(7), 52(7), 101
Ottinger, C., 257(74), 262

P

Papa, L.J., 110(9), 111(9), 133(9), 135(9), 204, 250, 261

Pardue, H. L., 110(10), 111(10), 116(10), 117, 127(10), 131(14), 143(21, 22), 145(21, 22), 146, 150(29, 30), 151(10), 152(10), 153(34, 35, 37), 156(38-40), 157(38-40), 158(40), 159(10), 172(29, 35), 173(48, 49), 174(49), 175(48), 177, 178(51), 179, 180(51), 184, 185(53)-187(53), 191, 192(14), 193(14), 195(56), 196(56), 197(59), 198(59), 199(59), 201, 202(37), 203, 205-207, 250(61), 261
Parker, R. A., 184, 185(53)-187(53), 207
Pausch, J. B., 196(57, 58), 198(58), 207
Pearson, R. G., 233(33), 260, 286(15), 292, 297(21), 298(26), 312
Perche, A., 233(32), 260
Perez, A., 233(32), 260
Peterson, N. C., 295(12), 295(13), 303(32), 304(32), 310(32), 311-313
Peterson, T. I., 294(2, 3), 311
Pinkel, D., 137(16), 188, 189(16), 205, 250(59), 261
Pitha, J., 40(8), 41
Platt, A. E., 233(28), 260
Plumpe, W. H., 250(60), 261
Plumpe, Jr., W. H., 143, 146(19), 147, 155, 156(19), 205
Pons, B. S., 44(12-14), 102
Prostak, A., 44(8, 15), 45, 49(19), 65(47), 67, 101-103
Pyun, C. W., 268(13), 292

R

Rabinovitch, B., 268(10), 292, 297(18), 312
Rechnitz, G. A., 109(1-3, 7), 110(7), 111(7), 117(7), 133(7), 135(7)-138(7), 197(7), 204

Reilley, C. N., 110(9), 111(9), 133(9), 135(9), 204, 250, 261
Repges, R., 211(16), 259
Riemann, J., 211(13), 244(13), 259
Rijnsdorp, J. E. R., 225(22), 259
Rodiguin, N. M., 297(23), 298, 312
Rodiguina, E. N., 297(23), 298, 312
Rodriguez, P. A., 143, 145(21), 146, 205
Roepke, H., 211(13), 244(13), 259
Roth, J. A., 240(41), 243(41), 260
Rubloff, G. W., 57(37), 103
Ryzhik, I. M., 303(33), 304(33), 313

S

Saemann, R., 217, 259
Sands, M., 48(31), 102
Sarmousakis, J. N., 251(65), 262
Savitsky, A., 40(9), 41
Schaad, L. J., 268(9), 287, 292, 297(17), 312
Schechter, R. S., 70(53), 103
Schiesser, W. E., 211(1), 258
Schmidt, A., 225(24), 227(24)-229(24), 259
Schmidt, O., 236(37), 239(38), 240(38), 260
Scholfield, C. R., 219, 220(21), 221(21), 259
Schreider, Y. A., 268(8), 292
Schulz, L. G., 44(2, 4), 45(22), 101, 102
Schwartz, M. A., 250(63), 261
Scott, G. D., 44(17), 57(38), 102, 103
Scott, N. W., 44(6), 101
Seifert, R. L., 70(54), 103
Sennett, R. S., 44(17), 102

Sharpe, L. H., 52(34), 103
Sherry, A. E., 167, 194(46), 195(46), 207
Shirahata, R., 57(38), 103
Shriver, D. F., 36(6), 41
Slifkin, M. A., 211(15), 225(23), 259
Smith, G. F., 196(57), 207
Smith, S. D., 58, 103
Snell, F. M., 245(47), 261
Sommerfeld, J. T., 256(66), 262
Sorokin, P. P., 295(11), 311
Spangler, R. A., 245(47), 261
Statescu, C., 45(20), 102
Stehl, R. H., 175, 176(50), 207
Stokes, A. R., 62(44), 103
Stone, E. W., 26(5), 41
Sugita, M., 247(53-55), 261
Suzuki, T., 217(19), 259

T

Tanimura, Y., 217(19), 259
Taylor, A. M., 46(24), 102
Toren, E. C., 167(45, 46), 190(45, 54, 55), 194(45), 195(46), 196(46), 200, 201(45), 206, 207
Torren, E. C., 250(56, 57), 261
Tucker, G. T., 244(43), 247(43), 260

V

Van Der Beek, H. J., 65, 103
van de Vusse, J. G., 225(22), 259
VanSwaay, M., 148(28), 206
Vichnevetsky, R., 225(22), 259
Visser, W. M., 46(26), 49(26), 102

W

Wajc, S. J., 214(18), 259
Walles, W. E., 233(28), 260
Walter, C., 233(34), 245(44, 48), 247(44), 260, 261
Watson, C. C., 294(1), 311
Watson, G. N., 309(34), 313
Weber, Jr, W. J., 52(33), 103
Weichselbaum, T. E., 143, 146(19), 147, 148(26), 149(26), 155, 156(19), 205, 250(60), 261
Weinstein, F. C., 57(39), 103
Wells, Jr., D. J., 148, 205
Whittaker, E. T., 309(34), 313
Williams, T. J., 254(69, 70), 262
Willis, B. G., 184, 185(53)-187(53), 195(56), 196(56), 197(59)-199(59), 207
Winograd, N., 44(9-11), 101, 102
Winstrom, L. O., 44(12, 14), 58(12), 102
Winterbottom, A. B., 70(52), 103
Wolf, E., 48(32), 102
Woodruff, W. H., 197(59)-199(59), 207

Y

Yamaguchi, T., 57(36), 103
Yang, C. C., 247, 261
Yano, M., 211(11), 259
Yatsimerskii, K. B., 109(8), 204
Yeager, E., 65(48), 103
Yokota, K., 245(50), 261
Yoshida, S., 57(36), 103

Z

Zalewski, E., 295(13), 312
Zuman, P., 239(38), 240(38), 260

SUBJECT INDEX

A

Absorbance, 51
 vs angle of incidence, 53, 55
Absorption
 drug, 244
 hydrogen, 254
Absorption coefficient, 51
Absorption intensity, 23
Absorptivity, 51
 molar, 51
Acetic acid, 236
Adiabatic heat transfer, 257
ADP, 247
Aldolase, 247
Alkaline phosphatase, 156, 179, 186, 192, 195, 201
Alkyl-automatic ring bond, 32
Alkyl bonds, 32
Alkyl iodides, 257
Amines, 230
p-Amino-N-diethylaniline, oxidation with persulfate, 233
Amperometry, 116, 141, 149
(+)-Amphetamine, 244, 245
(+)-Amphetamine sulfate, 247
Amplifier
 differential, 19
 logarithmic, 113, 116, 177
Analog domain, 138, 139
Analog operations, 212, 213
Analog signal, 15
Analog voltage, 19
Angle
 critical, see Critical angle
 of incidence, 47
Ascorbic acid, 159
Aspirin, 250, 251

ATP, 247
ATR, see Attenuated total reflection
Attenuated total reflection, 47, 49, 50
Attenuation index, 49
Automatic pipets, 116, 153
Automation, 152, 203

B

Baseline drift, 28
Baseline shift, 28
Batch processing, 33
Beer's law, 142
Benzene, 6
 chlorination, 233
Bessel functions, 287
 first kind, 298
 third kind, 298
Beta function, 298
Binary, 22
Binary coded decimal (BCD) information, 15, 16
Binary counter, synchronous four-stage, 164
"Biological clock," 245
Biological tissue, 244
Biphenyl, 28
Biphenyl radical anion, 28
Biphenyls, substituted, 11
Blood, 243
Blood serum, 156, 179, 186, 192, 195
Bromine, 236
Bronsted acid-base, 250
Buffer, 201
n-Butyl benzyl sulfide, 239
Butylcyclopentenes, 214
Butylhypochloride, 227

321

C

Calcium, 198
Calculations
  closed form, 268
  Monte Carlo, 267
  statistical analysis, 268
Cassette, magnetic tape, 14
Catalyst, 115
Catalytic hydrogenation, 219
  heterogeneous, 223
Catalytic reactions, 109
Cauchy function, 40
Cesium, 257
Chemical analysis, 109
Chemisorption, 251
Chloranil, 225
Chlorination
  with butylhypochlorite, 227
  with hypochlorite, 230
Chlorine, 230
Chromatography, 217
Clinical chemistry, 204
Clock, 164
  crystal, 184
CO_2, 232
Comparator circuit, 141
Compiler language, 32
Complex refractive index, 49
Computer languages
  high level, 14
  machine, 14
Concentration-time curves, 214, 233
Conductometry, 127
  bipolar pulse technique, 150
Continuous flow methods, 152
Copolymerization, anionic, 225
Coulometry, 45
  controlled current, 233
Counters, 15, 18
Coupler, 20
Coupling constant, 11, 22, 26, 28
  deuterium, 31

Critical angle, 47, 49, 50, 53, 54, 61, 67
Crystal oscillator, 128
Curve fitting, 25, 214
Cyclohexane isomerization, 214
Cystine, 157
Cytoplasm, 248

D

Data rate, 13
Data smoothing, 36
Data transmission, phone line, 14
Degeneracy, accidental, 23
Density, bulk vs thin film, 44, 45
Depolarization, 63
Depth of penetration, 49, 52
Derivative circuits, 121
4, 4'-Dialkylbiphenylide
  radical anions, 6, 29, 30, 31
Dielectric constant, 257
Differential amplifier, 15
Differential equations, 210
Digital differential analyzers, 210
Digital domain, 138, 139
Digital plotter, 34
Digital stepping recorder, 15, 19
Digital voltmeter, 123
Dihydroxyacetone phosphate, 246
Dimethyl terephthalate, 225
Diphenyl sulfide, 236
Disease diagnosis, 114
Distortion, 109
Di-o-tolylcarbonate ester,
  chlorination, 217
Dividers, 211
Domain conversion, inter and intra, 140
Domain converters, 116
Dosage, drug, 247
Double bonds, 220
Drift, 121, 127, 141, 146, 171
  long-term, 145
  short-term, 142

SUBJECT INDEX

Drugs, 244
Dynamic measurement, 109

E

E. coli, 248
Effective thickness, 52
Electric fields, 52, 53
  attenuated, 50
  incident plane wave, 48
  time averaged, 51
  total internal reflection, 49
Electrochemistry, 65, 86
Electrode, 141, 150
  gold film, 44
  optically transparent, 44
  platinum film, 44
  rotating platinum, 150
Electrode potentials, 122
Electromagnetic radiation, 5
Electromagnetic waves, 48
Electromodulation, see Electroreflectance effect
Electron microscopy, 57, 58
Electron nuclear double resonance (ENDOR), 11
Electron spin resonance, 6, 15
Electroreflectance effect, 65
Ellipsometry, 46, 58
Elouich equation, 251
Emission spectroscopy, 146, 148
ENDOR, see Electron nuclear double resonance
Energy partitioning, 257
Enzymatic systems, 245
Enzyme activity, 114, 119, 125, 192
Equilibrium, 109
  liquid-vapor, 278
Equilibrium techniques, 109
Error analysis, 78
Error coefficients, 78
ESR, see Electron spin resonance
ESR derivative signal, 19

ESR spectra, 7-10, 19, 21, 22
  first derivative, 19
  line half-width, 28
  temperature error, 32
ESR spectrometer, 6
Esterification, 225
Evanescent wave, 49, 52
Exchange function, 16
Excretion, 244
Exponential integral, 298
External reflection, 47
Extinction coefficient, 44-46;
  see also Absorption coefficient, refractive index
  definition, 49

F

Fast reactions, 233
Fatty acid esters, 223
  catalytic hydrogenation, 223
Fatty acids, 220
File marks, 20
Files, disk, 35
Filter (optical), 142
Filter photometer, 143
Flip-flop, 19
Fluorometry, 116, 141, 148
Fourier transform, 14
"Free pellet," 244
Free radical, 6
Frequency standard, 18
Fresnel equations, 54
  reflection coefficients, 56
Frobenius solution, 301, 307
Fructose-1, 6-diphosphate (FDP), 246
Fructose-1, 6-phosphate (F6P), 246
Function multiplier, 225

G

Galactose, 175
Gamma function, 298
Gastrointestinal tract, 245

Gaussian curves, 39
Gaussian function, 40
Gaussmeter, 18, 20
Glactose oxidase, 175
Glass/air system, 53-55
Glass/gold film/air system,
    53-55, 82 ff
  spectra of, 83
Glucose, 159, 165, 245-247
  analysis, 173
Glucose oxidase, 153, 165, 175
Glutamic oxaloacetic
    transaminase, 210
Glyceraldehyde-3-phosphate
    (GAP), 246
Gold
  films, 82 ff
  optical constants, 84
  reflection spectra, 54, 55, 83
Graphic display, 34
Gravimetry, 44

H

Hamiltonian, 23
Hansen equations, 48, 49, 54, 56, 60
Heat convection, 143
Heat transfer, 146, 257
"Heteropoly blue," 159, 183
Hindered rotation, 32
Hybrid computer, 210
Hydrozine reduction, 220
Hydrocarbons, 6, 233
Hydrogen gas, 254, 256
Hydrogen ion, 196
Hydrogen peroxide, 241
Hydrogenation of oils, 220
Hydroxylamine, 250
N-(2-Hydroxypropyl)
    imidazolidinone, 232
Hyperfine splitting, 6, 29
Hyperfine splitting constants,
    6, 23, 24
Hypergeometric function, 298
Hypochlorite, 230

I

Incubation period, 201
Infinite series, 295, 299
Infrared spectrum, 40
Inhibition, 233
Inhibitors, 269
Initial reaction period, 235
Instrument drift, 109
Integrated circuits, 108
Integrated rate laws, 294
Integrator, 210
Interface, 11
Interferometry, 44, 45
Internal reflection elements,
    61, 64
Internal reflection spectroscopy,
    47 ff
  applications, 52
Iodine, 248
Ionic strength, 110, 122
Ion-pair, 257
Iron (III), 196
IRS, see Internal reflection
    spectroscopy
Isolinoleate esters, 223

J

Johnson noise, see Noise

K

K_m, see Michaelis constant
K lines, 23
Ketone, 190, 250
Keypunch, 13
Kinetic analysis, 109, 110
  "derivative" method, 117, 121
  derivative technique, 154
  differential rate methods, 133
  fixed-time methods, 111,
    115, 121, 123, 127,
    159, 191
  graphical extrapolation method,
    133

SUBJECT INDEX 325

initial rate methods, 112, 115, 118, 126, 136; see also variable time and reactions, fixed-time methods, pseudo-reactions, first-order
method of proportional equations, 135, 137, 188, 197
multi-component methods, 110, 132
pseudo-fixed-time method, 191
"signal-state" method, 117
single-component methods, 110, 132
slope method, 116
variable-time methods, 111, 115, 121, 131, 172, 191
Kinetic separation, 109
Kinetics
  electrochemical, 210
  enzyme, 225
  first-order, 215
  heterogeneous, 210
  homogeneous, 211
  homogeneous isothermal, 214
  in vivo, 245
  ion-pair recombination, 257
  nonisothermal, 210
  pH effects, 233
  second order, 223
  steady state approximation, 268
  zero-order reactions, 257

L

Lactic dehydrogenase (LDH), 146, 195
"Lag time," 245
Lead (II), 196, 198
Least-squares, polynomial, 25
Least-squares analysis, 22, 27, 28, 197, 281
  nonlinear iterative, 22, 294
Least squares iteration, 25
Legendre function
  first kind, 298

second kind, 298
Light-source fluctuations, 142
Light-source intensity, 122, 171
Light-source stabilization, 142
Line voltage fluctuations, 142
Linear collision model, 257
Linear equations, 300
Linoleate esters, 223
Linoleic acid, 220
Linolenate esters, 223
Linolenic acid, 220
  hydrazine reduction, 220
Linseed oil, 220
"Living anions," 227
Log-antilog circuit, 211
log (K)-pH profile, 250
Lorentzian line shape, 26
Lorentzian line shape function, 22, 23

M

Machine unit (m.u.), 215
Machine variable, 223
Magnesium, 198
Magnetic field, 4-6, 18
Magnetic nuclei, 11
Magnetic permeability, 56
Magnetic resonance, 5
Magnetic tape, 13
Matrix, 25
  derivative, 25
Maxwell equations, 47, 54
Membranes, biological, 244
Memory core, 14
Metabolism
  amphetamine, 245, 246
  glucose, 245
  iodine, 248
Metabolites, 244
Methanol, 225
Methyl linoleate, 220
Methyl linolenate, catalytic hydrogenations, 219
Methyl oleate, 220

5-Methyl-2-oxazolidinone, thermal decomposition, 232
Michaelis constant, K_m, 114, 200
Michaelis-Menten mechanism, 114, 200; see also Reactions, enzyme
Microscopy, electron, see Electron microscopy
Microwave power levels, 6
2 MnO · Cr_2O_3, 254, 256
Mo(VI), 165
Modulation amplitude, 6
Modulation frequency, 6
Molar absorptivity, see absorption coefficient
Molecular orbitals, 6
Molecular ray, 44
Molecular streaming, 44
Molybdate, 159
12-Molybdophosphoric acid, 165
Monochromator, 142
Monomer, 225
Monomethylterephthalate, 225
Monostable multivibrator, 164
Monte Carlo method, 267, 297
Multiphase system, see n-phase system
Multiple reflections, 52
Multiplier, 210

N

n-phase system, 54
Newton's method, 70-72
p-Nitrophenylphosphate, 201
NMR, see Nuclear magnetic resonance
Noise, 108, 127, 159
　amplifier, 121
　detector, 142, 146
　60 Hz, 121
　Johnson, 146
　light-source fluctuations, 121
　low frequency, 109, 121
　shot, 121

Nonlinear equations, Riccati type, 299, 307
Nonlinear response, 113, 114
Nuclear magnetic resonance (NMR), 6
Numerical integration, 295

O

Oleate esters, 223
Oleic acid, 220
　hydrazine reduction, 220
Optical constants, 58
　bulk vs thin film, 58-60
　calculation, 64 ff
　　estimated error, 80
　　flowchart, 69, 77
　definition, 44
　and electrochemistry, 65, 86
　gold films, 84
　platinum, 59, 76, 79
　thin film
　　via photometry, 44
　　via polarimetry, 45
　variations, 58
Optical feedback, 143
Oscillator, 18
Oscilloscope, 19, 108, 198, 214, 251
Osmium, 192
OTE, see Electrodes, optically transparent

P

Paper tape, 13, 14
Parallel logic, 210
Parallel polarization, see Polarization
Parity, 20
Parity errors, 13
Pb(II), see Lead (II)
Perpendicular polarization, see Polarization
Perylene, 29

SUBJECT INDEX 327

Perylene radical cation, 29
pH, 110, 122, 137
pH-stat, 190, 250
Phenethicillin, 250
Phosphate, 159, 160
   analysis, 165, 172
Phosphofructokinase, 247
Photocurrent, 151
Photography, 35, 195
Photolysis, UV, 217
Photometry 44, 45; see also Spectrophotometry
Photomultiplier, 142, 146, 148, 171
Photomultiplier counter, 39
Photomultiplier tube, 113
Photon counting, 142, 152
Phototube, vacuum, 146
Platinum, see also Electrodes, optically transparent
   films, 62, 65 ff
   optical constants, 49, 76, 79
   spectra, 68, 81, 82
PL/I, 5
Plotter, CALCOMP, 22
Polarimetry, 46
Polarized light, 46, 52, 53
   parallel, 53, 56, 57, 61
   perpendicular, 53, 56
   polarizers, 62, 67
   transverse electric, 53
   transverse magnetic, 53
Polarizer, 62, 67
Potassium, 4, 4'-bis(trideuter-omethyl)biphenylide, 7
Potassium, 4, 4'-di-n-butylbiphenylide, 9
Potassium, 4, 4'-di-sec-butyl biphenylide, 10
Potassium, 4, 4'-di-tert-butylbiphenylide, 10
Potassium, 4, 4'-diethylbiphenylide, 8
Potassium, 4, 4'-diisopropyl biphenylide, 9

Potassium, 4, 4'-dimethylbiphenylide, 7, 11
Potassium 4-methyl, 4'-(methyl^{-13}C) biphenylide, 8
Potassium permanganate, spectrum, 65
Potentiometers, 214
Potentiometry, 116, 121, 127, 141, 149, 157
Proline, 225
Proline-chloranil complex, 225
Propagation vector, 48
Proton, 32
β-Proton, 11
Punched cards, 13

Q

δ-Quinolinol, 240

R

Radical, 241
Radical anion, 6
Radical cation, 29
Radioactivity, 271
Radiolysis, 240
   X-ray, 242
Raman spectroscopy, 36
Random events, 268
Rate equations
   differential form, 112
   integral form, 112
Rayleigh scattering, 39
Reaction half-life, 109
Reaction mechanism, metabolism, 247
Reactions
   anionic copolymerization, 225
   autocatalytic, 233, 241
   bimolecular, see Reactions, second-order
   branched chain, 225
   bromination, 236, 239, 240

catalytic heterogeneous
 hydrogenation, 223
catalytic hydrogenation, 219
catalyzed, 109, 115, 120, 126
Ce(IV)-As(III),
 chlorination, 227, 230, 233
competitive, 221
competitive and consecutive, 221
complex, 283
consecutive, 215, 277
consecutive first-order, 271
Cs with alkyl iodides, 257
decomposition, 269
enzyme, 110, 114, 119, 124,
 130, 243
esterification, 225
first-order, 110, 118, 123,
 128, 134, 196, 215, 269
greater than second-order, 225
hydrazine reduction, 220
hydrolytic degradation, 250
iodine-azide, 157
irreversible, 110
ketones, 250
metabolic, 243
monomolecular, 269
p-nitrophenyl phosphate
 hydrolysis, 201
oxidation, 233
oxygen-hydrogen, 233
pseudo-first-order, 110, 118,
 123, 128, 134, 136, 196,
 201, 219, 271
pseudo-zero-order, 136
radical, 241
radioactive decay, 271
radiolysis, 240
reversible, 223, 277
second-order, 223, 278
thermal decomposition, 232
Reduced thickness, 54
Reflection absorbance, 52
 of three-phase system, 56, 57
 vs angle of incidence, 53, 55
Reflection coefficient, 56

Reflectivity, 51
 of three-phase systems, 56
 vs angle of incidence, 53, 54, 61
Reflectometry, 58
Refractive index, 44-47, 49
 complex, 49
Refractometry, 45
Register, 15, 16
Register storage, 18
REP-OP mode, 210
Resistivity
 bulk vs thin film, 45
 gold, 45
Resonance absorption, 6
Riccuti equations, see Nonlinear
 equations
Runge-Kulta integration, 295

S

Safflower oil, 220
Saline solution, 192
Scaling
 concentration, 224
 time,
Scaling factors, 17, 214
Schmidt trigger circuit, 141
Serum, see Blood serum
Short noise, see Noise
Signal
 analog, 15
 frequency, 16, 19
 nonlinear, 17
 voltage, 17
Signal-to-noise ratio, 108, 109,
 127, 131, 146, 164, 191, 196,
 203
Silicon-control switch, 177
Simultaneous nonlinear equations,
 70 ff
Snell's law of refraction,
 47, 49, 50
Sodium hydroxide, 250
Solvent effects, 122, 137
Soybean oil, 220

SUBJECT INDEX

Spectral bandwidth, 142
Spectroelectrochemistry, 64, 65, 86
Spectrometers, 4
Spectrophotometer, 61
  calibration, 62
  Cary Model 14 modification, 14, 61-63
  dual-beam, 142
  grating, 145
  nonlinearity, 80
  single-beam, 142
  stopped-flow, 195
Spectrophotometry, 116, 121, 122, 141, 142, 146, 171, 179, 186, 217
Spectroscopy
  attenuated total reflectance, 47, 49, 50
  internal reflection, 47 ff
  transmission, 51
Spectrum
  derivative, 32
  double integral, 32
  ESR, 27
  simulation, 22
  theoretical, 22, 25
Spin quantum number, 23
Sr/Ca ratios, 198
Statistical analysis, 294
Statistical methods, 268
Steady-state approximation, 114
Stearate esters, 223
Stearic acid, 220
Stepping motor, 39
Stoichometry, 109
Stopped-flow methods, 153
Stopped-flow spectrophoto-meter, 165, 195
Storage devices, 13
Strontium, 196, 198
Substrate, 114, 243
Succinic acid, 217
Succinylcholine hydrolysis, 217, 219

Succinylmonocholine chloride, 217
Sulfuric acid, 29
Symmetry, 6
Syringe, 116, 122
Systematic error, 62, 63

T

Taylor series, nonlinear first-order, 24
Teletype, 15
Temperature control, 110, 146
Temperature controller, proportional thermistor bridge, 148
Temperature effects, 137, 142
Terephthalic acid, esterification, 225
Terminal, 13
Tetrahydrofuran, 6, 7
Thermal decomposition, 232
Thermostat, 146
Thermostated cell, 116, 133, 146
THF, see Tetrahydrofuran
Thick films, 66
  anomolous behavior, 82 ff
  refractive index, 45
Thin films, 66
  anisotropies, 66, 67
  applications, 44, 57
  electrically deposited, 45
  evaporated, 44
  gold, 45, 53, 54, 82 ff
  microstructure, 45, 58
  model, 54, 60, 62
  platinum, 62, 65 ff, 76, 79, 81
  resistivity, 45, 85
  thickness, 44-46, 55, 65
Thiocyanate, 196
Three-phase systems, 53, 54 ff, 64 ff; see also n-phase system
Thyroid gland, 248
Time
  computer, 216

real, 216
scaling, 216
Time base, 15
 external, 15
 internal, 15
Time-interval (Δt) domain, 139
p-Toluic acid, 241
Total internal reflection, 47, 50
Trans-1,2-diaminocyclohexane-N,N,N,N-tetraecetate (CyDTA^{4-}), 196, 198
Transducers, 121
Transfer function, 112
Transmittance, 51
Transverse electric field, see Polarization
Transverse magnetic field, see Polarization
Triose phosphate isomerase, 247
Triphenylsilylamine, chlorination, 227, 230
Tryphtophan synthetase, 248
Tunnel diode, 177
Two-phase systems, 47 ff; see also n-phase system

U

Urine, 245

V

Vacuum technology, 57
Variable angle IRS cell, 53, 60, 80
 alignment, 67
 calibration, 61
Voltage-to-frequency converter, 15, 17

W

Wavelength, 4, 49, 142
 in vacuum, 55
Word, 20

X

X-band spectrum, 6
X-ray, 242
p-Xylene, 240
p-Xylene hydroperoxide, 241